£109.00

Soil Nutrient
Bioavailability

SOIL NUTRIENT BIOAVAILABILITY

A Mechanistic Approach

STANLEY A. BARBER
John B. Peterson Distinguished
Professor Emeritus of Agronomy
Purdue University

JOHN WILEY & SONS, INC.

New York ■ Chichester ■ Brisbane ■ Toronto ■ Singapore

This text is printed on acid-free paper.

Copyright © 1995 by John Wiley & Sons, Inc.

All rights reserved. Published simultaneously in Canada.

Reproduction or translation of any part of this work beyond
that permitted by Section 107 or 108 of the 1976 United
States Copyright Act without the permission of the copyright
owner is unlawful. Requests for permission or further
information should be addressed to the Permissions Department,
John Wiley & Sons, Inc., 605 Third Avenue, New York, NY
10158-0012.

This publication is designed to provide accurate and
authoritative information in regard to the subject
matter covered. It is sold with the understanding that
the publisher is not engaged in rendering legal, accounting,
or other professional services. If legal advice or other
expert assistance is required, the services of a competent
professional person should be sought.

Library of Congress Cataloging in Publication Data:

Barber, Stanley A.
    Soil nutrient bioavailability : a mechanistic approach / Stanley
A. Barber. — 2nd ed.
        p.   cm.
    Includes bibliographical references and index.
    ISBN 0-471-58747-8 (cloth)
    1. Crops and soils—Mathematical models. 2. Crops—Nutrition—
Mathematical models. 3. Soil fertility—Mathematical models.
    I. Title.
    S596.7.B37   1995   94-22899
    631.4'22—dc20

10 9 8 7 6 5 4 3 2

To my wife
Marion
and my daughters
Darlene and Rebecca

Contents

Preface

Growing plants to produce adequate food and fiber depends, in part, on nutrient supply from the soil. Empirical methods have usually been used to measure soil nutrient bioavailability. This has entailed using numerous field experiments in order to obtain information on the relationship between an empirical soil measurement and the response to an applied nutrient for a variety of plant species growing in different soil types. Fortunately, our understanding of the process involved in nutrient flux at the soil–root interface has increased greatly in the last two decades. This has enabled us to use a basic approach in evaluating soil nutrient bioavailability and reduced the necessity of conducting numerous crop-fertilization experiments to gain information about each crop–soil combination. This basic approach involves understanding the mechanisms governing the flux of nutrients from the soil to the root.

Soil Nutrient Bioavailability arose from my involvement in developing a mechanistic approach for describing plant uptake of soil nutrients and from the belief that presenting current information would stimulate use of this approach in research, extension, and teaching. Our understanding of the mechanisms governing the supply of nutrients to the plant and their uptake has developed to the point that these mechanisms can be described mathematically; hence, they can be quantified. While the mathematical solution may be complex, it can be readily handled by a computer.

Nutrient uptake involves properties of the plant and the soil, so that measurements of both are combined to predict nutrient uptake for a particular soil–plant–climate system. Several mechanistic models have been developed, and the accuracy of their uptake predictions has been verified for a number of soil–plant–climate conditions. Hence, it is possible to use a mechanistic approach to describe soil nutrient flux to the root, and this approach increases our understanding of the soil nutrient–availability process. Such an understanding is useful for determining the cause of fertility problems in the field and for developing practices that will correct the problem. Since both soil and plant parameters are involved, with plant parameters varying with cultivar and soil parameters varying with soil, using this approach should be of value to both soil and crop scientists.

While specific research papers provide fragments of the total mechanistic approach, I have written this book in order to present a unified treatment of the research I have been developing over the past 30 years, and to provide information to stimulate adoption of the mechanistic approach. Interest in this approach is indicated by numerous requests for reprints of my papers and invitations to lecture in both the United States and throughout the world. However, since the mechanistic approach is a recent development, many scientists may not be familiar with it. I believe it is a significant step in the evolvement of soil-fertility research, and more people should be aware of it.

Written as both a reference book for those involved in research, teaching, and extension services as well as a textbook for senior- and graduate-level courses, *Soil Nutrient Bioavailability* differs from other books in that the mechanistic approach is the central theme. Beginning chapters outline the reactions of nutrients with the soil and plant root that are important in determining nutrient flux into the root. The mathematical model and its parameters are discussed and then used to assess the availability of each nutrient. I have tried to present material in such a way that this book can be used by a wide audience. Early chapters provide chemical and biological principles that help with the understanding of later chapters.

Preparing a book of this nature involves the cooperation and assistance of many people. I wish to specifically acknowledge my gratitude for the generous assistance of the following: Purdue University for providing the environment to develop the research described in this book; the many graduate students and postdoctoral research associates who did much of the research and contributed greatly to the development of the approach; my colleagues for their suggestions and encouragement; Robert Austin for his assistance in collecting research data; Janet Hancock for her assistance in drafting many of the figures and in conducting research; Cynthia Boone for putting the manuscript on the word processor and for patiently making what seemed a never-ending number of revisions; Dr. Brian McNeal for his detailed review of the manuscript; M. C. Drew, M. J. Goss, F. J. Stevenson and P. B. Tinker for providing photographs or illustrative material; and Wiley-Interscience for its encouragement and cooperation in publishing of *Soil Nutrient Bioavailability*.

STANLEY A. BARBER

West Lafayette, Indiana
February 1984

Preface to the Second Edition

This edition includes research results obtained since publication of the first edition. It retains the emphasis on basic principles of nutrient uptake so that application can be made to a wide variety of soils and plants. Since the first edition the Barber–Cushman mechanistic uptake model has been adapted for use on PC computers. Advances in computers and programming allow the model to run rapidly on the PC computer. Hence, the program is readily available on a floppy disk. It has been distributed widely throughout the world. Previously the program ran on a mainframe computer and it was difficult to adapt to other mainframe computers. This availability enables the program to be used by students in studying the principles discussed in this book. More research on nutrient uptake has been done as well as the development of new programs. The book is written for use by a wide audience.

Some parts of the book have been eliminated or reduced while others have been expanded. The basic chapter structure has been maintained. Chapter 21 has been changed to "Application of the Mechanistic Uptake Model," which illustrates various uses of the model so that readers can see how the uptake model can be used to solve their problems.

Revision of the book used the research information of my students, postdoctorals, and visiting scientists for which I am grateful. The assistance of Beverly Bratton in typing the revisions is gratefully appreciated.

STANLEY A. BARBER

March 1994

Soil Nutrient Bioavailability

CHAPTER 1

Introduction

Plant growth involves the interaction of soil and plant properties. Soil is the normal medium for plant root growth. The plant's roots absorb nutrients and water from the soil and are an anchor to support the shoot. Maximum plant growth depends on the soil having the biological, chemical, and physical conditions necessary for the root system to maximize the plant's required absorption of nutrients and water and to enable the biochemical reactions that occur in the root. The plant's rate of absorption of nutrients involves processes going on in both the plant's root and the soil. Each of these processes is important in providing nutrients for use by the shoot.

Processes within both the plant's shoot and root are important in determining the rate of plant growth. Photosynthate is produced in the leaf by the action of sunlight energy on CO_2 from the air and water absorbed from the soil. This photosynthate combines with minerals absorbed by the roots to form the compounds necessary for plant growth. Crop yields may be limited by an insufficient rate of supply of any of the reactants, whether they originate primarily from the shoot or the root. Modification of the root environment is generally easier than modification of the shoot environment. Hence, a basic understanding of the plant root system and the mechanisms of nutrient and water absorption from the soil is important for determining why crop yields are low and developing practices that will maximize crop yield.

Nutrient supply from the soil usually has been studied by the soil chemist. Root growth and absorption of nutrients by the root usually have been studied by the botanist and plant physiologist. Thus, growth of plants in soil represents the integration of many disciplines. In the past, considerably less effort has been placed on integrating the processes than on studying each separate discipline. It has been a scientific axiom that more basic information can be obtained by studying the smallest segment of a system. Consequently, the plant physiologist has studied the kinetics of ion absorption by using either large single cells or excised roots. Soil chemists have also gone to simpler systems using clays from relatively pure deposits, such as montmorillonite or kaolinite, to measure ion bonding and ion movement in order to be able to describe the forces involved more accurately. Research in soil fertility has often relied on empirical correlations between increases in shoot growth or nutrient content and chemical measurements on samples of the soil. Useful correlations may be obtained without knowing the mechanisms involved, but they are generally unsatisfactory for predictive purposes involving other plant–soil situations or for developing an understanding of the processes involved.

During the past three decades, greater emphasis on soil-plant root research has occurred. Many research developments have helped bring about this change in emphasis. The development of rhizotrons enabled viewing and measuring the growth of roots during the life cycle of the plant. Study of plant roots in undisturbed soils was facilitated by development of mechanized equipment to obtain cores from field plots, which enabled sampling the root system with soil depth. Mechanized methods for washing roots from soil led to separation of the roots, and line and grid intercept methods for measuring root length facilitated measurements after roots had been separated from the soil. Bohm (1979) has described such developments in his book on sampling and measuring root systems.

Because of these types of research, it is now possible to use a mechanistic approach to describe the chemistry and biology of plant nutrient bioavailability. This book, an interpretative approach to mechanisms for understanding the supply of soil nutrients to plants and their resultant effect on crop yield, complements other recent works by Carson (1972), Nye and Tinker (1977), Russell (1979), Harley and Russell (1979), Marschner (1986), Wild (1988), and Baligar and Duncan (1990) that treat various aspects of this subject.

Weaver (1926) reported over 60 years ago in the introduction to his book *Root Development of Field Crops* that study of roots was greatly neglected. In his book *Plant Root Systems, Their Function and Interaction with the Soil* Russell (1979) stated that "the same claim could have been made more recently." Although Bohm (1979) found ten thousand papers on root ecology, this figure is small compared with the number of papers on the above-ground parts of the plant. In addition, few of these publications consider nutrient flux at the soil–root surface interface. Roots are important determinants of crop yield in the field, as evidenced by the observation that crop growth varies with differences in soil conditions of various locations in a field where above-ground conditions for plant growth are essentially the same. Roots absorb nutrients and water, hold up the shoot, synthesize organic chemicals needed by the plant, and exude waste products into the soil. Roots must be able to complete these functions without limiting plant growth.

To provide a useful basis for investigating a plant's root growth and nutrient uptake, a mechanistic uptake model was developed describing the more significant processes involved in soil nutrient uptake by plants (Claassen and Barber, 1976; Barber and Cushman, 1981). As will be shown in the following chapters, predicted uptake by the Barber–Cushman model agrees closely with observed uptake when accurate data for all the necessary factors for each situation are included. The uptake model describes size and growth of the roots, kinetics of nutrient uptake by the root, and supply of nutrients from the soil to the root surface. The model has three parameters describing the root surface area: L_0, initial root length; r_0, mean root radius; and k, rate of root elongation. Three parameters describe the relation between nutrient influx and nutrient concentration in soil solution at the root surface, C_{lo}; I_{max}, maximum influx at high C_{lo} values; C_{min}, C_{lo} below which net influx ceases, and K_m, $C_{lo} - C_{min}$ where net influx is $1/2 I_{max}$. Five parameters influence rate of nutrient supply to the root surface from the soil, C_{li}, initial nutrient concentration in the soil solution; b, buffer power of the labile nutrient in the soil for C_{li}; D_e, effective diffusion coefficient for diffusion of the nutrient in the soil; v_0, water influx; and r_1, the mean half-distance between adjacent roots.

The book considers soil factors that influence C_{li}, b and D_e in Chapter 2. In Chapter 3, the kinetics of nutrient absorption from solution by plant roots is discussed; Chapter 4 discusses nutrient absorption from soil. Chapter 5 describes the mathematical model and its consequences. Initially, the model assumes roots to be smooth cylinders without either root hairs or mycorrhizae. Roots are assumed to be absorbing at uniform rates and to serve purely as sinks for ions moving toward the root by diffusion and mass flow. Absorption is assumed to be independent of root age but capable of change with plant age. Since other factors may also influence the root–soil interface under specific conditions, Chapter 6 discusses the effect of the root on rhizosphere soil and the effect of soil on root properties. Moreover, since roots are not always smooth cylinders devoid of microorganisms, Chapter 7 discusses the role of root hairs, microorganisims in the rhizosphere, and the influence of mycorrhizae. Since there appears to be considerable interest in the supply mechanisms for specific nutrients, 12 chapters discuss these properties for most of the more important nutrients and water. Finally, illustrations are given of the application of the model to solution of plant nutrition problems. There is much more information available for some nutrients than for others. It is hoped that researchers will be encouraged to conduct investigations with those nutrients for which I could find only limited information. Where I have made statements on very limited information, I have qualified them; however, since overqualification makes for tedious reading, I have tried to keep the need for qualifications to a minimum. Because of the broad nature of the subject, material presented in this book is confined to subject matter that relates directly to those mechanisms and processes that influence nutrient flux into plant roots growing in soil.

DEFINITIONS

Several terms used frequently in this book require definition, since they have a qualitative nature and have been used differently by different authors. Mathematical definition is used when possible.

ABSORPTION: Ion uptake into the plant root. Each ion passes through the plasma membrane into the cytoplasm of a root cell. This may occur in the epidermis, cortex, or endodermis.

ADSORPTION: Chemical or physical bonding of ions to surfaces. Adsorbed ions can be desorbed into solution so an equilibrium exists between the adsorbed and solution ions.

AVAILABLE NUTRIENT: Although many avoid the term, it is important in this book. An available or bioavailable nutrient is one that is present in a pool of ions in the soil and can move to the plant root during plant growth if the root is close enough. An available nutrient must also be in a form that can be absorbed by the root. It is a diffusable ion with a D_e in the soil larger than 10^{-12} cm²/s. The term brings up problems of handling such nutrients as nitrogen or sulfur that may be present in large amounts in organic matter, which can mineralize and release organically combined ions to inorganic forms. The present definition is restricted to inorganic forms present in the soil during plant growth and includes ions released from organic combination during this time period. The amount is designated by C_s, in units of mol/L or dm³ of soil.

BIOAVAILABILITY: This term restricts the term availability to the processes of supplying nutrients to biological organisms which, in the context of this book, is primarily to plants.

DESORPTION: Removal of ions from solid soil surfaces rather than the root. As ions are removed from solution by plant root's uptake, ions from the solid phase go into solution. Ions may also be desorbed from the soil by other ions.

DIFFICULTY AVAILABLE NUTRIENTS: Soil nutrients that may be removed only by intensive cropping. They are differentiated on the basis of rate of release. These nutrients are released slowly as compared with available forms which tend to be released in less than 24 hours.

DIFFUSION: Movement of ions by random kinetic motion of molecules, sometimes referred to as Brownian movement. When a concentration gradient exists, net movement occurs by diffusion from locations of high concentration to locations of low concentration.

DIFFUSION COEFFICIENT: A coefficient describing the rate of movement of an ion by diffusion through a uniform medium such as water.

EFFECTIVE DIFFUSION COEFFICIENT: A coefficient describing the rate of movement of an ion by diffusion through the total soil volume calculated as though the soil volume were a uniform system.

EXCHANGE IONS: Ions are held on the soil's surface by negative or positive charges present in the soil; they may be exchanged by added cations or anions. The exchange is usually rapid but may take up to 24 hs.

EXCHANGE ADSORPTION: Adsorption of ions on the solid phase in positions where ions can be exchanged by other ions of similar charge.

FIXED NUTRIENTS: Some nutrients added to the soil are adsorbed into positions where they are not readily released back into solution; these are termed fixed nutrients. They are not usually desorbed by a particular extracting agent.

LABILE IONS: Ions in the soil that may exchange interchangeably with ions of the same element in solution. They are usually measured by adding a carrier-free radioactive isotope of the ion to the soil solution and measuring the disappearance of the radioactive isotope from solution, due to its exchange with nonradioactive isotopes of the ion. Equilibrium is usually approached within 24 hs.

MASS FLOW: Movement of ions to the root in the convective flow of water to the root surface.

NUTRIENT-UPTAKE EFFICIENCY: Proportion of a nutrient added to the soil that is absorbed by a plant growing in the soil.

ROOT INTERCEPTION: Ions in the soil that are intercepted by the growth of the root through the soil and hence do not have to move to the root before absorption.

SPECIFIC ADSORPTION: Adsorption of ions by the soil's solid phase so that they cannot be released by exchange for other ions.

SYMBOLS

A	area of root surface, m^2
a	affinity parameter in Langmuir equation, L/mg
a_i	activity, mmol/L

b	buffer power of C_s for C_l
B	absorption maximum in Langmuir equation, mg/kg
C_l	ion concentration in soil solution or nutrient solution, mmol/L
C_{li}	initial ion concentration in soil solution before plant growth, mmol/L
C_{l0}	ion concentration in solution at the root's surface, mmol/L
C_{min}	value of C_{l0} where influx = efflux and In = 0, mmol/L.
C_s	concentration of labile ions in the soil, mmol/L or dm^3
C_{si}	initial concentration of labile ions in the soil, mmol/L or dm^3
d	mean distance between ions, cm
d_i	diameter of hydrated ion, cm
d_p	particle diameter, cm
D_l	diffusion coefficient in water, cm^2/s
D_e	effective diffusion coefficient in soil, cm^2/s
d	distance, cm
E	efflux of ions from roots into solution, nmol/m$^2 \cdot$ s
e^-	electron, a negative charge, mol
Ep	potential energy difference, J/mol
F	change in free energy, J/mol
f_i	impedance factor for ion diffusion through soil, resulting from tortuosity, water density, and surface charges; dimensionless
I	influx of ions into plant roots, nmol/m$^2 \cdot$ s
In	net influx of ions into plant roots, nmol/m$^2 \cdot$ s
I_{max}	maximal net influx of ions into roots, nmol/m$^2 \cdot$ s
J	ion flux to the root, nmol/m$^2 \cdot$ s
k	rate of root growth; units depend on whether rate is linear, exponential, or sigmoid
K_i	equilibrium constant
K_m	Michaelis–Menten constant; ion concentration in solution $C_l - C_{min}$, where In = 1/2 I_{max}, mmol/L
K_{sp}	solubility product constant; dimensionless
k_l	distribution coefficient between ions in solution and ions on the solid phase for the ions indicated
k_B	Boltzmann constant, 1.3805×10^{-16} erg/deg $\cdot$ molecule
L	liter
L_A	length of roots per unit of land surface area, cm/cm^2 or 1/cm
L_V	length of root per unit of soil volume, cm/cm^3 or l/cm^2
L_0	root length of plants for initial conditions in using mathematical model to predict nutrient uptake, cm
q	charge on the ion
Q/I	quantity/intensity relationship, where Q is the amount potentially available and I is the amount in solution immediately available to the root; dimensionless
p^+	proton, a positive charge, mol
r	radial distance from root axis, cm
r_0	root radius, cm

r_1 radial distance from root axis where competition for nutrients by an adjacent root begins; one-half the distance between root axes, cm

R molar gas constant, 1.987 cal/deg · mol

s   second

T absolute temperature

t time, s

U_t uptake of ions by a plant at time t, mol

V_{max} maximal velocity of enzyme reaction in Michaelis–Menten equation, nmol/m² · s

V velocity of enzyme reaction in Michaelis–Menten reaction according to level of C_1

v volume

v_0 flux of water into the root, cm³/cm² or cm

w weight, g

x/m amount adsorbed per unit mass of soil; an expression used in the Langmuir equation, mol/kg

η viscosity of a liquid, centipoise

θ volumetric soil water content; dimensionless

γ activity coefficient

μ ionic strength

Subscripts

1   at half-distance between root axes

0   at root surface

l   in soil's liquid phase

s   on soil's solid phase

i   initial state

i   ionic species

F.W.   fresh weight

SI UNIT CONVERSIONS

The international system of units (SI units) is used throughout this book. In a few cases, units are given per cm² or cm³ instead of m² or m³ and per g rather than kg for convenience of interpretation. Many of my references used different units, hence a list of the more common conversions is given here.

CATION EXCHANGE CAPACITY: me/100 g = cmol(p⁺)/kg.

ADSORPTION ENERGY: calories/mol × 4.19 gives joules (J)/mol.

INFLUX: pmol/cm² · s × 10 gives nmol/m² · s.

SOIL–BULK DENSITY: g/cm³ = Mg/m³.

ROOT DENSITY: cm/cm³ × 10 gives km/m³.

MINERAL SPACING: Å × 0.1 gives nm.

SOIL WATER POTENTIAL: bar × 100 gives kPa.

REFERENCES

Baligar, V. C., and R. R. Duncan. 1990. *Crops as Enhancers of Nutrient Use.* Academic Press, San Diego, CA.

Barber, S. A., and J. H. Cushman. 1981. Nitrogen uptake model for agronomic crops. In J. K. Iskandar, Ed. *Modeling Waste Water Renovation-Land Treatment.* John Wiley, New York.

Bohm, W. 1979. *Methods of Studying Root Systems.* Springer-Verlag, New York.

Carson, E. W., Ed. 1972. *The Plant Root and Its Environment.* University Press of Virginia, Charlottesville.

Claassen, N., and S. A. Barber. 1976. Simulation model for nutrient uptake from soil by a growing plant root system. *Agron. J.* **68**:961–964.

Harley, J. L., and R. Scott Russell. 1979. *The Soil-Root Interface.* Academic Press, New York.

Marschner, Horst. 1986. *Mineral Nutrition in Higher Plants.* Academic Press, New York.

Nye, P. H., and P. B. Tinker. 1977. *Solute Movement in the Soil-Root System.* Blackwell Scientific Publishers, Oxford, England.

Russell, R. Scott. 1979. *Plant Root Systems: Their Function and Interaction with the Soil.* McGraw-Hill, New York.

Schenk, M. K., and S. A. Barber. 1979. Root characteristics of corn genotypes as related to phosphorus uptake. *Agron. J.* **71**:921–924.

Weaver, J. E. 1926. *Root Development of Field Crops.* McGraw-Hill, New York.

Wild, A. (Ed.), 1988. *Russell's Soil Conditions and Plant Growth*, 11 ed. Longman Scientific and Technical, copublished by John Wiley, New York.

CHAPTER **2**

Chemistry of Soil–Nutrient Associations

Plant nutrient availability in soil depends greatly on the amount and nature of nutrients in the soil solution and their association with nutrients adsorbed by or contained within the solid phase of the soil.

This chapter briefly describes the constituents of the soil, their composition, and the properties that influence the effect of the relation of nutrients between the solid phase and solution. The various procedures that have been used to describe this relation will be discussed. The reaction of nutrients with soil surfaces is influenced by the chemical and physical natures of the surface, which in turn is affected by the nature of the inorganic and organic materials that make up the soil particle. The nature of the soil material results from hundreds to thousands of years of physical, chemical, and biological weathering of the uppermost portion of the earth's crust. Weathering rate is influenced in turn by factors including temperature, rainfall, slope, drainage, rate of water infiltration, minerals present, and plant growth. Plant roots absorb nutrients from the soil; they are translocated to the shoot and then returned to the soil's surface when the plant or plant part dies. Hence, nutrients are transferred from the subsoil to the surface, they may move through the soil profile once more with leaching waters. Soil material can be subdivided into crystalline inorganic, amorphous inorganic, and organic categories. The identity of crystalline minerals can be determined from the nature of the structural arrangement of the atoms, and amorphous minerals can be described from their chemical composition and morphology. The organic compounds can be described from their chemical composition and by their contents of various reactive groups, such as carboxyl, hydroxyl, and methyl.

The physical arrangement of soil components is important for determining the chemistry of nutrient supply to plant roots. Soil particles may be either rock fragments or combinations of inorganic and organic materials cemented together in a porous structure. The top 15-cm soil layer may contain 95% (w/w) inorganic and 5% organic matter. However, when expressed on a volume basis, which is important when considering nutrient flux, the same soil may be 38% (v/v) mineral, 12% organic matter, and 50% pore space. The pore space may contain 15 to 35% water (v/v), depending on the soil's moisture content. The balance of the pore space is filled with air. The soil's organic content can vary from almost 50% (v/v) for an organic soil, a Histisol, to less than 1% in a desert soil, an Aridisol.

In this chapter, the fabric of the soil is discussed briefly, then each component of the soil is discussed in terms of its properties that influence release of nutrients to the soil solution and adsorption or precipitation of applied nutrients from solution. The general mechanisms are discussed here; the specific effects for a particular nutrient are discussed in the chapter discussing the nutrient concerned. Nutrient release from the solid phase of the soil may result from processes such as exchange, decomposition, dissolution, or desorption. Rate of release as well as the capacity to release nutrients will be important in determining the significance of release in supplying nutrients for movement through the soil to the plant root.

SOIL PARTICLE SIZE

Soil minerals can be subdivided on the basis of particle size into sand, silt, and clay fractions; particle diameter ranges for each fraction are given in Table 2.1. Sand and silt particles may be similar mineralogically and differ only in particle size and relative abundance of particular minerals within the particle. Most clay particles, on the other hand, differ mineralogically from the sand and silt, though they generally have developed from weathering of silt and sand particles. Table 2.1 gives the average number of particles per gram and the average surface area per gram for each size fraction. The surface area of cubical or spherical particles may be calculated from the relation

TABLE 2.1   Soil Particle Size and Surface Area

Particle classification	Particle diameter (d_p, mm)	Particles/g[a]	Surface area[b] (cm^2/g)
Very coarse sand	2–1	112	15.4
Coarse sand	1–0.5	895	30.8
Medium sand	0.5–0.25	7.1×10^3	61.6
Fine sand	0.25–0.1	7.0×10^4	132
Very fine sand	0.1–0.05	89×10^5	308
Silt	0.05–0.002	2×10^7	888
Clay-swelling	<0.002	4×10^{11}	8×10^{6c}
Clay-nonswelling	<0.002	4×10^{11}	4×10^{5c}

[a] $1/(\text{mean } d_p^3 \times 2.65)$ except for clay.
[b] $2.31/d_p$ except for clay, median value of range used for calculation.
[c] Calculated as clay plates with a thickness of 1×10^{-7} cm for swelling clay and 2×10^{-6} cm for nonswelling clay.

surface area = $2.31/d_p$, where d_p is the particle diameter. However, clay particles are plate-shaped and have a much larger surface area per gram. Cihacek and Bremner (1979) measured the surface area of 33 soils with clay contents ranging from 1 to 72% (w/w) and organic carbon contents of 0.3 to 9.4% (w/w). The authors obtained an average value for the soils of 100 m^2/g, with a range of 9 to 297 m^2/g. If the particles had been spheres the average surface area would have been in the range of 1 to 3 m^2/g. Hence, the plate shape of clay particles increases surface area greatly, and the bulk of soil's surface area is due to the clay. In Cihacek and Bremner's study, surface area was correlated with clay content ($r = 0.80$), but not with organic carbon. Removing organic carbon before measurement had little effect on the surface area/g measured. Clay particles in these soils had surface areas averaging 300 m^2/g. Surface area of clay and the chemical and physical nature of its surface are important in determining the nature of adsorption and release of nutrients by the soil solid phase. The total surface area of the soil in the upper 20 cm of profile, assuming a surface area of 100 m^2/g and a bulk density of 1.3 Mg/m^3, would be 26×10^6 m^2/m^2 of soil surface. This large surface area plays a significant role in supplying nutrients to plant roots.

SOIL MINERALOGICAL COMPOSITION

The chemical composition of a mineral plays a role in supplying some nutrients, particularly in relation to the mineral's solubility and rate of solution. Lindsay (1979) has described equilibrium–solubility relations for a large number of minerals found in soils. In soils that have been highly weathered for hundreds or thousands of years, the minerals remaining in the soil have very low solubilities and only go into solution slowly. Soil minerals may also play a role in plant nutrition because of the nature of the adsorption and desorption of applied nutrient ions by their surfaces. This effect is influenced by the ionic structure of the mineral particle.

Soil Rocks

Minerals are combined together to form rock particles in the silt and sand fractions of the soil. The rocks present depend on the prior geologic history; they may be igneous, metamorphic, or sedimentary.

Igneous Rocks

These rocks result from cooling of the magna from the earth's core. Slow cooling of the magna allows separation according to the melting points of the mineral and gives coarse-grained or plutonic rocks. Fast cooling gives fine-grained volcanic rock or glass. Forms of plutonic rocks include granite, diorite, gabbro, and peridotite. Granite is a common igneous rock made up primarily of quartz and feldspars. Forms of volcanic rocks include basalt, andesite, docite, and rhyolite.

Metamorphic Rocks

These rocks occur as a result of heat and pressure applied to sedimentary and igneous rocks over long periods of time. Gneiss is a laminated rock similar mineralogically to granite. Mica schists contain laminated mica minerals, slates result from compression of shales. Marble is metamorphosed limestone, and quartzite is metamorphosed sandstone.

Sedimentary Rocks

These rocks result from weathering of igneous and metamorphic rocks and their deposition as sediments in lakes and seas. Sedimentary rocks cover three-fourths of the earth's land area and 80% of these are shales, 15% are sandstones, and 5% are limestones. Shales result from deposition and compression of silts and clays, sandstones result from deposition and compression of sands. Limestones result from deposition of $CaCO_3$ and $MgCO_3$, usually as shells, in marine deposits and their eventual compression into rock.

Soil Minerals

The total elemental composition of soil may reflect the minerals present; however, it does not indicate the amounts of nutrients that may become available for plant uptake. Nevertheless, it is useful to know how the composition varies among soils. The compositions of four different soils and the earth's crust are given in Table 2.2. Silica and aluminum are the dominant

TABLE 2.2 Percentage Composition of Four Different Soils and the Earth's Crust

Element	Mollisol Marshall silt loam A^a 0–25.4[b]	Mollisol Marshall silt loam B^a 25.4–76.2[b]	Aridisol Mohave loam A 0–15.21	Aridisol Mohave loam B 5.2–35.6	Ultisol Cecil clay loam A 0–15.2	Ultisol Cecil clay loam B 15.2–101.6	Oxisol Columbiana clay A 0–25.4	Oxisol Columbiana clay B 25.4–63.5	Earth's crust
SiO_2	72.63	72.79	67.44	68.27	58.33	49.95	19.8	20.4	59.08
Fe_2O_3	3.14	3.96	5.31	5.35	6.38	9.31	15.6	16.3	6.82
Al_2O_3	12.03	12.77	14.40	13.86	20.06	26.54	37.1	37.8	15.23
MnO	0.10	0.12	0.10	0.09	0.06	0.04	0.26	0.23	0.12
CaO	0.79	0.70	2.04	2.51	Trace	Trace	0.25	0.10	5.10
MgO	0.82	1.04	1.52	1.53	0.58	0.45	0.50	0.43	3.45
K_2O	2.23	2.08	2.72	2.59	2.20	1.43	0.14	0.14	3.11
Na_2O	1.36	1.32	1.73	1.86	0.45	0.40	0.23	0.22	3.71
P_2O_5	0.12	0.10	0.22	0.12	0.10	0.04	0.31	0.31	0.29
SO_3	0.12	0.08	0.10	0.11	0.07	0.07	0.24	0.15	0.15
Ignition loss	6.01	4.46	3.35	3.47	10.79	10.63	24.1	22.0	
N	0.17	0.09	0.03	0.01	0.08	0.03			

Source: Byers et al. (1935); Lawton (1955).
[a]Horizon sampled.
[b]Depth of sample in cm.

13

elements in the soil, while their oxides are dominant constituents of the silicate and aluminosilicate minerals that make up much of the mineral structure. Soil minerals may be either crystalline or amorphous; these forms differ in how they adsorb nutrients onto their surfaces.

Crystalline Soil Minerals

Much of the soil's mineral content is present as crystalline-structured minerals that can be identified by analytical techniques such as X-ray diffraction. These minerals are mainly oxides, silicates, and aluminosilicates. The most common soil minerals are discussed in this section.

Mineral Structure

Mineral structure affects the extent and chemistry of the exposed surfaces of mineral particles. Dissolution, precipitation, desorption, adsorption, or exchange of nutrients on mineral surfaces is greatly affected by the chemistry of the surface; hence mineral surface chemistry plays a large role in determining the rate of nutrient supply to plant roots growing in soil. Minerals with extensively exposed surfaces are likely to have the greatest effect on soil–plant root nutrient relations. This section contains a brief discussion of mineral structure as it pertains to mineral surface chemistry. More detailed discussions on mineral structure are given by Marshall (1964) and Dixon and Weed (1989).

The structural arrangement of the cations and anions in a mineral gives the mineral its individual characteristics. Oxygen is the largest ion and occupies much of the volume of many minerals. The diameters of ions commonly found in soil minerals are given in Table 2.3. Oxygen, with a diameter of 0.132 nm, dominates the structure of aluminosilicate minerals, since Al, with a diameter of 0.057 nm, and Si, with a diameter of 0.039 nm, are much smaller. The size of Si is such that it just fits into the cavity that results when four oxygens are fitted around the Si. This unit of Si with four oxygens fitting around it is called a silica tetrahedron; it is the building block of many soil minerals. Because four oxygens fit closely around the Si, Si is said to have fourfold coordination, or a coordination number of four. Only cations whose size is such that they have fourfold coordination can occupy such a site without causing undue strain on the crystal.

The Al ion is larger than Si and thus can accommodate six surrounding oxygens. Aluminum with six oxygens around it is in sixfold coordination. Actually, Al has a size such that it can also go into fourfold coordination and proxy for the Si in tetrahedral positions. Most crystalline soil minerals are made up of combinations of layers of silica tetrahedra combined with layers of Al octahedra or silica tetrahedra in a three-dimensional structure. The oxygens of the silica tetrahedron will be shared with an adjoining tetrahedron or octahedron, and any excess negative charge will be commonly balanced with

TABLE 2.3   Diameters and Coordination Numbers of Ions
            Found in Soil Minerals

Ion	Diameter (nm × 100)	Coordination number
Oxygen	13.2	–
Fluorine	13.3	–
Water	14.5	–
Chlorine	18.1	–
Lithium	7.8	6
Sodium	9.8	8+
Potassium	13.3	8+
Ammonium	14.3	8+
Magnesium	7.8	6
Calcium	10.6	8+
Barium	14.3	8+
Cobalt	8.2	6
Aluminum	5.7	4 or 6
Iron, 3+	6.7	6
Iron, 2+	8.3	6
Chromium	6.4	6
Manganese, 4+	9.1	4
Silicon	3.9	4
Titanium	6.4	6
Zirconium	8.7	6
Zinc	8.3	6

K, Na, Li, Ca, Mg, Fe, or Zn. The size of these ions is such that they are able to fit into the crystal lattice without distortion.

Quartz.   Quartz makes up the major portion of the sand and silt fraction of most soils. It is a three-dimensional mineral composed of silica tetrahedrons and has Si and O in the ratio of SiO_2. The various quartz minerals include chalcedonite, agate, flint, chert, and opal; they differ in the structural arrangement of Si and O_2 atoms. Quartz is a dominant soil mineral, because it weathers slowly if silt- or sand-sized.

Feldspars.   Feldspars are three-dimensional, anhydrous aluminosilicates of K, Na, and Ca. They constitute about 60% of the igneous rocks weathering to form soil. Feldspars are made of silica-oxygen tetrahedra linked by shared oxygens, with some of the negative charge balanced by K, Na, or Ca. Magnesium is not present, because its size does not give a stable structure.

Orthoclase and microcline ($KAlSi_3O_8$) are K feldspars containing 13.7% K (w/w). Plagioclase is actually a series of Na and Ca feldspars ranging in composition from $NaAlSi_3O_8$ to $CaAl_2Si_2O_8$.

Other Nonlayer Silicates.   Other silicates in the soil include amphibole and pyroxene, which are Ca Mg Fe silicates. Hornblende is a common amphibole, and augite is a common pyroxene. Olivene $(MgFe)_2SiO_4$ is a MgFe silicate. Apatite $Ca(OH)_2 \cdot 3Ca_3(PO_4)_2$ and fluoroapatite $CaF_2 \cdot 3Ca_3(PO_4)_2$ are the main P-bearing minerals in the soil. Tourmaline, a borosilicate, is the primary B-containing mineral.

Micas.   Micas occur as sheetlike minerals having two layers of silica tetrahedra sandwiching one layer of Al octahedra or brucite, $Mg_3(OH)_6$ in each sheet. The oxygens of the silica tetrahedron are shared with an adjoining tetrahedron or octahedron, and the excess negative charge is commonly balanced with K, Na, Li, Ca, Mg, Fe, or Zn, which are able to fit into the crystal lattice. Muscovite $K_2Al_2Si_6Al_4O_{20}(OH)_4$ and biotite $K_2Al_2Si_6(Fe^{2+},Mg)_6O_{20}(OH)_4$ are common micas, containing approximately 9.8 and 8.7% K (w/w), respectively.

Clay Minerals

Clay minerals are those mineral particles below 0.002 mm in diameter. They consist of sheetlike crystalline minerals as well as silica and amorphous minerals. Crystalline clay minerals have been studied intensively; many of the investigations have been with relatively pure minerals found in clay deposits rather than the mixtures of various clays usually found in soils. Because of their sheetlike structure, clays provide the majority of the particle surface area found in soils. Clays are made up of layers of Si tetrahedra and Al octahedra in either 1:1 or 2:1 arrangement. The 1:1 arrangement has one octahedral layer and one tetrahedral layer. The 2:1 arrangement has a tetrahedral layer on each side of an octahedral layer that is sandwiched in the middle.

Isomorphous Substitution.   Cations that have the same coordination number and are similar in size can proxy for one another in a mineral. In clay minerals, Al^{3+} can proxy for Si^{4+} in the tetrahedral layer; when this occurs, there is an excess negative charge, which is balanced by an exchangeable cation that is outside the mineral structure. This negative charge is part of the soil's cation-exchange capacity. The proxying of Al for Si occurred at the time of mineral formation.

The octahedral layer can also have Mg^{2+}, Fe^{2+}, Zn^{2+}, and Li^+ proxying for Al^{3+} in the octahedral position. This gives further negative charge to the octahedral layer, which is balanced by cations outside the clay sheet. The octahedral layer has six positions for aluminum within a unit cell (the smallest crystal unit that is repeated), though when aluminum is present, only four of these positions must be filled to balance the negative charge of the oxygens. When magnesium proxies completely for aluminum at the time of mineral formation, there could be six magnesium ions with all positions filled. This would also result in complete balancing of the negative charge, so there would be no residual cation-exchange capacity. In many clay minerals, there is only one magnesium proxying for each aluminum during isomorphous substitution, however: hence, a negative charge arises, which is balanced by exchangeable cations held outside the mineral structure.

When magnesium substitutes for a fraction of the aluminum, the number of octahedral spaces filled with either magnesium or aluminum is close to four; the average varies between 4.00 and 4.44. This type of clay is called dioctahedral. In other clays, when magnesium is the dominant ion, there may be a small number of other ions substituting for magnesium; they result in either a negative or positive charge arising from the octahedral layer. The total cations per unit cell in the structure is close to six; the range varies from 5.76 to 6.00. These clays are labeled trioctahedral.

Types of Clay.   Clays may be divided into 1:1 and 2:1 clays. The 1:1 clays have one silica tetrahedral layer and one alumina octahedral layer in each sheet. The 2:1 clays have one silica tetrahedral layer on each side of an alumina octahedral layer. These layers are bonded by sharing oxygens between layers.

The 1:1 clays have little if any isomorphous substitution and, hence, low negative charge. The charge holding exchangeable cations on these clays is believed to come mainly from broken bonds at crystal edges. The 2:1 clays usually have isomorphous substitution in either the tetrahedral or octahedral layers or both. The amounts of substitution can vary widely, and specific substitutions give clays different properties. Some of the common clays and their substitutions are given in Table 2.4. Some clays have the clay sheets bonded together so that they do not swell; for example, potassium bonds clay sheets of illite together. The potassium ion is similar in size to the natural cavity that exists in the surface of the tetrahedral layer. When sufficient potassium ions are present between the tetrahedral layers and the clay sheets have a sufficiently high negative charge, usually greater than 150 $cmol(p^+)/kg$, the clay sheets are held together by potassium ions.

Magnesium hydroxide (brucite) can also bind clay sheets together, as in the mineral, chlorite. Vermiculite has a layer of hydrated magnesium ions between the clay sheets, permitting only limited swelling of this species as well.

Negative Charge on Clays.   Clays have a negative charge that holds exchangeable cations, this charge can be due to (1) isomorphous substitution in the octahedral layer, (2) isomorphous substitution in the tetrahedral layer, (3) broken bonds at the crystal edges, and (4) pH-dependent charge on amorphous and oxide minerals. The total of these negative charges is

TABLE 2.4 Structure of Common Clay Minerals

Clay	Dominant isomorphous substitution	Additional substitution	Cation-exchange capacity cmol (p⁺)/kg
1:1 Structure			
Kaolinite	None	None	1 to 15
Halloysite	None	None	1 to 15
2:1 Structure (swelling)			
Pyrophyllite	None	None	1 to 15
Montmorillonite	Mg for Al	Al for Si, Fe for Al	80 to 150
Beidellite	Al for Si	Al for Si	70 to 100
Saponite	3 Mg for 2 Al		
Hectorite	Li for Al		
2:1 Structure (limited swelling)			
Vermiculite	Al for Si	Hydrated Mg between layers	140 to 200
2:1 Structure (nonswelling)			
Illite	Al for Si	Mg for Al	40 to 70
Chlorite	Al for Si	Brucite sheets between layers	30 to 40

measured as cation-exchange capacity; it is reported as cmol(p$^+$)/kg soil. Each source of charge tends to hold cations with a different strength. In general, bonding to negative charges resulting from isomorphous substitution in the tetrahedral layer is the strongest; that due to isomorphous substitution in the octahedral layer, the next strongest; that due to broken bonds at crystal edges and pH-dependent charges, the weakest. However, the cation involved may have some influence on differences in bonding strength between sources.

Types of Clay Minerals.   The types of clays commonly found in soils are shown in Table 2.4. These are designated according to isomorphous substitution and structural arrangement. The type of clay structure influences the reactivity of the clay surface; hence, it is important to determine the types of clay in a soil. One surface of a mineral may be a solid sheet of hydroxyl ions, as occurs on the surface of the octahedral layer exposed in 1:1-type minerals. On the other hand, both sides may have hexagonal rings of oxygens exposed, as occurs on the surfaces of tetrahedral layers in 2:1 clays. The reactivity of these surfaces differ; there are also obvious differences in the tendency for adjacent clay sheets to collapse and trap ions in the cavities on the surfaces. These differences are discussed in greater detail later in this chapter.

Carbonates, Oxides, and Sulfates.   Oxides and hydroxides of aluminum and iron are common in soil. They may have a crystalline structure or they may be amorphous. Gibbsite is a common aluminum hydroxide. Iron oxide minerals vary with both the oxidation state of iron and its degree of hydration. Some common iron oxides are hematite, Fe_2O_3; goethite, $Fe_2O_3 \cdot H_2O$; and magnetite, Fe_3O_4. The limestones are calcium and magnesium carbonates and may vary from almost pure $CaCO_3$, calcite, to equal-molar amounts of $CaCO_3$ and $MgCO_3$, dolomite. They persist in soils only with pH above approximately 7.0. Sulfates commonly occur as $CaSO_4$ because of the amounts of calcium in soil and the low solubility of $CaSO_4$.

Amorphous Soil Minerals

Many soils also contain minerals that do not have a definite crystal structure. Hence, they are described according to their chemical composition and the reactivity (including solubility) of their surfaces. Oxisols and soils developed from volcanic materials can have large amounts of amorphous minerals. Noncrystalline clay-sized minerals in the soil include (1) allophane, a hydrous aluminosilicate, (2) aluminum oxides and hydroxides, and (3) iron oxides and hydroxides. Some of these minerals are also present in crystalline form. Allophane is a dominant mineral in soils developed from volcanic material. Aluminum and iron oxides and hydroxides are present in Oxisols and Ultisols. However, small amounts of amorphous minerals may be found in most soils, often as coatings on crystalline minerals.

An average chemical composition for allophane is SiO_2, 28.2%; Al_2O_3, 40.3%; Fe_2O_3, 0.36%; MgO, 0.10%; CaO, 2.31% (Dixon and Weed, 1989). Allophane can possess a cation-exchange capacity of 20–50 cmol(p^+)/kg of mineral at pH 7.0, with the cation-exchange capacity being highly pH dependent. Gibbsite is present in Oxisols and Ultisols. Amorphous aluminum and iron oxides are present in at least small amounts in most soils.

The significance of amorphous soil minerals can be greater than suggested by the fractional amount present, since coatings of amorphous minerals can occur on surfaces of crystalline minerals. Because of the pH dependency of the cation-exchange capacity of amorphous minerals, their exchange capacity must be measured at a particular soil pH. Broken bonds at the edges of crystalline minerals exhibit pH-dependent cation-exchange capacity in the same manner as do amorphous minerals. Amorphous minerals may have different effects in the distribution of nutrient ions between the soil and the soil solution and hence a different effect on nutrient bioavailability.

SOIL ORGANIC MATTER

Soils contain organic matter in amounts varying from 0.1% in desert soils to over 50% (w/w) in Histisols (organic soils). The organic matter is amorphous, varies widely in composition, and a portion of it is in constant flux as microorganisms break it down as a source of energy. This process releases carbon into the air as carbon dioxide while plants and animals add new organic material to the system.

Organic matter can be divided into nonhumic and humic substances. The nonhumic are attacked readily by microorganisms and disappear rapidly. They consist of carbohydrates, proteins, amino acids, fats, waxes, alkanes, and low-molecular-weight organic acids (Schnitzer, 1978; Stevenson, 1982). Humic substances are the majority of organic matter, and they decompose slowly. They are chemically complex organic compounds with molecular weights from a few hundred to several thousand. The organic matter in a soil can be characterized by its chemical composition and the amounts of various reactive groups that are active in cation exchange.

The chemical composition of organic matter is approximately 50% C, 5% N, 0.5% P, 0.5% S, 39% O and 5% H (w/w); however, these values can fluctuate from soil to soil. The main functional groups in humic matter are carboxyl, phenolic hydroxyl, alcoholic hydroxyl, and carbonyls. The carboxyl and some of the phenolic hydroxyls provide exchange sites for cation exchange; the exchange capacity is strongly pH-dependent. Cation-exchange capacity at pH 7.0 ranges from 100 to 400 cmol(p$^+$)/kg; values of approximately 150 cmol(p$^+$)/kg are common.

In addition to adsorption of cations in readily exchangeable form, organic matter can adsorb multivalent cations as coordination complexes. These complexes are not readily exchangeable with monovalent cations and do not dissociate readily into soil solution. Such cations as Mn, Zn, Cu, and Fe can be adsorbed in these complexes. Walker and Barber (1960) measured both complexed and exchangeable manganese on 12 Indiana soils and found that there was almost as much manganese (12 mg/kg average) complexed as held in exchangeable form (18 mg/kg average).

Organic matter is also present in soil solution. Soluble organic compounds may increase the metal's cation concentration in solution. Hodgson et al. (1966) found that much of the copper and zinc in soil solution was present as complexed soluble organic compounds. They compared 20 Colorado soils with pH levels of 6.9 to 7.9 with 10 New York soils with pH's ranging from 4.1 to 8.1. The New York soils were A and B horizons from each of five soils, while the Colorado samples came from the top 18 cm of soil. The organic matter in solution ranged from 2 to 25 mmol(p$^+$)/L in Colorado soils and from 15 to 75 μmol(p$^+$)/L in New York soils. The Cu^{2+} concentration in solution was 0.31 mg/L for the New York soils and 0.009 μg/L for the Colorado soils. The concentration of metal ions in solution was influenced greatly by soluble organic complexes.

CATION ADSORPTION

Negatively charged soil particles normally hold cations so loosely that they can be readily exchanged with cations in the soil solution (Kelley, 1948). The quantity of each exchangeable cation on the soil is usually measured by displacing the cations with a salt, such as 1 mol/L of neutral ammonium acetate, and then measuring all cations in the displaced solution. The total number of exchange sites can be determined by summing the quantities of all exchangeable cations or by saturating the exchange sites with a common cation, such as ammonium, and then displacing and measuring it. This quantity, the cation-exchange capacity, varies with the pH of the displacing solution because of the presence of pH-dependent exchange sites. In soil, the cations balancing the negative charge on the exchange site are in equilibrium with cations in solution. The latter are balanced with soluble anions in solution. The rate of equilibration between these phases is usually very rapid (within minutes). The strength of bonding of cations by the exchange site depends on the nature of the source of the negative charge and on the valance, hydration status, and size of the cation. The relative strengths of binding of different cations determines the equilibrium solution concentration of the cations when more than one cation is present. For further detail, see Bohn et al. (1979).

Source of Negative Charge on Soil

Sources of cation-exchange sites include the following:

1. Permanent charge resulting from:
   a. Negative charge due to isomorphous substitution in the tetrahedral sheet of the clay mineral.
   b. Negative charge due to isomorphous substitution in the octahedral sheet of the clay mineral.
2. pH-dependent charge resulting from:
   a. Broken bonds at crystal edges.
   b. Dissociation from surfaces of amorphous minerals and hydrous oxides.
   c. Carboxyl groups on organic matter.
   d. Hydroxyl groups on organic matter.

The relative amounts of permanent and pH-dependent charge vary from soil to soil. Pratt (1961) measured the two types of charge on a diverse group of 15 California soils. Measurements of cation-exchange capacity were made at pH 8 and pH 3. The pH-dependent cation-exchange capacity (the difference between that at pH 8 and pH 3) was 46 ± 12% of total cation-exchange capacity. Of this percentage, 15 ± 4% came from the clay and the remainder from soil organic matter. Hence, pH-dependent charge was a significant part of the charge holding exchangeable cations in these soils.

Nature of Permanent Charge

The permanent negative charge on the clay is due to isomorphous substitution of a lower valency cation for the dominant cation Si or Al present in the center of the tetrahedral and octahedral oxygen sheets. The negative charge not satisfied by the substituting cation in the mineral structure is satisfied by an exchangeable cation held at the clay surface. The strength of the charge may vary due to the distance between the negative charge and the closest approach of the exchangeable cation. Charges originating in the octahedral layer will be farther from the balancing cation than charges originating in the tetrahedral layer. The strength of charge is given by Coulomb's law:

$$F_a = \frac{q_1 q_2}{d^2 D} \tag{2.1}$$

where F_a is the force of attraction, q_1 and q_2 are the electrical charges, d is the distance separating the charges, and D is the dielectric constant (78 for water at 25°C). Hence, the strength of bonding is directly related to the strength of the charges and inversely related to the square of the distance of their separation.

Nature of pH-Dependent Charge

The pH-dependent charge of amorphous minerals and hydrous oxides arises from dissociation of H^+. When an oxide particle is placed in water, water is adsorbed on its surface and H^+ and OH^- dissociate from the adsorbed water. If predominantly OH^- dissociates, the particle becomes positively charged, and if predominantly H^+ dissociates, the particle becomes negatively charged. The pH of the solution surrounding the particle influences the relative degree of OH^- versus H^+ dissociation. At a particular pH, there is equal dissociation, and the resultant particle has no charge. This is called the pH of zero point charge. As pH increases, H^+ dissociates from the surface, and the number of cation-exchange sites increases. Because of this, cation-exchange capacity is measured at a specific pH, usually 7.0 or 8.0. Measurements at pH 8.0 and 3.0 can be used to separate the cation-exchange capacity into that due to pH-dependent charge and that due to permanent charge.

Diffuse Double Layer

The exchangeable cations at soil surfaces tend to diffuse into the solution due to their kinetic energy until the counter potential restricts further movement away from the soil surface. This distribution is called a diffuse double layer because of the negative charge layer of the clay balanced by the diffuse layer of positively charged cations. The concentration distribution of the cations perpendicular to the surface can be described by the Boltzmann equation,

$$\frac{C_1}{C_2} = \exp\left(\frac{-\Delta E}{k_B T}\right) \tag{2.2}$$

where E ($E_1 - E_2$) is the difference in potential energy of ions at distances 1 and 2 from the particle surface, and the kinetic energy of the ion is defined by k_B, the Boltzmann constant, times T, the absolute temperature. C_1 and C_2 are the ion concentrations at distances 1 and 2. The extent of the double layer corresponds to the point where C_1 is no longer larger than the concentration in the bulk solution. The double layer is suppressed by addition of salt to the bulk solution. Because divalent ions are attracted more tightly than monovalent cations, the extent of the double layer is also less for divalent ions.

The extent of the double layer is as great as 50 nm in Na-saturated soil at low salt concentrations (10^{-5} mol/L) and as low as 0.5 to 1.0 nm when the soil is saturated with divalent cations at high salt levels (0.1 mol/L). The diffuse double layer may have an effect on the supply of nutrients to plant roots, due to interpenetration of diffuse layers from the soil with those on root surfaces and also the diffuse double layer may affect diffusion in the soil.

Ion Activities

The activity of an ion in solution is its effective concentration as measured thermodynamically. Activity is equal to concentration only at infinite dilution. When other ions or negative exchange sites are present, interaction with the ions generally causes activity to be less than the concentration. The activity a_i is equal to an activity coefficient γ_i times the concentration C_i as shown in Equation 2.3

$$a_i = \gamma_i C_i \tag{2.3}$$

The activity of ions in solution can be calculated from a knowledge of the ionic strength of the solution and by using the Debye–Hückel equation. Ionic strength μ is a measure of the intensity of the electric field in a solution; it can be calculated from

$$\mu = 1/2 \sum C_i Z_i^2 \tag{2.4}$$

where Z_i is the valence of the ion. The sum of C_i for all ions present times their respective Z_i^2 value is used to calculate μ.

The activity coefficient can be approximated from the Debye–Hückel equation,

$$\log \gamma = -\frac{AZ_i^2 \sqrt{\mu}}{1 + Bd_i \sqrt{\mu}} \tag{2.5}$$

where A and B are constants for a given solvent and temperature. For water at 25°C, A is 0.508 and B is 0.328×10^8. d_i is the effective diameter of the hydrated ion. For most natural waters, the greater μ, the lower γ_i, and, therefore, the smaller a_i.

Cation-activity studies in soils have involved using ion-selective electrodes (Marshall, 1964). The glass electrode, for example, is an electrode for selectively measuring hydrogen ion activity (pH). These electrodes estimate ion activity in solution on each side of a semipermeable membrane by using the Nernst equation (Equation 2.6). The inner solution is of standardized composition, so the potential of the outer solution relative to the inner solution indicates the ion activity in the outer solution $(a_i)_o$

$$E = \frac{RT}{Z_i F} \ln \left[\frac{(a_i)_i}{(a_i)_o} \right] \tag{2.6}$$

where R is the gas constant, T is the absolute temperature, and F the Faraday constant.

Ion-selective electrodes have been used to characterize the bonding of cations to clay exchange sites as the clay is increasingly saturated with the cation studied. Investigations of this type have been reported by Marshall (1964). Figure 2.1 illustrates the increase in K activity as K replaces H on the cation-exchange sites of montmorillonite.

Cation-activity measurements of cations such as Ca and K are similar to measurements of H ion activity made with the glass electrode. The same reservations regarding the inability to obtain strict thermodynamic measurement of single ions and problems with junction potential for measurements in clay suspensions apply. Hence, activity measurements serve their greatest purpose in measuring relative values as the degree of saturation of cation-exchange sites changes and complementary ions change.

While ion activities in soil solution are important in determining effects such as solubility, ion concentration is important for determining the amount of ions moving by mass flow and diffusion to the root.

Fraction Active

The term *fraction active* has been used with clay suspensions to express the fractional activity (with respect to concentration) of an ion in the system. It is analogous to the activity coefficient in solutions.

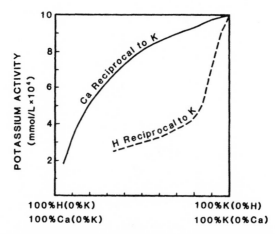

FIGURE   2.1   The effect of percent saturation of K and H- and Ca-montmorillonite on K activity of clay suspensions. Reproduced from McLean and Marshall (1948) by permission of Soil Science Society of America.

Cationic Bonding Energies

Cation-activity measurements can be used to calculate mean free-bonding energies for bonding cations by exchange sites on clays. The difference in cation activity between the cation in free solution and the cation in the clay suspension represents the reduction in activity due to bonding the cation by the clay. The change in free energy of the cation is the work done in moving the cation from its state in association with the clay to the condition of complete dissociation in solution. The change in free energy, ΔF, can be calculated according to Equation 2.7,

$$\Delta F = RT \ln C_i - RT \ln a_i$$
$$= RT \ln \left(\frac{C_i}{a_i} \right) \tag{2.7}$$

where C_i is the activity on complete dissociation in solution.

The effect on ΔF of the cation and its degree of saturation of the exchange sites has been measured for several clays (Marshall, 1964). Figure 2.2 represents the results of experiments with Putnam clay, a beidellite-type clay with a cation-exchange capacity of 70 cmol(p$^+$)/kg. Values are with H (and Al) as the complimentary cation. The bonding energy for divalent cations was about double that for monovalent cations. Increasing the amount on the clay initially increased the mean bonding energy and then decreased it as more exchange sites were occupied. The complementary ion also affected the results. Using a divalent complementary cation in place of a monovalent

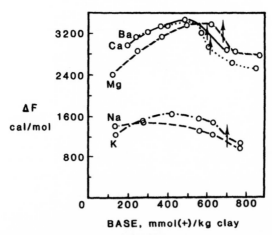

FIGURE 2.2 The effect of base saturation of Putnam clay (a beidellite) on bonding energies of several cations. Reprinted from Marshall (1964) by permission of John Wiley & Sons, Inc.

cation (i.e., using Ca in place of H) decreased the bonding energy of the monovalent cation.

Ease of Displacement of Exchangeable Cations

The ease of displacing exchangeable cations from soil can also be evaluated by partial displacement of exchangeable ions from addition of a salt of a cation not already present, such as NH_4^+. This partial displacement measures the relative strength of binding of the cation. Early research by Gieseking and Jenny (1936), Jenny (1932), and Schachtschabel (1940) indicated that relative amounts of displacement of monovalent cations followed their order of hydration; Li > Na > K > Rb > Cs, with the largest hydrated ion Li held the most weakly when measured on clays such as montmorillonite and kaolinite. The order for displacement of divalent cations was Mg > Ca > Sr > Ba. Because of their greater charge, divalent cations were held more tightly than monovalent cations.

The dissociation of cations from clays was investigated by Marshall (1964). Cation-sensitive membrane electrodes were used to measure activities of potassium and calcium in clay systems that were then used to calculate mean bonding energies of the cations to the clay. Divalent cations were held with two or more times the energy of monovalent cations. Sodium was usually held with less energy than potassium. However, the order of strength of bonding varied with the clay investigated and the degree of saturation of the exchange sites with the investigated cation. When clays were compared at equal saturation with a cation like potassium, the 2:1 clays held potassium more tightly than did the 1:1 clays. The latter, in turn, held the potassium more tightly than did pH-dependent sites, such as on organic matter.

Effect of Exchange Site

Juo and Barber (1970) investigated the relative bonding of Ca and Sr in soils by adding Ca-saturated soil samples to 0.05 mol/L $CaCl_2$ containing concentrations of $SrCl_2$ ranging from 10^{-4} to 10^{-3} mol/L. They measured the solution concentration of Ca and Sr after equilibration and found that Sr was adsorbed preferentially to Ca on predominantly permanent-charge exchange sites. Adsorption on the mineral sites was in accord with the hydrated size of the two ions. With organic exchange sites that are primarily pH-dependent, however, the field strength of the site was also important. In fact, a reversed sequence occurred, so that Ca was adsorbed more strongly than Sr. This adsorption was according to the relative *dehydrated* size of the cations. Hence, the relative adsorption strengths of cations on permanent-charge and pH-dependent exchange sites can vary for cations of the same valency.

McLean and Snyder (1969) investigated the effects of differing charge sites on bentonite and illite as they influence equilibrium levels of cations in solution (Figure 2.3). Bentonite had 11% of the exchange sites due to pH-dependent charge, and illite had 33%; the remainder were due to permanent

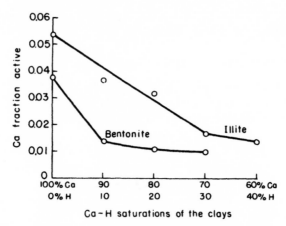

FIGURE   2.3   Calcium fractions active at various Ca–H saturations of bentonite (1.21% w/v) and illite clay (3.8% w/v) suspensions at equal levels of total exchangeable calcium. Reproduced from McLean and Snyder (1969) by permission of Soil Science Society of America.

charge. Ca-sensitive electrodes were used to determine the fraction of total exchangeable plus soluble Ca that was represented by measured Ca activity, with H as the counter ion. At Ca saturation below 70% on illite and below 90% on bentonite, the fraction of Ca active was below 0.02. However, increasing Ca saturation above these values caused a rapid increase in the fraction of Ca active. It appears that Ca is held mainly on permanent-charge sites when the clay is only partially saturated with Ca and dissociation from these sites is low. When all sites are filled with Ca, additional Ca is held on the pH-dependent charge sites and is not held as tightly, so that a higher proportion is dissociated. Hydrogen appears to be preferentially held on the pH-dependent charge sites over calcium. When the pH is below 5.2, aluminum will be present with hydrogen.

McLean and Bittencourt (1973) studied the effect of pH-dependent charge on potassium and calcium displacement from bentonite and illite clay suspensions. Potassium displacement was studied by saturating the clays with K to Ca [mol(p⁺)/L] ratios 0.05:0.95, 0.125:0.875, 0.25:0.75, 0.5:0.5, 0.75:0.25, 0.875:0.125, and 0.95:0.05. These clays were 100% saturated with potassium and calcium. A similar experiment with the same ratios was conducted where 20% of the bentonite sites and 40% of the illite sites were saturated with hydrogen. When illite was 100% base saturated, with 25% or less as potassium, the relative potassium displacement was much higher than when more than 25% of the clay was saturated with potassium. When illite was only 60% base saturated, degree of potassium saturation had little effect on potassium displacement by water. This indicates that hydrogen (or hydrogen and aluminum) was the primary cation on the pH-dependent sites and when there was enough hydrogen to saturate all pH-dependent sites, potassium was held entirely on the relatively uniform permanent-charge sites. Hence, degree of saturation had little effect on fraction displaced. The same type of relationship, though to a lesser degree, occurred with bentonite. There appeared to be some potassium going onto pH-dependent sites of the bentonite even when there was sufficient hydrogen to saturate them. Such research gives a general idea about the effect of type of exchange site on distribution of calcium, potassium, and hydrogen among exchange sites and how this, in turn, should affect the relative concentrations of these ions in solution. Additional information on the relative bonding of cations is given in chapters discussing individual cations.

Cationic Selectivity Coefficients

When two exchangeable cations of similar valence are present in a soil system, the cations will distribute themselves between exchange sites and solution according to the relative strength of bonding of the cation by the exchange site. This distribution can be characterized by a cation selectivity coefficient or distribution coefficient, calculated as shown in Equation 2.9 for the Rb and K system

$$\frac{[\text{Rb - soil}](\text{K}^+)}{[\text{K - soil}](\text{Rb}^+)} = k_{\text{Rb/K}} \tag{2.9}$$

where brackets indicate molar fractions of K and Rb on the exchange sites (mol/kg of soil) and the parentheses molar fractions in solutions (mol/L of soil solution). A value of $k_{\text{Rb/K}}$ greater than one indicates preferential adsorption of Rb over K by the exchange complex. Values of $k_{\text{Rb/K}}$ may vary from soil to soil because of the different nature of exchange sites and their effects on the relative adsorption of these two cations.

Comparisons can also be made of Rb-K selectivity where these cations are only a fraction of the total exchangeable cations present. Baligar and Barber (1978) measured the selectivity coefficient in four soils at two levels of potassium each and found that the selectivity coefficient decreased as the degree of potassium saturation of the soil increased (Figure 2.4). This indicated that exchange sites can vary in their relative strengths of bonding for K and Rb.

Khasawneh et al. (1968) investigated $k_{\text{Sr/Ca}}$ for samples of 63 Indiana soils. A molar ratio of Ca to Sr of 187:1 was used in the equilibrating solution. Values of $k_{\text{Sr/Ca}}$ varied from 0.61 to 1.51. Organic matter content was negatively correlated with $k_{\text{Sr/Ca}}$ ($r = 0.83$). The organic fraction of the soil adsorbed Ca preferentially to Sr, but the inorganic fraction in most instances adsorbed Sr preferentially.

Juo and Barber (1969) used Eisenman's exchange theory to explain the difference between $k_{\text{Sr/Ca}}$ on organic exchange sites and $k_{\text{Sr/Ca}}$ on inorganic exchange sites. Eisenman showed that cation-responsive characteristics of a

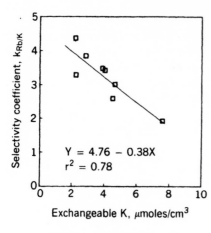

FIGURE 2.4 Effect of exchangeable potassium levels in the soil on the selectivity of adsorption for potassium and rubidium. Reproduced from Baligar and Barber (1978) by permission of Soil Science Society of America.

silicate glass is due to the substitution of Al^{3+} for Si^{4+} in the tetrahedral coordination. The resulting $(AlOSi)^-$ type exchange site has relatively low field strength. The selectivity rank order among the alkali-metal cations for the $(AlOSi)^-$ site was found to follow the lyotropic series. Exchange sites in bentonite and vermiculite arise mainly from the charge deficiency caused by isomorphous substitution in the octahedral and the tetrahedral sheets, respectively. Therefore, we may regard these clay minerals as the kind of exchanger that has a relatively low field strength at the exchange site.

The exchange sites on humic acid are predominantly pH-dependent carboxyl groups. They adsorb calcium preferentially to strontium, which is contrary to the finding for bentonite and vermiculite; Eisenman's theory can also be used to interpret this result. The carboxyl group exchange site is considered to have a small size and a high field strength (Reichenberg, 1966). Thus, under this circumstance, electrostatic interaction between the exchange site and the counterion becomes much more important than the hydration energetics of the counterions. Thus because of its smaller ionic size, the Ca^{2+} ion is adsorbed preferentially to Sr^{2+} ion by these exchange sites.

Cation-Exchange Equations

In soil systems at equilibrium, cation-exchange equations are helpful in predicting the distribution of ions between the adsorbed, or exchangeable, and solution phases of the soil as amounts of cations present are changed. In the simplest system with only two monovalent cations present, a mass-action equation can explain the results. When a soil saturated with potassium is placed in a solution of sodium chloride, the equilibrium shown by Equation 2.10 occurs:

$$\text{K-soil} + \text{NaCl} \rightleftharpoons \text{Na-soil} + \text{KCl} \qquad (2.10)$$

The exchange equation for this reaction is

$$\frac{[Na]}{[K]} \frac{(K)}{(Na)} = k_1 \qquad (2.11)$$

where brackets refer to ions on the exchange sites and parenthesis to activity of ions in solution. Values for k_1 will differ for different exchange materials, since the proportionate strength of adsorption of the two ions will vary with exchange site. Because several types of exchange sites usually occur in a soil, the value of k_1 may also vary with the ratio of potassium to sodium. Equation 2.11 is commonly used to describe exchange in homovalent systems.

The divalent-monovalent exchange system is more complex, but it more nearly represents the situation in the soil, where potassium, calcium, and magnesium are the dominant exchangeable cations. Equations have been developed by Kerr (1928), Vanselow (1932), Gapon (1933), and Krishnamoorthy et al. (1948); the first three are developed from mass-action equations, while the last is from statistical thermodynamics. The Gapon equation is widely used to describe the monovalent-divalent ion exchange; the Gapon equation for the potassium–calcium systems is

$$\frac{[K](Ca)^{1/2}}{[Ca](K)} = k_1 \qquad (2.12)$$

Verification of the Gapon equation has been attempted by several investigators, with variable results. Most have found that the value of k_1 decreased as K adsorption increased due to the addition of K. The decrease using the Gapon equation was usually less than for the other equations; hence, the Gapon equation was usually selected.

Cation-exchange equations are useful for determining the nature of the change in solution levels of divalent versus monovalent cations as solution anion concentrations change and ratios of divalent to monovalent cations in a specific soil change. Cation ratios have been used to evaluate the effect of sodium on plant growth.

The U.S. Soil Salinity Laboratory Riverside, California, developed an empirical equation to predict the effect of additions of $CaSO_4$ during sodic soil reclamation on the (Ca)/(Na) ratio in the soil. Their equation had the same form as the Gapon equation:

$$\frac{(Ca)^{1/2}[Na]}{(Na)[Ca]} = k_1 = 0.01475 \qquad (2.13)$$

The value for k_1 was determined from observations made on many western U.S. soils. Equation 2.13 is a useful relation for evaluating the status of soils high in sodium, and for predicting the amount of treatment necessary to alleviate a high sodium problem.

Complementary-Ion Effect

When an ion is adsorbed on a clay, the degree of dissociation of the ion into solution or its exchangeability with a second ion depends on the complementary ions associated with it on exchange sites. Jarusov (1937) was one of the first to show that the exchangeability of an ion was influenced by the degree of saturation and exchangeability of the accompanying ion on the exchange complex. Jenny and Ayres (1939) and Wiklander (1946) further investigated the complementary-ion effect. Starting with clay in solution, the procedure used was to add NH_4Cl in amounts equivalent to the total exchangeable cations on the clay. After equilibration and filtration, the solution was analyzed for the cations being studied. When a system initially had only potassium and hydrogen as exchangeable cations, replacement of hydrogen with calcium greatly increased the exchange of potassium from the clay. In this case, a more tightly held divalent cation displaced the monovalent potassium to weaker sites, where it is more readily exchanged.

When hydrogen in a Ca–H system was exchanged for potassium as the complementary ion, only a small and unpredictable effect on calcium exchangeability occurred. In this case, since one monovalent ion was substituted for another, neither one displaced calcium from the more strongly held sites, so there was little effect. The extent of the complementary-ion effect depends on the clay and the ions involved. A principal reason for the effect is probably due to differences in bonding strength between permanent-charge sites and pH-dependent charge sites. The divalent ions go to the permanent-charge sites, where they are tightly held. The monovalent ions, on the other hand, go to the pH-dependent sites that hold cations more loosely.

In addition to permanent and pH-dependent sites, soil particles have a wide range of types of cation-exchange sites, which bind cations with different strengths. Every soil will have a range of sites, with the relative number of each type varying with the soil. The strength of binding also varies with the cation of interest and the associated cations present. Because of this variation, it is difficult to predict behavior accurately for any one soil. However, general relationships can be used to show effects that occur when moisture or cation levels are changed, a useful concept in this regard is Donnan equilibrium.

Donnan Equilibrium

The effect on cation activities of salt addition to, or dilution of, a soil suspension can be evaluated by means of the Donnan equation. Donnan equilibrium expresses equilibrium activities of ions in solution in a system where a restriction is placed on the movement of anions in one part of the system. For the usual calculation, salts are assumed present in a system divided by a semipermeable membrane. All the cations can pass freely through the membrane, but an anion on one side cannot pass through the membrane. The system is illustrated in Figure 2.5, where R is the anion with restricted mobility. At equilibrium, the product of the activities of the anions and cations on side 1 must equal the product of the activities of the cations and anions on side 2. The nontransferable anion does not enter into the calculation. For a monovalent system, the equation is

$$C_1 \times A_1 = C_2 \times A_2 \qquad (2.14)$$

where C and A refer to cation and anion activities, respectively. Since there are additional nonmobile anions in side 1 to balance cations, the cation concentration in side 1 is higher than in side 2.

To relate this to a soil system, we use the exchange sites of the soil as the nondiffusable anion. The soil system differs, however, in that there is not a membrane between the solution and solid phase across which equilibrium can be calculated. Hence, there is not a clear-cut barrier to use in calculating the Donnan system, but cations are retained on the exchange sites as if a physical barrier *were* present.

Perhaps the most significant use of the Donnan equation is for predicting what occurs between monovalent and divalent ions. When monovalent and divalent ions are both present, a situation common to soils, we can predict the relative effect of dilution or concentration by salt addition on the levels of each type of ion in the soil solution. If we have two systems at equilibrium, both having only potassium and calcium and separated by a

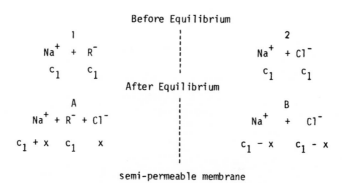

FIGURE   2.5   Diagram of a Donnan system.

semipermeable membrane, we can express a relation between the soil solutions as

$$[(K)/(Ca)^{1/2}]_A = [(K)/(Ca)^{1/2}]_B \tag{2.15}$$

A nondiffusable anion makes the difference between the levels in each system.

Equation 2.15 can also be used to predict what may happen when a soil is diluted with water. As the system is diluted, $K/(Ca)^{1/2}$ remains constant due to equilibration of the solution with exchangeable K and Ca if it follows Donnan equilibrium, but levels of potassium and calcium decrease. Calcium in the soil solution will decrease as the square of the potassium concentration decrease. Hence, dilution affects soil solution potassium concentrations much less than it does soil solution calcium concentrations. This is significant in evaluating potassium availability in soils since the proportion of K in solution determines K supply to the plant root.

Ratio Law

A more general description of the relation shown in Equation 2.15 as it applies to soil, termed the ratio law by Schofield (1947), states

> When cations in a solution are in equilibrium with a larger number of exchangeable ions, a change in the concentration of the solution will not disturb the equilibrium if the concentrations of all the monovalent ions are changed in one ratio, those of all the divalent ions in the square of that ratio and those of all the trivalent ions in the cube of that ratio.

Schofield verified the ratio law over a limited range of concentrations for K, Mg, Ca, Al, and H on Rothamstead soils. He tested the law by percolating salt solutions through columns of soil and measuring the cation concentrations in the percolate.

The Q/I Relation

Beckett (1964) investigated the activity ratio over a range of fractional potassium saturations of cation-exchange sites. He used this, in turn, to evaluate the potassium-absorbing characteristics of the soil. Each of series of 10-g samples of soil were shaken with 50 ml of 2 to 7×10^{-3} mol/L $CaCl_2$ to which graded amounts of potassium had been added. After equilibration, the samples were filtered and analyzed for potassium. Data were used to construct the graph in Figure 2.6 for the potassium–calcium relationships on a Lower Greensand soil. Characteristically, the relation was linear between change in exchangeable potassium level and activity ratio over most of the range. At low levels of potassium, however, the relation became curvilinear, which suggests that the last potassium held by the soil was held with

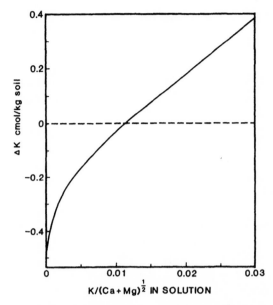

FIGURE   2.6   Q/I relation of potassium in a soil–solution system. Redrawn after Beckett (1964) by permission of Blackwell Scientific Publications Ltd.

increasing strength as the amount in solution decreased. McLean and Marshall (1948) and Barber and Marshall (1951) have observed the same relation at the lower end of such curves when using potassium sensitive electrodes.

The slope of the curve in Figure 2.6 was called the potential buffering capacity by Beckett. The point where the curve crossed the abscissa, the point of no change in exchangeable potassium, gives the AR_0 (activity ratio) for the untreated soil. The relation is called a Q/I relation, where Q refers to the quantity of exchangeable potassium and I, the intensity or concentration of potassium in the soil solution. The Q/I relation for soil potassium has been measured on diverse soils by many investigators, who have obtained curves similar to the one in Figure 2.6 except for variations in slope and intercept.

Exchangeable-Solution Relations

As will be discussed in more detail in later chapters, the relation of the ion concentration in solution to the total diffusible ion concentration in the soil, the buffer power, has a large effect on the bioavailability of the ion, since the quantity of the ion at the root surface depends on the initial concentration in the soil solution, C_{li} and the amount it is buffered by the ions in the soil, C_{si}, that equilibrate with those in solution. The relation between C_i and C_s and their change with additions of ions such as phosphorus or potassium vary with soil due largely to the effects that have been discussed previously in this chapter.

The effect of soil on the relation between exchangeable soil, C_{ex}, and soil solution potassium among samples from the 0- to 20-cm layer of 33 widely varying soils is shown in Figure 2.7 where four soils are used to give the range of values found. The curves were described by fitting the potassium data to the expression $C_{ex} = m(C_{li})^n$. Values for m give the slope of the linear relation, which varied from 7.3 to 235.1, while values of n, the curvilinearity, varied from 0.29 to 0.82. The relation between C_{ex} and potassium added was linear while that of C_{li} to potassium added was curvilinear (Kovar and Barber, 1990). The differences among soils reflect differences in strengths of bonding of potassium adsorption sites. The range of values obtained was used to give potassium application rates at the higher levels that would occur where all the applied potassium was concentrated in bands in the soil. The values obtained may differ from those for potential buffering power from the Q/I relation of Beckett (1964), which used the activity ratio of $K/(Ca + Mg)^{1/2}$, where Ca and Mg levels were constant to show the change in solution K with K additions. In Chapter 10 more detailed information on potassium will be given. The data in Figure 2.7 were given to show an example of wide differences among soils in the relative values of C_{li} and C_{ex} for potassium. Values for C_{si} are the sum of both C_{li} and C_{lx} when corrected for volumetric fraction of soil water.

Measurement of C_{li}, C_{si}, and b

This book is about the use of the mechanistic nutrient uptake model to predict nutrient uptake. The soil supply part of the model involves measuring soil

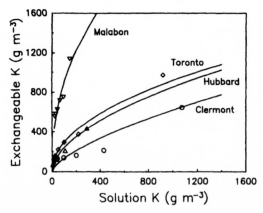

FIGURE   2.7   Relation between exchangeable potassium and solution potassium for four soils representing the range of values found in 33 soils. Curves fit with a Freundlich equation. Reproduced from Kovar and Barber (1988) by permission of Soil Science Society of America.

solution concentration, C_{li}, and labile nutrient concentration in the soil, C_{si}. The relation $\Delta C_s/\Delta C_l$ is the buffer power. Values of C_{li} and the buffer power of C_s for C_l are the most important since b also influences the size of D_e.

Determination of C_{li}

The initial concentration of an ion in soil solution may be obtained by displacing the soil solution and measuring C_{li} in the displaced solution. In my laboratory a procedure similar to that of Adams (1974) was used. A 400-g sample (oven dry basis) of moist soil that had been incubated at approximately −30 kPa water tension and 25°C for 3 weeks was placed in a 7.5-cm-diameter Plexiglas column. The soil surface was covered with filter paper. Deionized water was added at 4 to 8 mL/h until the soil reached field capacity, the soil was equilibrated for 24 h, then 40 ml (20 ml for sands) was added at 4 mL/h and the displaced soil solution filtered and analyzed for the respective ions. The ion concentration in the soil solution is C_{li}.

Determination of C_{si}

Cations.   Cations can be displaced from the soil by shaking for 10 min a sample of moist soil with 1 mol NH_4OAc/L, pH 7.0, with a soil:solution ratio of 1:10, separating the solution by filtration or centrifugation, and analyzing. The displaced cations can then be calculated per unit soil volume to obtain C_{si}. This value contains both exchangeable cations and the cations in solution.

Anions.   Anions such as phosphorus may be separated from the soil by using an anion-exchange resin procedure. In the procedure used in my laboratory a 1.0-g sample (oven-dry basis) of moist soil, 1.0 g of Dowex 2 × 8 Cl-saturated exchange resin > 0.425 mm particle size, and 100 mL of deionized water were placed in a 400-mL plastic bottle and shaken for 24 h at 25°C to desorb phosphorus from the soil. The resin was separated from the soil by washing through a 0.425-mm opening sieve. It was assumed any sand particles retained with the resin did not absorb appreciable phosphorus. The phosphorus was extracted from the resin by shaking with 50 mL of 1.0 mol/L NaCl (warmed to 60°C) for 6 h. The solution was separated from the resin and analyzed for phosphorus (Murphy and Riley, 1962) and C_{si} calculated.

Different procedures may be needed for other ions. The procedure should measure the amount of ion in the soil that equilibrates with the ion in solution in a time period related to the rate of uptake vs. root age.

Determination of b

The movement of ions from the soil to replace ions removed by plant roots from solution at the root surface is an important aspect of plant nutrition.

The buffer power of the soil for a particular ion is the concentration of labile ion per unit soil volume (including those in solution) divided by the concentration in solution per unit soil volume. Hence if you have the labile ion per kilogram of soil this concentration has to be multiplied by bulk density to get the labile concentration per unit volume. The concentration in displaced soil solution must be multiplied by the volumetric water, θ, to give solution ions per unit soil volume.

Kovar and Barber (1988) used these procedures in a study where 33 diverse soils were incubated for 3 weeks at −30 kPa water tension before measurement of C_{li} and C_{si}. There was a wide range of relations of C_{li} vs. C_{si} among the 33 soils. Figure 9.4 shows the relation of anion exchangeable phosphorus to solution phosphorus for four soils representing the range of values obtained. The relation between C_{si} for phosphorus and P added was linear and the slope ranged from 0.86 to 0.34; the values indicating the fraction of added P that was anion exchangeable. Hence when phosphorus is added to moist soils that are incubated moist, part of the added phosphorus is not extracted with the anion-exchange resin; the proportion not extracted varied with soil. More detail is given in Chapter 9. The relation between C_l and C_s could be described with a Freundlich adsorption equation

$$C_s = mC_l^n \tag{2.16}$$

where m and n are regression constants.

Studies by Holford and Mattingley (1976) and Syers et al. (1973) have shown the relation between soil phosphorus and C_{li} for phosphorus fit the Langmuir equation where all the added phosphorus was assumed to remain in equilibrium with phosphorus in solution. The Langmuir equation is

$$q = \frac{aBC_l}{1 + aC_l} \tag{2.17}$$

where q is the moles of a substance, phosphorus in this case, adsorbed per unit mass of solid, C_l is the concentration of phosphorus in solution, a is an affinity parameter related to bonding energy, and B is the adsorption maximum.

ANION ADSORPTION

Anions such as phosphate, sulfate, and borate are frequently adsorbed by or precipitated on soil surfaces. Because of the complex nature of soil surfaces, it is difficult to distinguish the exact mechanisms that occur. In addition, reactions are frequently so slow that equilibrium conditions are not reached. Hence, disappearance of added phosphate from a soil solution can be evaluated on the basis of surface adsorption reactions as well as solubility reactions. It is difficult to verify which type of reaction occurs or even whether both types are involved; thus, both mechanisms will be discussed here.

Surface Adsorption

When phosphate, borate, or sulfate anions are added to a soil or clay suspension, part of these anions disappear from solution over time. By diluting the system with water or adding anion-exchange resin, desorption or dissolution of the anion can be demonstrated. Adsorption is usually exponential with time, so that major adsorption occurs within 1 to 15 days, depending on temperature and soil.

When a series of increasing rates of phosphate are added to a soil suspension, the relation between phosphorus adsorbed and phosphorus remaining in 0.01 mol/L of $CaCl_2$ solution after 14 days equilibration at 25°C is of the type shown in Figure 2.8. The curvilinear relation is not evidence of a particular mechanism of removal from solution, such as adsorption or precipitation. The adsorption concept is frequently evaluated by testing the goodness of fit of the data to either the Langmuir or the Freundlich adsorption equations.

Specific Adsorption

On particles with pH-dependent charge, the overall charge on the particle is positive at pH values below that giving the zero point charge. Adsorption of anions on these particles due to the presence of positive charge is regarded as nonspecific, and such ions are readily exchangeable. However, there are anions such as phosphate, sulfate, and fluoride which are adsorbed in amounts much greater than attributable to nonspecific adsorption alone. Hence, they are considered to be adsorbed specifically. Hingston et al.

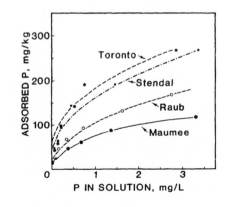

FIGURE 2.8 Relation between solution phosphorus and adsorbed phosphorus for four soils (Adepetu, 1975). The soils were Maumee sandy loam (Typic Haplaquolls), Raub silt loam (Aquic Argiudolls), Stendal silt loam (Aeric Fluvoquents), and Toronto (Udollic Ochroqualfs).

(1969) developed a theory to explain this adsorption; they studied adsorption of anions on goethite, FeOOH, an iron oxide commonly found in soil. The surface of goethite consists of a layer of Fe^{3+} ions octahedrally coordinated with OH^- ions and H_2O molecules. The authors found that specific adsorption varied with pH: At any specific pH, there was a point of maximum adsorption as the anion concentration in solution was increased. When the maxima were plotted against pH, different curves were obtained for each of the various anions investigated. Breaks in slopes of the curve corresponded with pK values for the acid forms of each anion. Hingston et al. developed the concept that the surface is able to donate or accept a proton. Donating a proton to the anion that is to be specifically adsorbed makes it more positive and causes adsorption. The anion then would be coordinated with Fe^{3+}. Displacement of specifically adsorbed anions by another anion first requires adsorption of the ion to increase the negative charge on the surface, which causes displacement of the initial specifically adsorbed anion. Anions that are not specifically adsorbed themselves do not, in turn, displace specifically adsorbed anions. This research with goethite needs investigation so that application of the theory to soils can be evaluated.

Precipitation

Slightly soluble compounds in solution dissociate as indicated in Equation 2.18.

$$C^+ + A^- \rightleftharpoons CA \qquad (2.18)$$

The solubility of this compound is characterized by the relation where K_i is the equilibrium constant.

$$\frac{(C^+)(A^-)}{(CA)} = K_i \qquad (2.19)$$

Since CA is a precipitate, its activity can be considered as unity (by traditional convention), and we have the relation

$$(C^+)(A^-) = K_{sp} \qquad (2.20)$$

where K_{sp} is the solubility product.

In a saturated system, when C^+ or A^- or both are added to the soil, they will precipitate until they eventually reach equilibrium, at which time $(C^+)(A^-)$ again is equal to K_{sp}. Values of K_{sp} for a large number of compounds have been determined and are reported in chemical handbooks, values for the equilibrium constants commonly used for soil systems are given by Lindsay (1979).

If we assume that P concentrations or activity in soil solution is solely due to solubility of slightly soluble compounds, we can have Ca, Fe, and Al compounds each controlling the phosphorus level in solution. The amount of P in solution will then depend on the levels of Ca, Al, and Fe in solution as well as on the solubility product constant of the mineral forms that are precipitating. In soil, the level of Fe and Al in solution will vary with soil pH, because the solubility of Fe and Al oxides is greatly affected by the pH of the system. Hence, the Fe and Al concentrations in a soil solution are usually calculated from the K_{sp} for $Fe(OH)_3$ and $Al(OH)_3$, respectively. The phosphorus concentration in solution is then calculated from the resulting Fe and Al levels and the solubility product constants for those Fe and Al phosphate compounds that are assumed to form. Calculations can also be made for several Ca–P compounds. The level of Ca in solution is frequently not controlled by a precipitate, except at high pH where $CaCO_3$ solubility may control soluble Ca levels. The partial pressure of CO_2 will in turn affect $CaCO_3$ solubility. Where $CaCO_3$ does not affect Ca solubility, a value of 1 mmol/L of Ca in solution is commonly assumed. The relation between pH and log $H_2PO_4^-$ or log HPO_4^{2-} in solution where various compounds control solubility is shown in Figure 2.9. Experimental values of P concentration in solution as related to pH can be plotted on such a graph with their correspondence to the solubility data used to indicate which compound appears to be controlling P concentrations in solution. The fact that P concentration in solution versus pH for any given soil falls along any particular line is evidence for, but not ironclad proof, that solubility of that particular compound is controlling the P concentration in solution in that soil.

The use of the solubility-product principle for estimating the concentration of specific nutrients in solution is discussed further in specific nutrient chapters.

SOIL SOLUTION

Plant roots absorb nutrients from the soil solution, so the equilibrium level of nutrients in the soil solution at the time absorption begins is a factor in determining the rate at which plant roots can absorb nutrients. Previous sections have discussed the role of such processes as solubility, adsorption and desorption, and exchangeability in influencing concentrations of ions in the soil solution; an additional factor is the salt level in the soil. Adding fertilizers containing soluble salt, such as the chloride in KCl, the nitrate in nitrogen fertilizers and sulfates, can affect the level of salt concentration. In Aridisols and Mollisols, where rainfall may not be sufficient to leach soluble salts from the soil profile, soluble salts can accumulate to much higher levels than in soils from high rainfall areas. The concentrations of ions in the displaced solution of the Ap layer from several types of soil are shown in Table 2.5. Values for the Aridisol are generally much higher in Ca, Mg, and

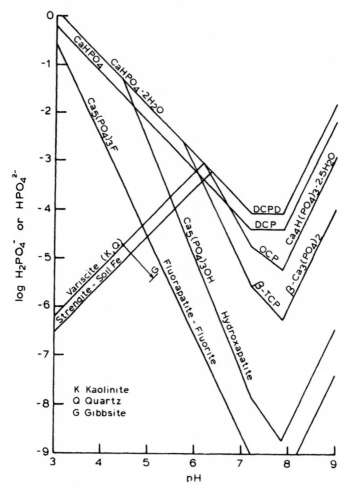

FIGURE 2.9 Solubility relation for Ca–P compounds in solid-solution systems. Reproduced from Lindsay (1979) with permission of John Wiley & Sons, Inc.

Na than values for soils that have been leached; where salts have accumulated due to evaporation, values can increase even more markedly.

Ion Activities in Solution

It is not known to date whether ion activity or concentration determines the uptake rate of an ion. While we might expect activity to be important for ion uptake where uptake is by active (energy-dependent) processes, it may not be as important where uptake is passive. It is difficult to determine the significance of activity for many ions, because uptake is usually from dilute solution where activity is nearly the same as concentration. In this

TABLE 2.5   Relative Concentration of Ions in Displaced Soil Solution from Three Types of Soil (μmol/L)

Ion	Alfisols and Mollisols[a]	Ultisol[b]	Aridisol[c] Normal	Aridisol[c] Saline
Ca	1500	1650	3,300	37,000
Mg	2500	500	1,940	34,000
K	150	220	700	400
Na	–	–	12,200	79,000
PO$_4$	1.6	1.0	–	–
SO$_4$	–	270	4,930	47,000

[a]Barber et al. (1962), mode of 134 values.
[b]Adams et al. (1980), Decatur silty clay loam.
[c]U.S. Dept. of Agric. Handbook 60.

TABLE 2.6   Calculated Ionic and Ion-Pair Concentrations and Ionic Activities of a Displaced Soil Solution and One-Fifth Hoagland Solution (mmol/L)

Ion or Ion pair	Soil solution[a] Conc.	Soil solution[a] Activity	¹/₅ Hoagland Conc.	¹/₅ Hoagland Activity
Ca^{2+}	15.22	5.42	2.35	1.54
Mg^{2+}	5.47	2.22	0.94	0.64
NH$_4^+$	68.96	48.85	0.003	0.003
K$^+$	9.08	6.54	1.09	0.98
SO$_4^{2-}$	10.99	3.29	0.79	0.51
Cl$^-$	95.00	68.37	1.02	0.91
NO$_3^-$	0	0	5.00	4.45
H$_2$PO$_4^-$	0.08	0.06	0.097	0.087
HPO$_4^{2-}$	0.027	0.008	0.0	0.0
CaSO$_4^0$	3.40		0.15	
MgSO$_4^0$	1.24		0.05	
NH$_4$SO$_4^-$	2.03		0.0	
KSO$_4^-$	0.22			

[a]Reproduced from Adams (1971) by permission of Soil Science Society of America.

case, uptake will be almost the same whether concentration or activity are controlling the uptake. When an ion becomes sufficiently concentrated, that activity is much *less* than concentration, the curve for rate of uptake versus concentration has usually reached a maximum. Thus, uptake will be similar whether calculated from activity or concentration in this case, as well. However, concentration of ions other than the one investigated can affect activity appreciably. Relations between concentration and activity have been calculated by Adams (1971). Both ionic strength and the formation of ion pairs can influence the effect of ion concentration on ion activity. (Some cations and anions form pairs in solution and are not ionized.) An example of the result of calculations made by Adams for a soil solution and a nutrient solution used for the culture of plants is shown in Table 2.6. The difference between activity and concentration was greatest for Ca^{2+}, Mg^{2+}, and SO_4^{2-}, which form the greatest concentrations of ion pairs in solution.

While data indicate that potassium activity was only 0.72 of concentration, in the case of this soil, salt concentration solution was unusually high. Calculations using 1/5 Hoagland solution, which has a salt concentration closer to that usually found in soils, gave an activity for potassium equal to 0.90 of the concentration.

Strongly Adsorbed Ions

When buffer power is evaluated from desorption curves, it is generally assumed that ions being removed from solution are replaced by ions formerly adsorbed by the soil. However, some of the adsorbed ions may be adsorbed so strongly that they are not readily desorbed. They therefore should not be used in calculating buffer power, since they will not influence ion flux in the soil appreciably and the value for buffer power will be less than if they were included. It is possible that a value for the buffer power of strongly adsorbed ions should be included where plants grow for long periods in the soil. The highly variable nature of adsorption and desorption usually makes measurement of b rather arbitrary, so we have to measure a value of C_s that involves equilibrating the soil for a length of time comparable to the average time that much of the ion will be diffusing to the root.

Slow Release of Ions from the Solid Phase

While buffer power is measured where there is rapid exchange between C_i and C_s there is also generally a slower rate of release that varies with ion and soil and helps maintain the C_s level in the soil. Such release occurs whenever levels of C_i and C_s have been reduced, whether or not plants are growing. Much research in this area has centered around rates of potassium release. Two types of release may occur; one is diffusion of ions out of positions between the plates in micalike minerals; the second type involves the slow dissolution of compounds within the soil. Each type of release occurs after concentration in solution has been reduced by plant uptake. Experimental values for release of these types are discussed in subsequent chapters on potassium and phosphorus.

Rates of release that are great enough to significantly contribute to the rate of supply from soil to the root depend on the specific nature of the soil minerals. Highest rates usually occur in soils where little weathering and leaching has occurred.

THE SOIL PROFILE

Up to this point, soil has been discussed as a uniform entity. Plants in a field are usually growing in soil that has developed over hundreds to thousands of years. The development of soil and its characterization by horizons is discussed in detail by Buol et al. (1980). Only a few statements are included here to emphasize the vast differences that can occur between soils and even horizons of the same soil. The soil profile is divided into *A, B,* and *C* horizons, which are characterized by different effects of soil-developing processes. The *A* horizon has an accumulation of organic matter as a result of plant growth; it has usually lost clay, iron, or aluminum by weathering and as a result is higher in quartz. The *B* horizon has an accumulation of clay, iron, or aluminum from the *A* horizon. Material in the *A* and *B* horizons are altered from the original material; the *C* horizon is the unaltered material from which the soil is presumed to have developed.

Chemical and physical properties of soil horizons frequently vary widely. The *A* horizon becomes an Ap horizon with tillage, fertilization, and cropping of the soil. The plant's root system may extend into all three horizons, a fact that must be considered when studying the nutrient flux to plant roots.

The *A* horizon usually is more fertile than *B* or *C* horizons, because of nutrient recycling by plants during soil development and fertilizer and lime additions. With plant growth, plant roots absorb nutrients from the *B* and *C* horizons and transport them to the shoot. The shoot then may die and release its nutrients into the *A* horizon. The organic matter content of the *A* horizon is due to organic matter accumulated from previous plant residues. The low organic matter content in the *B* horizon comes from movement of soluble fractions into the *B* horizon and decomposition of plant roots. The proportion of roots growing in each horizon will affect the total nutrient supply to the plant. Schenk and Barber (1980) found that less than 1% of the phosphorus and less than 10% of the potassium absorbed by corn plants were supplied by roots growing in the *B* and *C* horizons of the soil (an Aquic Argiudoll) that they studied, even though over half of the root system was in these horizons. Available phosphorus and potassium levels in the *B* and *C* horizons were much lower than in the *A* horizon. Nutrient supply to plants depends on both the proportions of roots growing in each soil horizon and relative levels of nutrients in these horizons.

Soil Structure

Soil structure refers to the physical arrangement of soil particles; it is greatly influenced by the relative proportion of sand, silt, clay, and organic matter present. Individual soil particles are cemented into soil aggregates. The nature of the soil aggregates determines the porosity of the soil through which the roots grow. Pores have to be larger than root tips if roots are to enter and expand. Soil porosity and its nature also regulate the flow of water and the diffusion of ions to the plant root. The effect of soil structure on root growth, as well as the influence of the root in changing soil structure near the root, will be discussed in Chapter 6. The chemistry of the soil is influenced by soil structure, due to its influence on air–water relations in the soil. Plant root growth and morphology is affected by soil structure, which in turn affects the ability of the plant to obtain nutrients.

REFERENCES

Adams, F. 1971. Ionic concentrations and activities in soil solutions. *Soil Sci. Soc. Am. Proc.* **35**:420–426.

Adams, F., C. Burmester, N. V. Hue, and F. L. Long. 1980. A comparison of column displacement and centrifuge methods for obtaining soil solutions. *Soil Sci. Soc. Am. J.* **44**:733–735.

Adams, F. 1974. Soil Solution. In Carson E.W. ed. The Plant Root and Its Environment. Univ. of Virginia Press, Charlottesville VA, 441–482.

Adepetu, J. A. 1975. Evaluation of the Kinetic Process Involved in Phosphorus Availability to Plant Root in Soil. Ph.D. diss. Purdue Univ.

Baligar, V. C., and S. A. Barber. 1978. Potassium and rubidium adsorption and diffusion in soil. *Soil Sci. Soc. Am. J.* **42**:251–254.

Barber, S. A., J. M. Walker, and E. H. Vasey. 1962. Principles of ion movement through the soil to the plant root. *Proc. Int. Soil Conf.*, New Zealand. 121–124.

Barber, S. A., and C. E. Marshall. 1951. Ionization of soils and soil colloids. II. Potassium-calcium relationships in montmorillonite group clays and in attapulgite. *Soil Sci.* **72**:373–385.

Beckett, P. H. T. 1964. Studies in soil potassium. II. The immediate Q/I relations of labile potassium in the soil. *J. Soil Sci.* **15**:9–23.

Bohn, H., B. L. McNeal, and G. A. O'Connor. 1979. *Soil Chemistry.* John Wiley, New York.

Buol, S. W., F. D. Hole, and R. J, McCracken. 1980. *Soil Genesis and Classification*, 2d ed. Iowa State University Press, Ames.

Byers, H. G., L. T. Alexander, and R. S. Holmes. 1935. The composition and constitution of the colloids of certain of the great groups of soils. *USDA Tech. Bull.* 484.

Cihacek, L. J. and J. M. Bremner. 1979. A simplified ethylene glycol monoethyl ether procedure for assessment of soil surface area. *Soil Sci. Soc. Am. J.* **43**:821–822.

Dixon, J. B. and S. B. Weed. 1989. *Minerals in Soil Environments.* Soil Science Society of America. Madison, WI.

Gapon, E. N. 1933. Theory of exchange adsorption in soils. *J. Gen. Chem. USSR* **3**:144–152.

Gieseking, J. E., and H. Jenny. 1936. Behavior of polyvalent cations in base exchange. *Soil Sci.* **42**:273–280.

Harter, R., and D. E. Baker. 1977. Applications and misapplications of the Langmuir equation to soil adsorption phenomena. *Soil Sci. Soc. Am. J.* **41**:1077–1080.

Hillel, D. 1980. *Fundamentals of Soil Physics.* Academic Press, New York.

Hingston, F. J., R. J. Atkinson, A. M. Posner, and J. P. Quirk. 1969. Specific adsorption of anions on goethite. *Ninth Int. Conf. Soil Sci. Trans.* **1**:669–678.

Holford, I. C. R. 1980. Adsorbed phosphate in soils and sediments. *Soil Sci. Soc. Am. J.* **44**:441–442.

Holford, I. C. R. and G. E. G. Mattingley. 1976. Phosphorus adsorption and plant availability of P. *Plant Soil* **44**:377–389.

Hodgson, J. F., W. L. Lindsay, and J. E. Trierweiler. 1966. Micronutrient cation complexing in soil solution. II. Complexing of zinc and copper in displaced solution from calcareous soils. *Soil Sci. Soc. Am. J.* **30**:723–726.

Jarusov, S. S. 1937. Mobility of exchangeable cations in the soil. *Soil Sci.* **43**:285–303.

Jenny, H. 1932. Studies on the mechanism of ionic exchange in colloidal aluminum silicates. *J. Phys. Chem.* **36**:2217–2221.

Jenny, H., and A. D. Ayres. 1939. The influence of the degree of saturation of soil colloids on the nutrient intake by roots. *Soil Sci.* **48**:443–459.

Juo, A. S. R., and S. A. Barber. 1969. An explanation for the variability in Sr-Ca exchange selectivity of soils, clays, and humic acid. *Soil Sci. Soc. Am. Proc.* **33**:360–363.

Khasawneh, F. E., A. S. R. Juo, and S. A. Barber. 1968. Soil properties influencing differential Ca to Sr adsorption. *Soil Sci. Soc. Am. Proc.* **32**:209–211.

Kovar, J. L., and S. A. Barber. 1988. Phosphorus supply characteristics of 33 soils as influenced by seven rates of P addition. *Soil Sci. Soc. Am. J.* **52**:160–165.

Kovar, J. L., and S. A. Barber. 1990. Potassium supply characteristics of 33 soils as influenced by seven rates of potassium addition. *Soil Sci. Soc. Am. J.* **54**:1356–1361.

Kerr, H. W. 1928. The identification and composition of the soil alumino-silicate active in base exchange and soil acidity. *Soil Sci.* **26**:385–398.

Krishnamoorthy, C., L. E. Davis, and R. Overstreet. 1948. Ionic exchange equations derived from statistical thermodynamics. *Science* **108**:439–440.

Lindsay, W. L. 1979. *Chemical Equilibria in Soils.* John Wiley, New York.

Marbut, C. F. 1935. Soils of the United States. In O. E. Baker, Ed. *Atlas of American Agriculture.* U. S. Dept. of Agric., Washington.

Marshall, C. E. 1964. *The Physical Chemistry and Mineralogy of Soils.* John Wiley, New York.

McLean, E. O., and V. C. Bittencourt. 1973. Complementary ion effects on potassium, sodium, and calcium displacement from bi-ionic bentonite and illite systems as affected by pH-dependent charges. *Soil Sci. Soc. Am. Proc.* **37**:375–379.

McLean, E. O., and C. E. Marshall. 1948. Reciprocal effects of calcium and potassium as shown by their cationic activities in montmorillonite. *Soil Sci. Soc. Am. Proc.* **13**:179–182.

McLean, E. O., and G. H. Snyder. 1969. Interaction of pH-dependent and permanent charges of clays: I. Use of specific ion electrodes for measuring Ca and Rb activities in bentonite and illite suspensions. *Soil Sci. Soc. Am. Proc.* **33**:388–392.

Murphy, J., and J. P. Riley. 1962. A modified single solution method for the determination of phosphate in natural waters. *Anal. Chim. Acta* **27**:31–36.

Pratt, P. F. 1961. Effect on pH on the cation-exchange capacity of surface soils. *Soil Sci. Soc. Am. Proc.* **25**:96–98.

Reichenberg, D. 1966. Ion exchange selectivity. In J. A. Marinsky, Ed. *Ion Exchange.* Marcel Decker, New York. Pp. 227–274.

Schachtschabel, P. 1940. Untersuchungen uber die sorption der tonminerallen und organischen boden-kollide. *Kolloid Beihefte* **51**:199.

Schenk, M. K., and S. A. Barber. 1980. Potassium and phosphorus uptake by corn genotypes grown in the field as influenced by root characteristics. *Plant Soil* **54**:65–76.

Schnitzer, M. 1978. Humic substances: chemistry and reactions. In M. Schnitzer and S. U. Khan, Eds. *Soil Organic Matter.* Elsevier, New York, Pp. 14–17.

Schofield, R. K. 1947. A ratio law governing the equilibrium of cations in solution. *Proc. Eleventh Int. Cong. Pure Appl. Chem., London* **3**:257–261.

Stevenson, F. J. 1982. *Nitrogen in Agricultural Soils.* American Society of Agronomy, Madison, WI.

Syers, J. K., M. G. Browman, G. W. Smillie, and R. B. Corey. 1973. Phosphate sorption by soils evaluated by the Langmuir adsorption equation. *Soil Sci. Soc. Am. Proc.* **37**:358–363.

Vanselow, A. P. 1932. Equilibria of the base exchange reactions of bentonites, permutites, soil colloids, and zeolites. *Soil Sci.* **33**:95–113.

U.S. Salinity Laboratory, 1954. Diagnosis and improvement of saline and alkaline soils. *U.S.D.A. Handbook 60.*

Walker, J. M., and S. A. Barber. 1960. The availability of chelated Mn to millet and its equilibria with other forms of Mn in the soil. *Soil Sci. Soc. Am. Proc.* **24**:485–488.

Wiklander, L. 1946. Studies on ionic exchange with special reference to the conditions in soils. *Ann. R. Agr. Coll., Sweden* **14**:1–171.

CHAPTER **3**

Nutrient Absorption by Plant Roots

Almost all nutrients absorbed by plants are in an inorganic form; organic forms of nutrients in the soil solution are usually mineralized to the inorganic form before absorption. Inorganic ions are primarily supplied to the plant by absorption through the root system. Carbon dioxide, oxygen, sulfur dioxide, and small amounts of other ions may be absorbed from the air through the leaves. Foliar sprays may be used to apply small amounts of nutrients that can be adsorbed through the leaves without injuring leaf cells. However, absorption through the root system is the primary supply mechanism. The kinetics of nutrient absorption into the root cells, and their translocation to other parts of the plant where they can be used, is one of the important processes in plant growth. The nature of absorption kinetics at the root of a particular plant species will affect the sufficiency of nutrient supply to the plant. In this chapter, I discuss root morphology as it is related to mechanisms of nutrient absorption, kinetics of nutrient absorption, and factors influencing the kinetics of nutrient absorption for specific plant species.

Plant roots are also important for anchoring the plant, synthesis of growth regulators, water absorption, and metabolizing photosynthate for root growth. However, discussion in this chapter is confined to those properties affecting nutrient influx.

ROOT MORPHOLOGY

When viewed as a crosssection, most roots can be subdivided into epidermis, cortex, endodermis, and stele (Figure 3.1). The stele contains the xylem, through which ions are translocated to the shoot, and the phloem, through which photosynthate is supplied from the shoot to the root. Ions move through the epidermis, cortex, endodermis, and stele and empty into the xylem, where they are transported to other parts of the plant.

Each cell consists of a cell wall that is permeable to water and ions. The outer epidermal cell wall may have a cuticle layer. Inside the cell wall is a plasma membrane, which is the barrier to passive movement of ions into and out of the cell. Inside the plasma membrane are the cell contents of cytoplasm, vacuole, and nucleus. The membrane between the cytoplasm and vacuole is the tonoplast. The vacuole serves for storage of ions. The cytoplasm of adjoining cells is connected by plasmadesma, so ions can flow in the cytoplasm from one cell to the cytoplasm of an adjoining cell. The number of rows of cortical cells between the epidermis and endodermis varies with plant species but is usually in the range of 5 to 10. Endodermal cells have a suberized band around them, known as the Casperian strip. The Casperian strip is a barrier to ion movement from the cell wall space of the cortex into

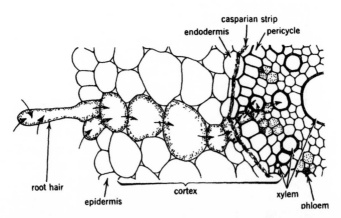

FIGURE   3.1   Root cross section of cells and tissues involved in ion absorption. Arrows indicate direction of ion movement through a selected series of cells. Reprinted from Esau (1965) by permission of John Wiley & Sons, Inc., copyright 1965.

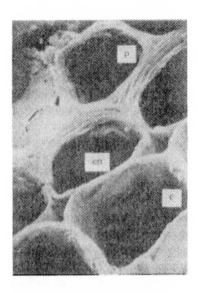

FIGURE 3.2 Scanning-electron micrograph of the endodermis of a maize root showing plasmadesma; en, endodermis; c, cortical cell; p, pericycle. Bar length, 10 μm. Reprinted with permission from Drew (1979). Copyright: Academic Press Inc. (London) Ltd.

the cell wall space of the stele. However, the cytoplasm of the cortical cells is connected through intercellular plasmadesma with the cytoplasm of the cells of the stele. Hence, ions can move freely through the cytoplasm from the epidermal cells and through the cortex and endodermal cells to the cells of the stele. The cytoplasm and plasmadesma that connect the cytoplasm from cell to cell are called the symplasmic pathway for ion movement from the root surface to the stele. A scanning electron microscope's view of the endodermis region of a maize root is shown in Figure 3.2. The pits in the cell walls show the plasmadesma connections between cells. The longitudinal morphology of the root system varies with plant species and environment for root growth.

Types of Root Systems

Type of root system, in terms of both range of root radii and spacial distribution of roots in the soil, varies greatly with environment and species. Kutschera (1960) has presented the distribution for many species diagrammatically. Because soil environment and climate so markedly affect the root system, only examples of differences between soils can be given. Angiospermous plants have two subclasses, monocotyledons and dicotyledons, according to their number of cotyledons. The initial root system developing from the seed differs for these plant subclasses.

Monocotyledons

Monocotyledonous plants include grasses and cereals. The root system of one of these, a corn (*Zea mays* L.) seedling, has one radical and several seminal roots, usually about four, that emerge from the seed and supply the seedling with water and nutrients during early growth but die after adventitious roots develop from transitionary nodes on the coleoptile. The adventitious roots form the main root system of the mature plant. These roots commonly branch into several orders, although, on some plants such as onion (*Allium cepa*) the roots do not branch. Adventitious roots may sometimes develop from nodes above ground; brace roots of corn plants are an example. Grasses form a crown at the root surface from which new tillers as well as new roots develop.

Dicotyledons

The seed of the dicotyledon sends out a single tap root that subsequently has laterals. The hypocotyl shoves the cotyledons above the soil's surface, and an epicotyl grows from the cotyledons to eventually form the shoot. The root system of the dicotyledon differs from the root system of the monocotyledon. In many plants, the tap root becomes very large and may be more than half the root weight, even though the tap root has only a small part of the root surface. Hence, root:shoot ratios on a weight basis may have little significance for evaluating nutrient uptake; root surface area is more important in determining nutrient uptake rate.

Monocotyledons usually have a smaller range in root diameters than dicotyledons. Depth of rooting is frequently greater for dicotyledons, although this is not always the case, Barber (1978) found that corn, a monocotyledon, had more of its root system in the subsoil than did soybeans (*Glycine max* L.), a dicotyledon. However, if alfalfa (*Medicago sativa* L.) is compared with bromegrass (*Bromus inermus* L.), the dicotyledon has a much greater proportion of roots in the subsoil. The tap root of alfalfa extends deep into the soil unless physical or chemical conditions in the soil restrict its penetration. Grasses, such as bromegrass, usually have a large part of the root system in the top 15 cm of soil.

Root Branching

Root branching varies with species. Branch roots arise in the pericycle at the periphery of the vascular cylinder and push through the endodermis, cortex, and epidermis. Later, the xylem, phloem, endodermis, and cortex of the branch roots become continuous with those of the main root. Stimuli for root branching is not known, but when the main root meristem is removed by cutting off the root tip, most plant species initiate root branches. Root branching is also initiated when the root tip encounters resistant soil layers, so that root extension is inhibited.

Root Hairs

Root hairs, which arise as papillae on epidermal cells in the root zone behind the zone of active cell division, are extensions of the epidermal cells. The cytoplasm concentrates in the root hair, and the nucleus migrates toward the tip. Root hairs may persist for days or weeks, so they are present along the root and not only in the zone behind the root tip. Root hairs are more numerous where there are pore spaces into which they can grow, and they often grow profusely in humid air. Roots growing into soils low in available phosphate have more and longer root hairs than roots growing into high-phosphate soils (Powell, 1974). Cormack (1962) has discussed mechanisms for root-hair development. Cormack believes that root hairs develop as a result of retardation of vertical elongation of epidermal cells. Retardation increases internal pressure, causing root hairs to develop from weaker portions of unequally hardened cell walls. The hardening process is believed to be due to incorporation of calcium into the outer pectic wall layer. There is a wide variation between species in the number and length of root hairs. A detailed discussion of the significance of root hairs in nutrient uptake is given in Chapter 7.

Amount and Distribution of Roots

Plants growing without competition have large root lengths and root surface areas per plant (Dittmer, 1937). Measurement of amount of roots per unit of either soil surface or soil volume, where plants are grown at densities that maximize shoot growth or grain yield per hectare, are more realistic for evaluating root growth under natural conditions. The amount of root length below each unit of soil area is termed L_A and has units of cm/cm^2. The amount per unit volume of soil at any location in the soil is given by L_V and has units of cm/cm^3.

Removal of plant roots from the soil and their measurement has been simplified over methods used in early work, where blocks of soil were excavated and roots were removed by carefully washing away the soil. Various measurement procedures may be used (Böhm, 1979). A common procedure is to wash the roots carefully from 5- to 15-cm diameter cores of soil taken as subsamples of the root system. The core can be subdivided according to either depth or soil-profile changes. Roots can be washed from the soil using mechanical procedures, where the soil is dispersed and washed through a screen, or roots can be separated from the soil by a combination of flotation and sieving (Smucker et al., 1982). Once the roots are separated from the soil, their length can be determined by line-intersect procedures, as evaluated by Tennant (1975). In this procedure, roots are spread over grid squares and the number of intersections of roots with horizontal and vertical grid lines counted. When a grid dimension is 1 cm, the number of intersects times 11/14 gives the root length in cm. Mean root radius can be calculated by assuming that fresh roots have a density of 1 Mg/m^3. Then root radius can be calculated from root volume and length using the relation $r_0 = (\text{root volume}/L\pi)^{1/2}$.

Root density varies with species. Root lengths for several plant species reported by different researchers are given in Table 3.1. Soybeans had the lowest root density, while annual monocotyledons, corn, and small grains had similar medium values. Perennial grasses had much larger root densities, with a large part in the upper 15 cm of soil.

The variation of L_A with plant age for corn, given in Figure 3.3 (Mengel and Barber, 1974b), shows that the maximum length of corn roots occurred when the corn reached 50% tassel. Root length increased exponentially with time during the first three weeks of seedling growth and then increased linearly until the plant changed from vegetative to reproductive growth at 80 days. Net root length was constant for two weeks during tassel and silk development then decreased rapidly. At maturity, root length was only one-third the maximum root length (Mengel and Barber, 1974a). McGonigle and Miller (1993) found the same general relation between corn root length and plant age in Ontario, Canada as Mengel and Barber (1974b). Barber (1978) investigated the relation between plant age and root growth for soybeans. Amsoy-71, the cultivar of soybeans used, was an indeterminate cultivar that did not have a specific stage where growth changed from vegetative to reproductive. As a consequence, the change in soybean root length with plant age did not show so sharp a maximum peak as had been found with corn. The medium L_A value for corn was approximately double that for soybeans.

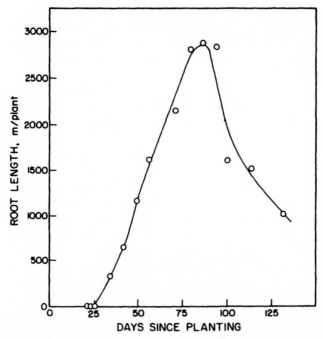

FIGURE   3.3   Relationship between L_A and plant age for field grown corn. Plants started to tassel at 75 days. Reproduced from Mengel and Barber (1974b) by permission of American Society of Agronomy.

TABLE 3.1 Root Length per Unit Soil Surface Area L_A and Unit Volume L_V for Several Plant Species Growing in the Field

Species	Plant Age (days)	L_V (0–15 cm; cm/cm^3)	L_A (cm/cm^2)	Source
Zea mays L. (corn)	79	4.1	145	Mengel and Barber (1974*a*)
Glycine max L. (soybean)	68	3.5	170	Schenk and Barber (1980)
Festuca arundinacea L. (Fescue)	85–92	2.0	80	Barber (unpublished)
Triticum aestivum L. (wheat)	400+	50	930	Johnson (1981)
Avena sativa L. (oat)	94	3.3	113	Welbank et al. (1974)
Hordeum vulgare L. (barley)	94	3.4	113	Welbank et al. (1974)
Phalaris arundinacea L. (Reed canary grass)	94	4.2	126	Welbank et al. (1974)
	400+	100	2500	Barber and Cushman (1981)
Oryza sativa L. (rice)	50	5	—	Slaton et al. (1990)
Gossypium hirsutum L. (cotton)	128	(0–10) 0.6 (10–20) 1.7	—	Brouder and Cassman (1990)

55

Although grown in different years, these crops were grown on the same plot area, so interacting effects of soil type on root growth should have been minimal.

The distribution of corn roots with soil depth as reported by Mengel and Barber (1974a) is shown in Figure 3.4. Consolidated soil below 75 cm restricted deeper root growth. More than half of the corn roots were in the top 15 cm of soil profile at plant age of 30 days; this fraction dropped to 30% by plant maturity. Root growth with depth was different for soybeans: In this case 31% of the roots were in the top 15 cm of soil at 60 days, and this value increased to 59% when soybean plants were 117 days old. With tap-rooted soybeans, branch roots grew out from the tap root into surface soil layers and then proliferated in the surface soil rather than the subsoil. For corn, branch roots continued to extend downward into the soil, so a progressively larger portion of the roots was below 15 cm as plant growth progressed. Roots of Reed canary grass (Johnson, 1981) were concentrated near the soil surface, and measurements showed that many of the roots were in the 0- to 2-cm layer and the amount decreased exponentially with soil depth. Root density in the 0- to 5-cm layer was greater than 25 cm/cm^3, but only 2.5 cm/cm^3 at the 15- to 35-cm depth. In addition to differences due to species, root distri-

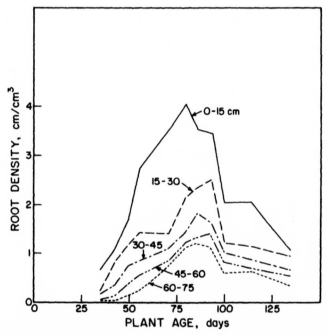

FIGURE   3.4   Relation between corn root density at selected soil depths and plant age. Reproduced from Mengel and Barber (1974a) by permission of American Society of Agronomy.

bution is affected by differences between surface soil and subsoil chemical and physical properties. Some subsoils may be so compact that roots cannot penetrate them (Taylor, 1974), or they may be so acid and high in exchangeable aluminum that roots cannot grow because of chemical factors (Pearson, 1974). Relative moisture levels in the surface soil and subsoil may also affect root distribution in the soil profile.

Kuchenbuch and Barber (1981) found maize root distribution with soil depth for 9 years on a rotation-fertility experiment varied with year. Weekly measurements made on five 15-cm depth layers of soil showed a significant correlation ($r = 0.79$) between the accumulated precipitation for 3 weeks prior to silking and the root density in the 0- to 15-cm soil layer at silking (Figure 3.5). This indicates that more than 60% of the year to year variation in root density in the 0- to 15-cm layer could be attributed to precipitation during the 9- to 12-week growth period.

Addition of phosphorus has been found to stimulate root growth in the phosphate-fertilized soil as compared to root growth in the unfertilized soil. The degree of root growth stimulation depends on the level of available phosphorus, C_{si}, in the fertilized soil as compared to that in the unfertilized zone. Zhang and Barber (1992) found the relation $y = 1.20 + 2.74 \log x$ ($r^2 = 0.97$) where y is the ratio of the root density in the fertilized zone to that in the unfertilized zone and x is the ratio of the C_{si} level in the fertilized soil

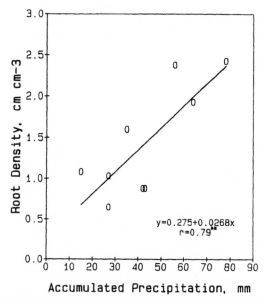

FIGURE   3.5   Relation between root density in the 0- to 15-cm soil layer and accumulated precipitation for 3 weeks prior to silking. Reproduced from Kuchenbuch and Barber (1988) by permission of Martinus Nijhoff Publishers B.V.

to that in the unfertilized soil (see Figure 21.4). Hence the greater the difference in C_{si} levels between the phosphate-fertilized and unfertilized soil, the greater the difference in root densities. This explains the large stimulation in root growth with placed phosphate in a situation where most of the roots are growing in solution that has no available phosphate. In addition, nitrogen distribution affects corn root distribution; however, since nitrate moves readily in the soil, root distribution is generally not influenced so much by initial nitrogen placement in the soil. Potassium variability has been found to have no effect on corn root distribution (Claassen and Barber, 1977).

Total amount of roots can also be influenced by uniformly changing the fertility level of the soil. In general, when nutrient levels are not so low that they reduce yield by more than about 20%, adding nitrogen or phosphorus will reduce the total yield of roots even though shoot growth is increased. In samples taken from fertility experiments at the Purdue University Agronomy Farm, Barber (unpublished data) found four-year average root lengths L_A in corn plots not receiving phosphorus to be 100 cm/cm², with an average grain yield of 8150 kg/ha. For plots receiving 50 kg P/ha per year, the average L_A value was 90 cm/cm², and the average grain had increased to 8720 kg/ha. Plots not receiving potassium had root lengths similar to those receiving potassium. In a different experiment, corn plots receiving 0 kg N/ha had an L_A value of 308 cm/cm² and a plant weight of 14,170 kg/ha; plots receiving 200 kg N/ha had an L_A value of 272 cm/cm² and a plant weight of 16,200 kg/ha (Edwards et al., 1974). Adding nitrogen to the total root system consistently resulted in less root growth. When a plant is deficient in nitrogen or phosphorus, it apparently diverts relatively more photosynthate to the roots and thus obtains greater root length, which, in turn, aids the plant in obtaining more nitrogen or phosphorus.

I wish to emphasize that these effects occur only when the corn without added nitrogen (or phosphorus) had enough available soil nitrogen (or phosphorus) that yield would be 80% of that when fertilized.

Soils where root growth is difficult because of poor physical conditions tend to produce plants with thicker, more irregularly shaped roots. Peterson and Barber (1981) grew soybean roots in sand continuously leached with a nutrient solution and compared them with roots grown in nutrient solution. The roots of 18-day-old soybean plants grown in sand had an average diameter of 0.49 mm, whereas roots grown in solution had an average diameter of 0.34 mm. The total length of roots was not significantly different for the two treatments, 125 versus 134 m/pot, so plants grown in sand provided more photosynthate to the roots than plants grown in stirred nutrient solution. The increase in diameter occurred primarily in the cortex; the stele had the same diameter in both treatments (Figure 6.7). There were the same number of cortical cells in a cross section of roots from both treatments, so the expanded cortex was due to larger diameter but shorter cells. This does not occur for all species nor in all soil situations.

When soil bulk density is larger than in the case just reported, root growth can be greatly restricted. Roots become larger in diameter, and total plant growth is reduced. Taylor and Ratliff (1969) showed a close relation between soil resistance measurements made with a penetrometer and the distance to which roots of cotton (*Gossypium hirsutum* L.) and peanuts (*Arachis hypogea* L.) penetrated the soil. Goss and Reid (1981) used a flexible container to apply different pressures to glass ballotini into which barley roots grew. They found root length to be progressively reduced as pressure was increased (Figure 6.6); even the first increment of applied pressure reduced root length. Russell (1977) reported that shorter roots, grown with increased pressure on the roots, had volumes of cortical cells similar to those of longer roots grown with less pressure. The cells were shorter in length but larger in cross section, giving a similar cell volume. This result is similar to that obtained for soybeans by Peterson and Barber (1981).

Root/Shoot Ratios

Green plants have leaves to absorb carbon dioxide and assimilate solar energy and roots to supply water and nutrients. Stems provide internal transportation and support for the leaves so that they may compete for light. There is a functional equilibrium between different plant parts that varies with species and cultivars. The nutrient supply to the shoot depends on the morphology and physiology of the roots in relation to shoot size. The shoot provides assimilate for root growth and nutrient uptake and a sink for absorbed nutrients.

As a seed germinates, root growth has priority, since the seed supplies this organ with internal energy and nutrients. Growth of the shoot must follow closely, however, so that it can provide a supply of photosynthate once the energy supply from the seed has been expended. While initial growth favors the root, subsequent growth favors the shoot. The relative growth of each plant part has been the subject of three hypotheses (Troughton, 1977) that portray the plant as composed of sources and sinks. The hypotheses are (1) the competitive hypothesis, stating that root growth is largely limited by lack of photosynthate from the shoot and shoot growth is limited by supply of nutrients from the roots; (2) the excess-carbohydrate hypothesis, which postulates that root growth depends on excess carbohydrate that cannot be used by the shoot; (3) the size-of-sink hypothesis, which postulates that root growth depends on the size of the sink for using carbohydrates. None of these hypotheses completely accounts for relative growth rates of shoots and roots that occur in nature, but each is useful for explaining relative root growth rates that are observed. Reduction in root growth when shoot growth changes from vegetative to reproductive (Mengel and Barber, 1974a) could be explained by the competitive hypothesis. Development of the grain competes with the roots for photosynthate. Increased root growth when nitrogen becomes deficient may be an example of the excess-carbohydrate hypothesis: Since there is not enough nitrogen to react with all of the photosynthate in the shoot, the excess goes to the root.

When roots are trimmed, photosynthate is diverted predominantly to the roots, and root growth is faster than shoot growth until the pretrimming shoot/root ratio is attained. This may indicate that reduced nutrient uptake causes surplus photosynthate to be available for use by the roots. There is also the hypothesis that photosynthate goes primarily to the nearest sink; in the case of grain filling, this sink may be physically closer than the root sink. Actual control of growth in the plant is probably much more complex than suggested by any of these simple hypotheses. There is little information concerning controls on root growth or determinants of whether roots will be long with a small radius or short with a large radius.

Klepper and Rickman (1990) reviewed the research on modeling root development. Their discussion is primarily for wheat and cotton, two crops for which considerable research has been done. Roots have a series of vertical axes and their branches. Their development depends on environmental soil conditions and on shoot growth. The detail that can be included in a model depends on the research data available.

ION-ABSORPTION KINETICS

Ion absorption is one of the principal functions of plant roots, and the kinetics of ion absorption affects nutrient supply and plant growth. A brief outline of current ion-absorption concepts is given here; for more details, refer to Marschner (1986). Many studies on ion-absorption kinetics have been conducted using excised, low-salt roots. Commonly, the roots of barley grown for 8 days in 0.01 mol/L $CaSO_4$ solution are excised, and nutrient absorption by roots for 10 to 20 min is used to measure ion-influx kinetics. Short time periods must be used, since the limited absorption capacity of the root and the lack of photosynthate supply may result in reduced uptake if the time for uptake were extended. Using radioactive tracers facilitates measuring of rate of uptake over short time intervals. Using excised roots removes the influence of translocation from the uptake process. Uptake studies have also been made with intact plants, for both short time and periods as long as several days. While excised-root studies are useful for studying uptake mechanisms, intact-plant studies are needed when studying uptake by plants growing in the plant–soil systems. Three types of ion-influx kinetics have been recognized: (1) passive ion movement of nutrients into the plant, which is independent of respiration energy; (2) passive ion uptake along an electrochemical gradient dependent on respiration energy; and (3) active ion uptake against an electrochemical gradient and requiring respiration energy. The effect of respiration energy on uptake is studied by measuring the effect on uptake of inhibiting respiration either with low temperature, for example, 1°C, or after adding respiration inhibitors to the system.

The membrane that provides the barrier to active uptake is the plasma membrane on the inside of the cell wall. Ions are actively moved across the plasma membrane into the cytoplasm. Plasmadesma connect the cytoplasm of adjoining cells so that ions can move across the cells of the cortex through the endodermis into the stele via the cytoplasm of each cell and the interconnecting plasmadesma (Figure 3.1).

Two pathways for ion movement from solution to xylem, the symplasmic and the apoplasmic, are proposed. In the apoplasmic pathway, ions move through the free space in the cell walls of the cortex and are thus free to move to the endodermis. The endodermal cells contain hydrophobic bands of suberin deposited in the radial walls (Casparian strip), which restrict movement from the free space of the cortical cells to the free space of the stele (Clarkson and Robards, 1975). Water and ions must pass into the symplasm to cross the endodermis; they may then move back into the free space of the stele, or they may continue via the symplasmic pathway to the xylem. In the symplasmic pathway, ions move across the plasma membrane of the epidermal and cortical cells. Once in the cytoplasm, water and ions can move to the stele and be secreted into the xylem for transport to the shoot. Calcium has been shown to move by the apoplasmic pathway, whereas phosphorus and potassium move by the symplasmic pathway.

The free space, or apoplasm, of the root represents 10 to 15% of the root volume; it occurs in the cell walls and intercellular spaces. The cell walls possess cation-exchange properties, probably due to carboxyl groups on the pectic cell wall matrix (Lauchli, 1976). The portion of the solution in the apoplasm where the solution's ion concentration is affected by exchangeable cations is called Donnan free space; the portion unaffected by exchangeable cations is termed the water-free space. Donnan effects on ion distribution were discussed in Chapter 2.

The absorption of ions for uptake through the symplasmic pathway occurs at the plasma membrane of the epidermal and cortical cells. Ions must first move through the cell wall before reaching the plasma membrane. Cell walls of young corn roots are about 1-μm thick. Ions diffuse through the apoplasm to the plasma membrane before they are actively moved across the plasma membrane; ions may diffuse through the apoplasm to all of the cortical cells. The inner barrier is then at the endodermis. Since the free space allows cortial cells to take up nutrients directly from the external solution, the total plasma membrane surface for nutrient absorption is presumably the combined surface of the plasma membranes of the epidermal and cortical cells. The plasma membrane surface area is not readily measureable, nor is the ion influx per unit of plasma membrane readily measured except where studies are made with large single cells. Transport through the cytoplasmic pathway by cytoplasmic streaming and diffusion from cell to cell through the plasmadesma to the stele, are not believed to be rate limiting for ion uptake, because many plasmadesma occur on the cell walls. Uptake rate has primarily been measured in terms of root surface area, root weight, or root length. However, an increase in absorption rate per m² of root surface as root radius increases indicates that the additional area of plasma membrane in larger roots may be important for ion uptake.

Carrier-Mediated Active Uptake Mechanisms

Active or carrier-mediated uptake is greatly reduced by adding respiration inhibitors or reducing temperature to reduce respiration, which suggests that energy is involved. Active uptake is also against a concentration gradient, so energy is needed. Uptake is also selective: Ions are not simply absorbed according to their ratios in solution. Uptake rate does generally increase as solution ion concentration increases, everything else being constant, as illustrated in Figure 3.6. Maximum rate, I_{max}, is approached as concentration increases. The shape of the curve can be described by Michaelis–Menten kinetics, which were developed to describe enzyme reaction rates. The Michaelis–Menten equation is

$$V = \frac{V_{max}C_i}{K_m + C_i} \tag{3.1}$$

where V_{max} is the maximum velocity, and K_m is the concentration where the reaction velocity is $\frac{1}{2} V_{max}$. Epstein (1966) found that Equation 3.1 fit the relation of net K uptake versus concentration over the solution K concentra-

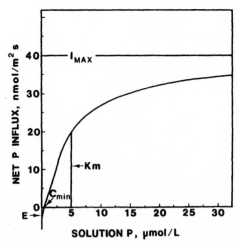

FIGURE   3.6   Phosphorus influx by 18-day-old corn roots as related to solution phosphorus concentration. Illustration of the use of E, C_{min}, K_m, and I_{max}, for describing the absorption isotherm for ion uptake by plant roots (Barber, unpublished).

tion range 0 to 1 mmol/L. Above 1 mmol/L, new values of K_m and V_{max} were obtained, which described uptake over the concentration range 1 to 10 mmol of K/L. This was believed due to separate mechanisms of uptake.

Energy-dependant uptake may be cation movement along an electrochemical gradient maintained by electrogenic proton (H^+) pumps. These pumps are mainly membrane-bound ATPases that extrude H^+ from the cytoplasm into the bathing solution. They cause the cytoplasm to have a negative electrical potential of the order of -100 meV with reference to the outside solution. Selectivity of cation now into the cytoplasm is theorized to be due to differences in cation bonding to the cell walls due to electrical strength of the cation-binding sites (Eisenman, 1962). The proton pump moves H^+ into the outside solution in relation to the movement of cations into the cytoplasm so that the negative electrical potential is maintained. This system represents indirect use of respiration energy to move cations from the external solution into the cytoplasm.

Since the interior of the cell is negative, respiration energy is necessary to move the negative anion against the electrochemical gradient and this requires use of respiration energy carriers, believed to be organic molecules; either ATP or molecules connected to ATP are located in the plasma membrane. The carrier combines with the ion on the outside, transports the ion through the membrane, then releases the ion into the cytoplasm. The process releases OH^- into the exterior solution. Respiration energy is required to operate the carrier. A detailed review of the process involved is given by Hodges (1975) and by Marschner (1986). In addition, the movement of a cation and anion together may be another mechanism for uptake and some cations, particularly K^+, may move into the cytoplasm with carriers.

The balance between cation and anion uptake will influence the pH of the external solution since the difference between the total molar amounts of cations and anions in solution will determine whether more H^+ than OH^- is released or vice versa.

The relation between absorption rates and ion concentration in the external solution over a wide concentration range suggests that there are several different carrier sites on the root that depend on ion concentration. Epstein (1966) suggested low- and high-concentration kinetics for potassium uptake. Nissen (1973) has looked in detail at isotherms and described them using a series of changing values for V_{max} and K_m as concentration in solution increases. Hodges (1973) suggested that the experimental results could also be explained by assuming a single carrier whose ion-binding sites interact as solution concentration increases, which would give different values of V_{max} and K_m. Sabater (1982) proposed a molecular mechanism for unitary multiphasic uptake, in which the carrier has two binding sites for the substrate. The first site would bind $n - 1$ molecules, with one additional molecule bound at the second binding site. Only the molecule bound at the second site would be transported by the carrier. Active transport mechanisms are prevented from operating at high substrate concentrations, thus preventing waste of energy by the cells.

Many ions, such as P, K, Cl, and S, appear to be absorbed by the symplasmic pathway. Calcium absorption, at least by some plants, is not affected by temperature; its uptake is believed to be by the apoplasmic pathway, and, hence, uptake is passive. Research on magnesium has given mixed results; Ferguson and Clarkson (1975) suggest that uptake is by the apoplasmic pathway, but Leggett and Gilbert (1969) found evidence for energy-dependent uptake of magnesium.

A nonstirred water film occurs at the surface of roots even when grown in stirred solution. Ions must diffuse through this water layer before they can move into the apoplast. The water film varies with conditions but is generally about 10-μm or more thick. The water in the film will move into the root at a rate governed by the rate of water uptake. The water film may have more significance in interpreting uptake from solution culture than it does for uptake from soil, because stirred solution does not exist in soil. Diffusion of ions through the water film into the root will only be affected by this water if the water density is different from that of normal water.

Ion Uptake by Intact Plants

The main emphasis of this book is ion uptake by intact plants growing in soil; therefore, the research results reported are primarily from intact-root uptake experiments. Excised-root experiments are most useful in studying uptake mechanisms, since uptake and translocation can be separated by this technique.

Ion-absorption isotherms using intact plants can be obtained in one of three ways. (1) Plants can be grown in flowing culture solutions maintained at a series of concentration levels, with influx calculated from plant uptake over several days; Asher et al. (1965) pioneered this technique. (2) Short-term (a few hours or less) uptake of radioactively labeled nutrients can be measured for a series of solutions varying in concentration level. This period of uptake is short enough so that solution concentration does not change appreciably. (3) Solution nutrient concentration change with time that occurs when plants reduce the nutrient concentration in solution can be used to calculate rate of net influx vs. nutrient concentration in solution (Olsen, 1950; Claassen and Barber, 1974; Nielsen and Barber, 1978). The relation can be used to calculate I_{max}, K_m, and C_{min} and then used to develop a curve such as that in Figure 3.6. The most common soil-solution potassium concentrations vary from 25 to 500 μmol/L. The concentration range for the first phase of uptake is 0 to 1 mmol/L. Because much of the potassium reaching the root surface may be from diffusion, the potassium concentration at the root is often reduced further to only 10 to 20% of initial potassium concentration of the soil solution, since a concentration gradient extends from the actively absorbing root.

The most common soil-solution phosphorus concentrations are between 1 and 20 μmol/L. This is in the range of the lowest concentration phase,

0 to 50 μmol/L, of the parameters describing phosphorus uptake by the Michaelis–Menten equation.

Influx Characteristics of Plant Roots

The absorption isotherm has been described in terms of I_{max}, K_m, and E, the rate of efflux from the root, by Claassen and Barber (1976). A typical absorption isotherm constructed from data obtained from a solution-depletion experiment (Claassen and Barber, 1974) is shown in Figure 3.6. Since the uptake rate was zero at a concentration above zero, the line was extended to the ordinate and the negative uptake at zero potassium concentration, termed the efflux from the root. Furthermore, since we are only using the Michaelis–Menten equation to describe the absorption isotherm and do not actually know if enzyme actions are responsible, the term V_{max} was changed to I_{max}. This is a more descriptive term for the maximum influx by the root. Hence, for the remainder of this book, the term I_{max} will be used in place of V_{max} wherever this enzyme kinetic equation is used to describe the maximum influx of nutrients per m^2 of root surface. Nielsen and Barber (1978) made a further modification by using the concentration in solution where the net influx reaches zero, rather than E as the third point describing the lower end of the absorption curve. This value was termed C_{min}. Hence, net ion influx In can be described by either Equation 3.2 or 3.3:

$$In = \frac{I_{max} C_l}{K_m + C_l} - E \qquad (3.2)$$

$$In = \frac{I_{max}(C_l - C_{min})}{K_m + C_l - C_{min}} \qquad (3.3)$$

Table 3.2 gives values for I_{max}, K_m, and C_{min}, for several nutrients and plant species at specific plant ages and temperatures; values are averaged for the complete root system. Values may vary with cultivar, root age, plant age, temperature, nutrient status of the plant, root morphology, and possibly other factors as described in the following sections.

Plant Demand and Nutrient Influx

Plants grown with their roots in stirred nutrient culture have a uniform concentration of nutrients around all roots. Plants grown in soil, on the other hand, generally have a variable supply of nutrients to various parts of the root system. The surface soil horizon is usually higher in fertility than subsurface horizons. Adding fertilizers and manures to surface soil may make fertility variable even within that horizon. Since evaluation of the kinetics of

TABLE 3.2 Values of I_{max}, K_m, and C_{min} for Uptake of Several Nutrients by Several Plant Species Using Intact Plants

Plant species	Plant age (days)	Nutrient	I_{max} (μmol/cm²·s)	K_m (μmol/L)	C_{min} (μmol/L)	Source
Zea mays L. (corn)	18–22	NO₃	10	10	4	Edwards and Barber (1976a)
	14–28	P	4	3	0.2	Jungk and Barber (1975)
	18	K	40	16	1	
Glycine max L. (soybean)	18–24	P	0.8	2	0.1	Edwards and Barber (1976b)
Triticum vulgare L. (wheat)	20–38	P	1.4	6	—	Anghinoni et al. (1981)
	20–40	K	7.0	7	—	Barber (unpublished data)
Festuca arundinacea L. (fescue)	200+	P	0.01	5	1	Barber (unpublished data)
		K	0.1	10	4	
Phalaris arundinacea L. (reed canary grass)		NH₄	2.0	50	4	Barber and Cushman (1981)
	100+	P	0.03	4	1	Barber (unpublished data)
	100+	K	0.6	10	5	Barber (unpublished data)
Robinia pseudoacacia L. (Black locust)	200+	K	1.7	1.8	0.7	Gillespie and Pope (1990)
Pinus elliotii Engelm (Slash Pine)	365	K	3.6	29	1	Van Rees et al. (1990)
Oryza sativa L. (rice)		NH₄	475	62	3	Teo et al. (1992)
Pinus taeda L. (Loblolly pine)	365	P	0.27	16	0.6	Kelly et al. (1992)

the root system in solution culture is under uniform conditions, it is also important to know the kinetics of influx under variable conditions. If only the demand for nutrients by the shoot determined influx, total nutrient uptake would be similar under variable conditions to that obtained under uniform conditions; this is not what has been found experimentally.

Phosphorus influx as related to the proportion of 12-day-old corn plant roots supplied with phosphorus was studied by Jungk and Barber (1975). A split-root experimental design was used for part of the treatments, with some roots grown in solution containing phosphorus and some in solution without phosphorus. The remaining treatments had portions of the root system removed by trimming the previous day so that similar shoots had varying amounts of root. The trimmed roots were placed in the phosphorus-containing solution. Uptake of ^{32}P over a 4-h period was measured; results are shown in Figure 3.7.

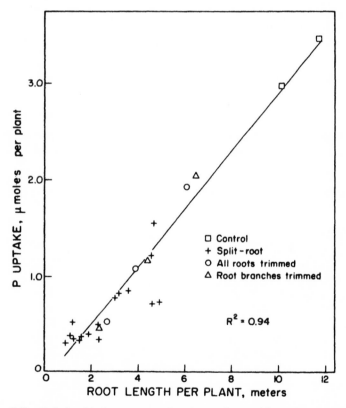

FIGURE   3.7   Relationship between root length per plant and P uptake over a 4-h period for 13-day-old corn plants having root length varied by split-root and root-trimming treatments. Reproduced from Jungk and Barber (1974) by permission of American Society of Agronomy.

Uptake was according to the amount of roots exposed to phosphorus. Increased shoot demand per unit of root exposed, where only a small portion of the roots were exposed to phosphorus, did not cause an increase in phosphorus influx. While shoot demand was responsible for creating a sink for phosphorus uptake by the roots, the uptake mechanism of the root had a finite maximal rate of influx. This was not surpassed by increased shoot demand per unit of exposed root.

When uptake per plant is reduced because only a small number of roots are supplied with nutrients, plant nutrient concentration decreases when compared with the situation where all roots are supplied. The fewer roots supplied, the greater the decrease in nutrient concentration in the plant. Claassen and Barber (1977) studied the effect of reducing potassium concentration on the maximal rate of potassium influx for 18-day-old corn plants. The results are shown in Figure 3.8. Shoot composition was varied by a variety of methods ranging from differences in solution potassium concentration during growth, removing potassium from the nutrient solution for varying periods of the 18-day growth period, and trimming roots to reduce potassium uptake. After these treatments, I_{max} for potassium was measured on each set of plants, and a correlation between potassium concentration in the

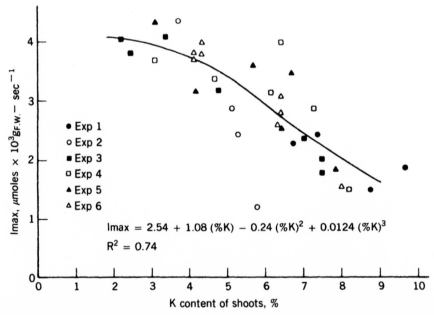

FIGURE   3.8   Relation between K concentration in the corn shoot and I_{max} for K uptake by corn roots. Reproduced from Claassen and Barber (1977) by permission of American Society of Agronomy.

shoot and the I_{max} value for potassium influx was noted. Influx increased almost threefold after potassium concentration in the shoot had been reduced from 9 to 2%. Although I_{max} increases when the plant lacks potassium, this adaptive procedure by the plant only aids in preventing the deficiency from becoming more severe where plants were grown in solution.

Edwards and Barber (1976a) investigated the effect of root trimming on nitrate uptake by corn plants; two experiments were conducted. In one, roots were split between N-containing and N-deficient nutrient solutions and plants were grown from 5 to 18 days. Then the nitrogen uptake rate was measured. In the second, roots were trimmed 2 days before uptake rate was measured. Results of these experiments are shown in Table 3.3. For the trimmed plants, root length as a proportion of root length for the untrimmed plants was used to express the proportion of roots exposed to nitrogen and hence the plant demand for nitrogen per unit of root surface area.

Uptake per plant decreased more when a smaller number of roots was absorbing nitrogen in the trimmed-root experiment than in the split-root experiment. Nitrogen concentrations in shoots and roots were reduced in the split-root experiment, whereas there was no time for appreciable reduction in nitrogen concentration in trimmed-root plants. Apparently when only part of the root system was exposed to nitrogen from the beginning of growth, reduction of nitrogen concentration in the plant stimulated the rate of nitrogen uptake. When shoot demand per unit of root was increased by root trimming

TABLE 3.3   Effect of Plant Demand for Nitrogen per Unit of Root Surface on Nitrogen Uptake Rate by Corn Plants Grown in Solution

Mean proportion of roots in N	N-uptake rate (pmol/m · s)	N-uptake/plant nmol/s	N concentration Shoot (%)	Root (%)
Trimmed roots				
1.00[a]	619a	17.3	3.74a[b]	2.58a
0.87	638a	15.5	3.74a	2.58a
0.80	697a	15.5	3.61ab	2.61a
0.40	728a	8.2	3.58b	2.48a
Split roots				
1.00	335a	18.2	3.77a	2.12a
0.68	317a	15.5	2.47b	2.13a
0.45	552b	16.2	2.38b	2.12a
0.28	1021c	11.8	2.63b	2.42a

Source: Reproduced from Edwards and Barber (1976a) by permission of American Society of Agronomy.
[a]Root length of trimmed plants as a proportion of root length of untrimmed plants.
[b]Values within columns followed by the same letter are not significantly different at the 5% level.

and the plant's nitrogen concentration was not affected, the trimmed roots did not increase their uptake rate appreciably in the two day period between trimming and uptake measurements. These results are in agreement with those of Jungk (1974) for phosphorus uptake by tomato roots. When roots are growing in soil, differences in root exudates (H+, etc.) also affect observed rates of soil nutrient supply to the root surface and hence nutrient uptake.

Data reported for nitrogen, phosphorus, and potassium uptake by corn roots indicated that a large proportion of the root system must absorb nutrients in order to supply the nutrient demand of the shoot.

Plant Age and Nutrient Influx

Previous data for corn or soybean plants represented only a single age period of 12- to 18-day-old plants. Since these plants were grown in a growth chamber, they were much larger than they would be after a similar period of growth at lower temperatures in the field. Jungk and Barber (1975) investigated phosphorus-influx characteristics of corn plants having both untrimmed and trimmed roots at stages of growth ranging from 12 days old to two weeks past tasseling. The corn was grown in an aerated nutrient solution. Influx kinetics were measured by the procedure of Claassen and Barber (1974). The uptake rate of phosphorus per cm of root in relationship to plant age and root trimming is shown in Figure 3.9. There was no significant difference between trimmed and untrimmed roots, uptake reached a maximum at 25 days, then declined rapidly with age thereafter. The decrease with age is similar to observations for field-grown corn by Mengel and Barber (1974b), where nutrient uptake rates were calculated from differences between nutrient contents and root lengths at each sequential harvest during crop growth in the field. Data are given in Table 3.4. Mean uptake rates for all nutrients decreased rapidly with increased plant age. The trimmed root experiments shown in Figure 3.9 indicate that the reduced rate with age is not just a reduction in plant demand per unit of root; it also reflects a reduction in the rate at which roots can absorb nutrients. It appears that a young root on an older plant does not absorb nutrients at the same rate as a root of the same age on a young plant. For modeling purposes, it will be assumed that all roots on a plant of a given age absorb at the same rate regardless of root age. In addition, uptake kinetics can be adjusted according to known effects of plant age.

Root Age and Nutrient Influx

Root age increases along the length of the root, with the root tip having the youngest cells. Measurement of rates of uptake along the root actually represents measurement of uptake with root age. Two procedures have been used to measure the rates of uptake and translocation by different segments of the root. Bowen and Rovira (1969) measured absorption of radioactive isotopes along the root using a radio-chromatogram scanner. Using a 15-min nutrient-absorption time, followed by a five-minute washing to remove

TABLE 3.4 Average Nutrient-Uptake Rate as Influenced by Corn Plant Age for Corn Grown in the Field

Plant age (days)	Calculated nutrient uptake									
	N	P	K	Ca	Mg	B	Cu	Mn	Zn	Fe
			(μmol/m per day)					(μmol/m per day $\times 10^2$)		
20	226.9	11.3	52.9	14.4	13.8	98.1	27.0	89.0	109.8	571
30	32.4	0.9	12.4	5.2	1.61	9.00	1.85	11.09	5.78	64.8
40	18.5	0.86	8.00	0.56	0.90	5.29	1.22	7.35	3.47	46.1
50	11.2	0.66	4.75	0.37	0.78	2.55	1.10	4.61	1.97	21.4
60	5.7	0.37	1.63	0.20	0.56	0.64	0.73	2.07	0.76	13.0
70	1.2	0.17	0.15	0.047	0.28	-0.20	0.41	0.06	-0.04	11.5
80	0.46	0.08	0.06	0.060	0.19	-0.53	0.32	-0.85	-0.24	-29.3
90	2.0	0.10	0.37	0063	0.17	-0.52	0.52	-0.39	0.50	-1.7
100	4.2	0.23	0.16	0.075	0.29	-0.41	1.03	0.51	2.19	6.7

Source: Reproduced from Mengel and Barber (1974*b*) by permission of the American Society of Agronomy.

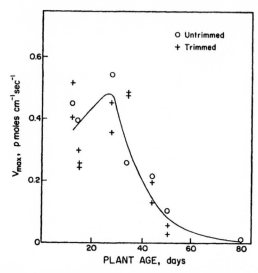

FIGURE   3.9   Effect of plant age and root trimming on maximum P uptake by corn roots grown in solution culture. Reprinted from Jungk and Barber (1975) by permission of Martinus Nijhoff Publishers B.V.

nonabsorbed ions, the authors found that little ion translocation into the shoot had occurred. The radioactivity along the root was a measure of the rate of absorption at each point for each nutrient studied.

Russell and Sanderson (1967) sealed off separate segments of the root in order to study uptake rate. Segments of root approximately 3 mm long were isolated from the remainder of the root system by sealing a root across the diameter of a polyethylene tube. Nutrient solution was then pumped through the tube. The nutrient solution inside the tube was the same as that outside except that the nutrient being studied was labeled with a radioactive isotope. Uptake by the segment was measured over a 24-h period. An example of the results obtained using this procedure are shown in Table 3.5 for absorption and translocation of labeled phosphate by segments of intact corn seedling roots (Ferguson and Clarkson, 1975). Uptake was not related to root age, since older roots continued to absorb phosphate. Hence, the assumption that all corn roots are absorbing phosphate at approximately the same rate is reasonable. Clarkson and Sanderson (1971) found that both phosphate and potassium uptake and translocation occurred for barley at all locations along the root. Ferguson and Clarkson (1975) observed that calcium uptake was restricted when the endodermis became suberized, although phosphate uptake continued. The difference between calcium and phosphate uptake is believed due to the fact that calcium uptake follows the apoplasmic pathway. Hence, its uptake is restricted when the endodermis becomes suberized. Phosphate uptake, on the other hand, follows the symplasmic pathway and is

TABLE 3.5 Absorption and Translocation of Labeled Phosphate by Segments of Intact Corn Seedling Roots

Distance from root tip (cm)	Amount translocated	Amount in treated segment (nmol H_2PO_4/mm^3 · ± SE)	Total uptake
1	0.75 ± 0.06	1.86 ± 0.17	2.61
4	0.94 ± 0.06	2.53 ± 0.35	3.47
8	0.54 ± 0.12	2.42 ± 0.15	2.96
12	0.64 ± 0.09	2.98 ± 0.57	3.62
20	0.53 ± 0.14	4.17 ± 0.42	4.70
28	0.57 ± 0.05	2.76 ± 0.58	3.33
30 (base)	0.13 ± 0.05	6.38 ± 0.76	6.51

Source: From Ferguson and Clarkson (1975) with permission of the New Phytologist Trust.

not restricted by endodermal suberization. Using the radio-chromatogram scanner, Rovira and Bowen (1970), observed that wheat (*Triticum vulgare* L.) roots absorbed phosphate at similar rates along the length of the root. Hence, influx was not appreciably affected by root age.

Root Exchange Capacity and Ion Uptake

Haynes (1980) has made an extensive review of research on the relation of root ion-exchange properties to ion accumulation by plants. Roots possess an appreciable cation-exchange property and even a small anion-exchange property. The measured cation-exchange properties are located within the apoplasm. Since these exchange sites are in equilibrium with ion levels in solution in the apoplasm, it is theorized that these sites may have an effect on the movement of ions through the apoplasm to the plasma membrane where active absorption occurs. Cell walls (Lauchli, 1976) are believed to consist of a framework of cellulose strands embedded in other materials such as hemicelluloses, pectins, and proteins. The origin of the cation-exchange properties is believed to be mainly due to the presence of carboxyl groups. Root's cation-exchange properties are theorized to vary with density of charge.

Some investigators (Nye and Tinker, 1977; Epstein, 1972; Bange, 1973) believe that the cation-exchange capacity of the root has no relation to ion uptake. Others (see review by Haynes, 1980) have reported that the relative absorption of calcium and magnesium increased with an increase in root cation-exchange capacity. These effects were explained on the basis of Donnan theory. Absorption of potassium by species with low cation-exchange capacity roots was greater at low levels of soil potassium in the soil than by species with high exchange capacity roots. This indicates low cation-exchange capacity roots absorb potassium more readily. The difficulty in analyzing the observed results lies in determining whether the cation-exchange properties themselves affect ion uptake or whether some other property of the root is responsible and simply correlates with root cation-exchange capacity. Comparisons are usually among species where other properties could also be important.

Some ions such as ammonium, potassium, and phosphate are absorbed through the symplasmic pathway. Calcium and possibly magnesium are absorbed through the apoplasmic pathway. The effect of exchangeable ions held on surfaces of the apoplasmic pathway on the movement of ions through the apoplasm to the endodermis, or through the apoplasm to the plasma membrane, determines how different levels of exchange capacity affect nutrient uptake. Divalent and trivalent cations dissociate much less from exchange sites than monovalent cations. With higher valency ions on the exchange sites, there is a larger solution volume in the apoplasmic pathway through which ions can diffuse to the plasma membrane or the endodermis; this may allow an increase in monovalent-ion uptake rate. It may also be the reason for increased uptake in the presence of calcium and other polyvalent cations (Hayries, 1980). Heavy metals have been found to accumulate in the apoplasm, where they would tend to reduce the charge on the cell wall. This would theoretically aid in increasing the relative absorption of monovalent nutrients. Whether there is a causal relation between the cation-exchange capacity of plant roots and the relative absorption of monovalent versus divalent cations remains unresolved.

Temperature Effects on Ion Absorption

Temperature response has been used as a method of differentiating between passive and active uptake or apoplasmic and symplasmic pathways for absorption of ions. Active absorption through the symplasmic pathway requires expediting respiration energy to transfer ions across the plasma membrane. Hence, when absorption is measured at 2°C, only passive uptake occurs since temperature is too low for respiration. However, increases in uptake rate with increased temperature can be due to either passive *or* active uptake processes.

When the effect of temperature on ion uptake by intact plants is considered, the indirect effect of temperature on many plant processes may be reflected in the rate of ion uptake. External factors that may influence the effect of temperature on ion influx include solution aeration, solution pH, and solute concentration. Plant factors that may interact with temperature are plant's nutrient status, growth rate, translocation rate, transpiration rate, and root respiration rate. Since temperature may affect each of these plant processes, the specific reason for the effect of temperature on ion uptake cannot be readily predicted. The interaction of temperature with these factors may also vary with plant species.

Ion uptake involves physical, chemical and biological reactions. The effect of temperature on individual reactions can be determined from measurements of the effect of temperature on the distribution of kinetic energy among molecules of the system at equilibrium (Nobel, 1974). For most reactions the molecule must have sufficient kinetic energy for the reaction to occur. As temperature is increased, a larger proportion of the molecules have energies above this energy barrier. The distribution of kinetic energy of the molecules is expressed by the Boltzmann energy equation

$$\frac{N(E)}{N_{\text{Total}}} = e^{-E/k_B T} \tag{3.4}$$

where $N(E)$ is the number of molecules with a kinetic energy above E, N_{Total} is the total number of molecules, k_B is the Boltzmann constant, and T is the absolute temperature.

The number of molecules with kinetic energy greater than U_B, an energy sufficient for activating a particular reaction, is proportional to $\sqrt{T} \cdot e^{-U_B/k_B T}$.

The rate at a temperature 10°C higher would be $\sqrt{T+10} \cdot e^{-U_B/[k_B(T+10)]}$. The ratio of these two values describes the increase in reaction rate with temperature. The change in rate with a 10°C rise in temperature is called Q_{10}. Nobel (1974) estimates a Q_{10} value of approximately 2 for diffusion through a membrane where an energy barrier U_B, exists. If no energy barrier to ion transfer exists, a membrane temperature will have little effect on influx, and the Q_{10} will approximate 1. For temperature changes in the range of 10° to 40°C, physical processes usually have a Q_{10} value in the range of 1.2 to 1.3. However, chemical and biological reactions have Q_{10} values between 2 and 4 (Precht et al., 1973; Nobel, 1974).

Changing the temperature of the root system affects plant growth rate as well as nutrient influx. Investigations have been made of the effect of temperature on the influx of ions into both intact and excised roots. With intact roots, the effect of increasing temperature on increased water use may affect influx of some ions. Figure 3.10 illustrates the effect of temperature on ion influx. As temperature was increased above 5°C, influx increased until a maximum was reached; at higher temperatures, influx decreased once more. The temperature for maximum influx varied with plant species. Temperature for maximum influx may also vary with type of nutrient. Low respiration rates may inhibit uptake at low temperatures. Breakdown of membrane structure may be responsible for decreased uptake at high temperature.

Rates of water influx and root respiration generally parallel the increase in ion influx with increased temperature. However, water influx usually does not decrease so markedly when temperatures increase beyond the tempera-

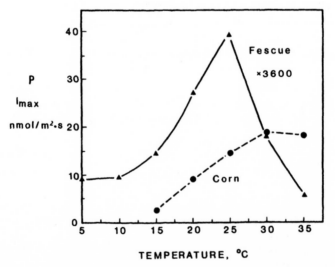

FIGURE   3.10   Relation between ion influx and temperature for corn and fescue.

ture for maximum ion influx. Some plants reduce transpiration rate at higher temperature by wilting. Values for the effect of temperature on influx have to be measured for each species and ion combination, since they cannot readily be predicted.

Balance of Nutrients between Shoots and Roots

The relation between ion translocation and ion absorption determines whether there is a disparity in ion concentrations between the shoot and the root. Such ions as potassium are very mobile within the plant, and the composition of the root and shoot are similar. Such ions as phosphorus, which are constituents of organic compounds found primarily in the shoot, have higher ion concentrations in the shoot than the root. Older and younger parts of the shoot may also vary in nutrient concentration according to the mobility of the nutrient within the plant, since heavy metal ions may accumulate in the root. It is sometimes difficult to determine if the metal is precipitated in the apoplasm, is in the symplasm within the plasma membrane of the root cortical cells, or is in the vacuole.

Collander (1941) investigated the nutrient distribution between roots and shoots in 16 species. He found potassium, rubidium, cesium, and magnesium to be present in about the same concentration in the root as in the shoot, while sodium and manganese were present at markedly higher concentrations in the roots of most species. Calcium, silicon, and lithium were more concentrated in the shoots than in the roots of nearly all species. Warncke and Barber (1974) measured ion concentrations of solution-grown corn plants at five stages of growth, from the four-leaf stage to 2 weeks past silking. Shoot concentrations of nitrogen, phosphorus, and potassium decreased with age, so that composition of the final sampling averaged one-half or less that of the first. Root nitrogen, phosphorus, and potassium composition decreased less dramatically, with the final sampling averaging a composition approximately 0.75 of initial. During early growth, concentrations in the shoot exceeded those in the root, but the root's composition was highest by the final harvest. Calcium and magnesium concentrations in both root and shoot changed little with plant age, but both were higher in the shoot than in the root.

Shoot Growth Rate and Ion Influx

Since the shoot requires nutrients to produce new cells and the shoot produces photosynthate used for active ion uptake, it is reasonable to assume a relation between shoot growth rate and ion influx. Pitman (1972) used length of photoperiod to alter relative growth rate and showed that there was a relation between relative growth rate and uptake rate of potassium per unit of barley root. When growth was limited by length of photoperiod, reduced sugar level in the roots appeared to control the export of potassium to the shoot. Claassen and Barber (1974) showed that potassium uptake rate decreased during the dark period. However, they suggested that part of this effect may have been due to the effect of transpiration rate on potassium uptake, since uptake increased to a rate greater than the initial rate immediately after the lights were turned on.

Using split-root techniques to supply a nutrient to only part of the root system is one way of altering the ratio of relative growth rate to the amount of root supplied with the nutrient. Jungk and Barber (1975) found that phosphorus influx by corn roots was not increased by increasing this ratio. Hence, the relation between relative growth rate and ion influx is complex, with the parameters directly controlling influx not yet identified. The ability of some plant species to grow at very low soluble-nutrient concentrations could be due to the low relative growth rate of the species and, hence, to the low rate of demand for nutrients. Rorisen (1968) compared two plant species with high growth rates with two species with low growth rates to determine if growth rate affected the optimum level of solution phosphorus needed for growth; he did not find a clear-cut relation. Hence, it appears that relative growth rate is not the only factor influencing optimum solution phosphorus levels.

Ion Competition for Absorption

When several nutrients are present in solution, an ion's rate of absorption may be influenced by the level of a second ion, since ions may compete directly for the same absorption site, or one ion may affect absorption of a second by its effect on other processes within the plant. Ion competition has been investigated in excised-root studies (Epstein, 1972). The presence of calcium increases the uptake of many ions. This effect, termed the Viets effect because it was first shown by Viets (1944), is believed to be due to the need for calcium to maintain the integrity of the cell membrane. Using the concept of separate carriers for moving different groups of ions across the plasma membrane, it is possible to determine through ion competition studies which ions are absorbed by the same carrier and which do not compete for the same carrier site. Such investigations indicate that H, K, NH₄, Rb, and Cs compete for the same carrier. Among the divalent cations, Ca, Sr, and Ba compete for the same carrier.

Some ions that are closely related in size and charge can proxy for one another in the plant, so competition for uptake is considerable, and there may even be no selectivity between ions with respect to ion uptake. These ion groups include K^+ and Rb^+, Ca^{2+} and Sr^{2+}, Cl^- and Br^-, and SO_4^{2-} and SeO_4^{2-}. However, one ion from each of these groups usually has little significance in plant nutrition and is normally present in only small amounts in the soil. Hence, in these cases, the competitive effect is more an academic interest than a practical nutrition problem.

Competitive effects of ion uptake by intact roots involves competition in translocation as well as absorption. These effects may involve competition for carrier sites, or they may be indirect effects where an excess of one ion interferes in some fashion with the translocation or absorption of a second ion. Some of the pairs of ions where interference in uptake occur are discussed in the following paragraphs.

Effect of Potassium on Magnesium Uptake.   Increasing the level of potassium will usually depress the rate of magnesium absorption, particularly where magnesium levels are relatively low. While the effect of potassium on magnesium uptake is large, the reciprocal effect of magnesium level on potassium uptake is small or negligible. Ammonium also depresses magnesium uptake to the same degree as potassium (Claassen and Wilcox, 1974). In a split-root experiment, Claassen and Barber (1977) showed the effect of the presence or absence of potassium on the rate of uptake of NO_3^-, $H_2PO_4^-$, and Mg^{2+}. Data in Table 3.6 show that magnesium uptake rate was reduced by one-half in the presence of potassium, while the nitrate uptake was increased, and phosphate uptake was not affected.

Effect of Potassium on Nitrate Uptake.   The level of potassium in the plant influences the assimilation of nitrate into protein (Koch and Mengel, 1974), which in turn is believed to influence the uptake of nitrate. Claassen and Barber (1977) showed with split-root studies that the presence of potassium in the nutrient media had a great effect on the rate of nitrate uptake (Table 3.6). It may be that these ions cooperate in each other's transport to

TABLE 3.6   Influence of the Presence of Potassium on the Uptake of Nitrogen, Phosphorus, and Magnesium by Corn Plants Grown in Split-Root Systems

Proportion of root (%)	Presence of K	Average uptake rate $(mol/g_{F.W.} \cdot s \times 10^4)$		
		NO_3^-	$H_2PO_4^-$	Mg^{2+}
50	+	32.3	5.30	1.82
50	+	35.2	4.48	1.90
75	+	28.9	4.33	1.33
25	−	26.8	5.46	2.89
50	+	39.6	5.01	1.89
50	−	25.2	5.38	3.34
25	+	50.0	4.84	2.00
75	−	25.5	5.35	3.88
15	+	39.2	6.87	1.29
85	−	18.4	5.05	3.37

Source: Reproduced from Claassen and Barber (1977) by permission from the American Society of Agronomy.

the shoot. In field experiments, Barber (unpublished data) has also shown that increasing nitrogen fertilization increases potassium uptake by corn plants. Increasing nitrogen from 62 to 212 kg/ha caused an increase in ear-leaf potassium concentration from 1.53 to 1.94% K.

Effect of Potassium on Calcium Uptake.   Adding potassium will reduce calcium uptake. However, the reduction in calcium uptake is only about one-half the effect of potassium on reducing magnesium uptake. As with magnesium, there is little evidence that increasing calcium concentration affects potassium uptake.

Effect of Calcium on Magnesium Uptake.   Little information is available on the direct effect of increasing calcium on magnesium uptake rate. In field experiments, however, increasing calcium levels reduced magnesium uptake (see Chapter 12 for details).

Effect of Phosphorus on Zinc Uptake.   Many investigators have shown that high levels of phosphate will depress the rate of zinc uptake to the extent that zinc deficiency may occur. Safaya (1976) showed that the effect of phosphorus on zinc uptake by corn was due to a reduction of zinc flux into the roots. Not *all* investigators have found that increasing the phosphorus level decreased zinc absorption; some (Watanabe et al. 1965) found the opposite. The effect appears somewhat dependent on age and species of the plant and levels of phosphorus and zinc used in the investigations. Phosphorus may also reduce the translocation of zinc within the plant.

Effect of Nitrogen on Phosphorus Uptake.   Adding nitrogen, particularly NH_4^+, can increase phosphorus uptake. When plants are grown in soil, this effect can be due to decreased pH in the rhizosphere soil (Figure 6.2). Miller (1974) reviewed available evidence for the effect of nitrogen on phosphorus uptake from solution by plant roots. Presence of nitrogen did not directly increase the absorption process but instead increased the rate of translocation of phosphorus to the shoot, which indirectly influenced the absorption rate.

SUMMARY

Many types of root systems exist. They allow plants to adapt to soil conditions so that they are able to get the nutrients and water needed for growth; hence the species growing the best may vary with soil. Root growth of annual plants usually increases until anthesis and then the amount of roots present may decline rapidly with time as the plant approaches maturity. Amount of plant root surface is important in determining the rate of nutrient uptake by the plant. Root growth varies with soil depth and may be enhanced in the soil depth that has the greatest nutrient supply, particularly nitrogen and phosphorus. The kinetics of nutrient uptake by the root has been the subject of much research. Factors that influence net nutrient influx include nutrient concentration in the soil solution, temperature, plant age, root age, shoot demand (influx kinetics), competition among nutrients, species, and cultivar. Information on the relation of these factors to net nutrient influx, using the mechanistic nutrient uptake model, can be used to predict their relative influence on nutrient uptake and plant growth.

REFERENCES

Anghinoni, I., V. C. Baligar, and S. A. Barber. 1981. Growth and uptake rates of P, K, Ca, and Mg in wheat. *J. Plant Nutr.* **3**:923–933.

Asher, C. J., P. G. Ozanne and J. F. Loneragan. 1965. A method for controlling the ionic environment of plant roots. *Soil Sci.* **100**:149–156.

Baligar, V. C., and S. A. Barber. 1978. Use of K/Rb ratio to characterize potassium uptake by plant roots growing in soil. *Soil Sci. Soc. Am. J.* **42**:575–579.

Bange, G. G. J. 1973. Diffusion and absorption of ions in plant tissue. III. The role of the root cortex cells in ion absorption. *Acta Bot. Neerl.* **22**:529–542.

Barber, S. A. 1978. Growth and nutrient uptake of soybeans under field conditions. *Agron. J.* **70**:457–461.

Barber, S. A., and J. H. Cushman. 1981. Nitrogen uptake model for agronomic crops. In I. K. Iskandar, Ed. *Modeling Wastewater Renovation-Land Treatment.* Wiley-Interscience, New York. Pp. 382–409.

Bohm, W. 1979. *Methods of Studying Root Systems.* Springer-Verlag, Berlin.

Bowen, G. D. and A. D. Rovira. 1969. New techniques to study nutrient relations in plants. *Atomic Energy Aust.* **12**:2–7.

Brouder, S. M., and K. G. Cassman. 1990. Root development of two cotton cultivars in relation to potassium uptake and plant growth in a vermiculitic soil. *Field Crops Res.* **23**:187–203.

Claassen, M. E., and G. E. Wilcox. 1974. Comparative reduction of calcium and magnesium composition of corn tissue by NH_4-N and K fertilization. *Agron. J.* **66**:521–522.

Claassen, N., and S. A. Barber. 1974. A method for characterizing the relation between nutrient concentration and the flux into roots of intact plants. *Plant Physiol.* **54**:564–568.

Claassen, N., and S. A. Barber. 1976. Simulation model for nutrient uptake from soil by a growing plant root system. *Agron. J.* **68**:961–964.

Claassen, N., and S. A. Barber. 1977. Potassium influx characteristics of corn roots and interaction with N, P, Ca, and Mg influx. *Agron. J.* **69**:860–864.

Clarkson, D. T., and J. Sanderson. 1971. Relationship between anatomy of cereal roots and the absorption of nutrients and water. *Agric. Res. Council Letcombe Laboratory Report,* Wantage, England. Pp. 16.

Clarkson, D. T., and A. W. Robards. 1975. The endodermis, its structural development and physiological role. In J. G. Torrey and D. T. Clarkson, Eds. *The Development and Function of Roots*. Academic Press, London. Pp. 415–436.

Collander, R. 1941. The distribution of different cations between root and shoot. *Acta Bot. Fenn.* **29**:4–12.

Cormack, R. G. H. 1962. Development of root hairs in angiosperms. *Bot. Rev.* **28**:446–464.

Dittmer, H. J. 1937. A quantitative study of the roots and root hairs of a winter rye plant *Secale cereal. Am. J. Bot.* **24**:417–420.

Drew, M. C. 1979. Properties of roots which influence rates of absorption. In J. L. Harley and R. S. Russell, Eds. *The Soil-Root Interface*. Academic Press, New York. Pp. 21–38.

Edwards, J. H., and S. A. Barber. 1976a. Nitrogen flux into corn roots as influenced by shoot requirement. *Agron. J.* **68**:471–473.

Edwards, J. H., and S. A. Barber. 1976b. Phosphorus uptake rate of soybean roots as influenced by plant age, root trimming, and solution P concentration. *Agron. J.* **68**:973–975.

Edwards, J. H., D. D. Warncke, S. A. Barber, and D. W. Nelson. 1974. Nitrogen uptake efficiency by four plant species in the field and growth chamber. *Water Resources Res. Tech. Rep. 40.* Purdue University, W. Lafayette, Ind.

Eisenman, G. 1962. Cation selective glass electrode and their mode of operation. *Biophys. J. Suppl.* **2**:259–323.

Epstein, E. 1966. Dual pattern of ion absorption by plant cells and by plants. *Nature (London)* **212**:1324–1327.

Epstein, E. 1972. *Mineral Nutrition of Plants. Principles and Perspectives*. John Wiley, New York.

Esau, K. 1965. *Plant Anatomy*, 2nd ed. John Wiley, New York.

Ferguson, I. B., and D. T. Clarkson. 1975. Ion transport and endodermal suberization in the roots of *Zea mays. New Phytol.* **75**:69–79.

Gillespie, A. R., and P. E. Pope. 1990. Rhizosphere acidification increases phosphorus recovery of black locust. I. Induced acidification and soil response. *Soil Sci. Soc. Am. J.* **54**:533–537.

Goss, M. J., and J. B. Reid. 1981. *Interaction between Crop Roots and Soil Structure. M.A.F.F. Ref. Book 341*. Her Majesties Stationery Office, London. Pp. 34–48.

Haynes, R. J. 1980. Ion exchange properties of root and ionic interactions within the root apoplasm: Their role in ion accumulation by plants. *Bot. Rev.* **46**:75–99.

Hodges, T. K. 1973. Ion absorption by plant roots. *Adv. Agron.* **25**:163–207.

Johnson, K. D. 1981. Tall fescue (*Festuca arundinacea* Schreb) root relationships. Ph.D. dissertation, Purdue University.

Jungk, A. 1974. Phosphate uptake characteristics of intact root systems in nutrient solution as affected by plant species, age and P supply. *Plant Analysis and Fertilizer Problems, Proc. Seventh Int. Coll.* **1**:185–196.

Jungk, A., and S. A. Barber. 1974. Phosphate uptake rate of corn roots as related to the proportion of the roots exposed to phosphate. *Agron. J.* **66**:554–557.

Jungk, A., and S. A. Barber. 1975. Plant age and the phosphorus uptake character-istics of trimmed and untrimmed corn root systems. *Plant Soil* **42**:227–239.

Kelly, J. M., S. A. Barber, and G. S. Edwards. 1992. Modeling magnesium, phos-phorus and potassium uptake by loblolly pine seedlings using a Barber-Cushman approach. *Plant Soil* **139**:209–218.

Klepper, B., and R. W. Rickman. 1990. Modeling crop root growth and function. *Adv. Agron.* **44**:113–132.

Koch, K., and K. Mengel. 1974. The influence of potassium nutritional status on the absorption and incorporation of nitrate nitrogen. *Plant Analysis and Fertilizer Problems, Proc. Seventh Int. Coll.* **1**:209–218.

Kuchenbuch, R. O., and S. A. Barber. 1988. Significance of temperature and pre-cipitation for maize root distribution in the field. *Plant Soil* **106**:9–14.

Kutschera, L. 1960. *Wurzelatlas, mitteleuropaischer Ackerunkrauter und Kulturpflanzen.* DLG-Verlag-GmbH, Frankfort.

Lauchli, A. 1976. Apoplasmic transport in tissues. In U. Luttge and M. G. Pitman, Eds. *Transport in Plants.* Springer-Verlag, New York. Pp. 22–29.

Leggett, J. E., and W. A. Gilbert. 1969. Magnesium uptake by soybeans. *Plant Physiol.* **44**:1182–1186.

Marschner, H. 1986. *Mineral Nutrition in Higher Plants.* Academic Press, New York.

McGonigle, T. P., and M. H. Miller. 1993. Mycorrhizal development and phospho-rus absorption in maize under conventional and reduced tillage. *Soil Sci. Soc. Am. J.* **57**:1002–1006.

Mengel, D. B., and S. A. Barber. 1974a. Development and distribution of the corn root system under field conditions. *Agron. J.* **66**:341–344.

Mengel, D. B., and S. A. Barber. 1974b. Rate of nutrient uptake per unit of corn under field conditions. *Agron. J.* **66**:399–402.

Miller, M. H. 1974. Effects of nitrogen on phosphorus absorption by plants. In E. W. Carson, Ed. *The Plant and Its Environment.* University Press of Virginia, Charlottesville, Pp. 643–668.

Nielsen, N. E., and S. A. Barber. 1978. Differences among genotypes of corn in the kinetics of phosphorus uptake. *Agron. J.* **70**:695–698.

Nissen, P. 1973. Multiphasic ion uptake in roots. In W. P. Anderson, Ed., *Ion Transport in Plants.* Academic Press, New York, Pp. 539–553.

Nobel, P. S. 1974. *Introduction to Biophysical Plant Physiology.* W. H. Freeman, San Francisco.

Nye, P. H., and P. B. Tinker, 1977. *Solute Movement in the Soil-Root System.* Blackwell Scientific Publishers, Oxford.

Olsen, C. 1950. The significance of concentration for the rate of ion absorption by higher plants in water culture. *Physiol. Plant.* **3**:152–164.

Pearson, R. W. 1974. Significance of rooting pattern to crop production and some problems of root research. In E. W. Carson, Ed. *The Plant Root and Its Environment.* University Press of Virginia, Charlottesville. Pp. 247–270.

Peterson, W. R., and S. A. Barber. 1981. Soybean rot morphology and K uptake. *Agron. J.* **73**:316–319.

Powell, C. L. 1974. Effect of P fertilizer on root morphology and P uptake by *Carex coriacea*. *Plant Soil* **41**:661–667.

Precht, H., J. Christopherson, H. Hensel, and W. Larcher. 1973. *Temperature and Life*. Springer-Verlag, New York.

Riley, D., and S. A. Barber. 1971. Effect of ammonium and nitrate fertilization on phosphorus uptake as related to root-induced pH changes at the root-soil interface. *Soil Sci. Soc. Am. Proc.* **35**:301–306.

Rorison, I. H. 1968. The response to P of some ecologically distinct plant species. I. Growth rates and P absorption. *New Phytol.* **67**:913–923.

Rovira, A. D., and G. D. Bowen. 1970. Translocation and loss of phosphate along roots of wheat seedlings. *Planta* **93**:15–25.

Russell, R. S. 1977. *Plant Root Systems, Their Function and Interaction with Soil*. McGraw-Hill, London.

Russell, R. S. and J. Sanderson. 1967. Nutrient uptake by different parts of the intact roots of plants. *J. Exp. Bot.* **18**:491–508.

Sabater, B. 1982. A mechanism for multiphasic uptake of solutes in plants. *Physiol. Plant.* **55**:121–128.

Safaya, N. M. 1976. Phosphorus-zinc interaction in relation to absorption rates of phosphorus, zinc, copper, manganese, and iron in corn. *Soil Sci. Soc. Am. J.* **40**:719–722.

Schenk, M. K., and S. A. Barber. 1980. Potassium and phosphorus uptake by corn genotypes grown in the field as influenced by root characteristics. *Plant Soil* **54**:65–76.

Slaton, N. A., C. A. Beyrouty, B. R. Wells, R. J. Norman, and E. E. Gbur. 1990. Root growth and distribution of two short-season rice genotypes. *Plant Soil* **21**:269–278.

Smucker, A. J. M., S. L. Burney, and A. K. Sirvastava. 1982. Quantitative separation of roots from compacted soil profiles by a hydropneumatic elutriation system. *Agron. J.* **74**:500–503.

Taylor, H. M. 1974. Root behavior as affected by soil structure and strength. In E. W. Carson, Ed. *The Plant Root and Its Environment*. University Press of Virginia, Charlottesville. Pp. 271–291.

Taylor, H. M., and L. F. Ratliff. 1969. Root elongation rates of cotton and peanuts as a function of soil strength and soil water content. *Soil Sci.* **108**:113–119.

Tennant, D. 1975. A test of a modified line intersect method of estimating root length. *J. Ecol.* **63**:995–1001.

Teo, Y. H., C. A. Beyrouty, and E. E. Gbur. 1992. Nitrogen, phosphorus, and potassium influx kinetic parameters of three rice cultivars. *J. Plant Nutr.* **15**:435–444.

Van Rees, K. C. J., N. B. Comerford, and W. W. McFee. 1990. Modeling potassium uptake by slash pine seedlings from low-potassium-supplying soils. *Soil Sci. Soc. Am. J.* **54**:1413–1421.

Zhang, J., and S. A. Barber. 1992. Maize root distribution between phosphorus-fertilized and unfertilized soil. *Soil Sci. Soc. Am. J.* **56**:819–822.

CHAPTER **4**

Nutrient Uptake by Plant Roots Growing in Soil

Chapter 3 discussed nutrient-uptake processes where roots were grown primarily in stirred aerated solution. In this chapter, we will consider the more complex system of soil as the medium that supplies nutrients for plant growth. When plants grow in nutrient solution, a uniform nutrient concentration is maintained at the root surface by mixing the solution. In soil, the nutrient concentration at the root changes with time as nutrients and water are absorbed at rates differing from the rates of supply by the soil. These changes must be considered in measuring nutrient flux from the soil into the plant root. The soil's initial nutrient level varies across soil pores and particles when viewed on a microscale. However, uptake by a root traversing this variability will tend to average the system. Presumably, average ion influx will approach a value similar to what would occur for uniform average concentration. During laboratory soil extraction, we also average the variability of the microscale, thus obtaining averaged measurements to use in evaluating nutrient availability.

The uptake of nutrients by plant roots growing in soil depends on both nutrient-influx characteristics of the root (Chapter 3) and the nutrient-supply characteristics of the soil; which are described in this chapter. These characteristics are influenced in turn by the nutrient-soil associations discussed in Chapter 2. Soil nutrient supply has frequently been evaluated by arbitrary empirical measurements that have been related by regression analysis to nutrient uptake by plants growing in the soil or to growth of the plants. While satisfactory correlations may occur for studies on similar soils, they frequently are not satisfactory when a wide range of soil types is included in the study. Hanway (1973) reports that "crop response to fertilizer applications has exhibited a generally large variability not explained by the laboratory index of nutrient availability in the soil."

Progress in developing an understanding of this area lies in studying processes that determine the rate of nutrient flow from the soil into the plant root. Nutrient flux into the root is a dynamic process, so kinetics is usually more important than thermodynamics in describing this system.

NUTRIENT SUPPLY

Roots have soil immediately adjacent to their surfaces as they grow through pores; by expansion, they compress the soil immediately adjacent to the root, so that it has fewer air and water-filled pores. The amount of disturbance of the soil depends on the nature of the soil's pore space and the diameter and degree of expansion of the root with growth. The effect of a cotton root on particle distribution in adjacent soil is illustrated in Figure 4.1. This figure shows a thin section across the root-soil interface, indicating that there was less pore space and, hence, a higher bulk density of soil nearer the root surface. Soil particles and soil solution at the root surface provide nutrients for initial absorption by the root. Because of a higher soil bulk density, initial nutrient concentration per cm³ in this soil will be hig her than the average for the soil.

Since the soil system is not stirred as in a solution, the quantity of nutrients immediately available at the root surface is limited. Albrecht et al. (1942) calculated the quantities of nutrients that would be present at the root surface if the root surface were completely covered with soil clay. Even though they used values from 1 to 25 cm² of root per cm³ of soil, the authors found that only 0.1 to 0.3% of the clay present in the soil would actually contact the roots. Hence, the proportion of soil nutrients in direct contact with the root as it grows through the soil is small.

Interception

Barber et al. (1963) used the term *root-interception* to describe soil nutrients at the root interface that do not have to move to the interface to be positionally available for absorption. They added to this term the quantity of ions

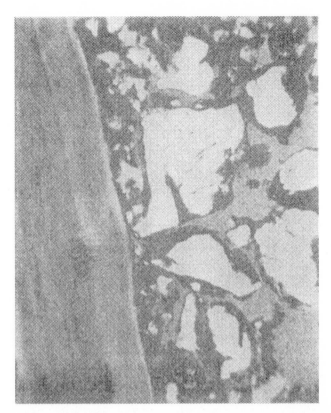

FIGURE   4.1   Thin section of a cotton root–soil interface showing the effect of the root on the distribution of soil particles. The white areas are sand and silt quartz grains. The dark areas are clay masses. The gray areas are pore spaces with a root magnification about 40×. Reproduced from Lund and Beals (1965) by permission of Soil Science Society of America.

displaced by root growth. The quantity of nutrients supplied by root interception was taken as the quantity present in a volume of soil equal to the root volume. While this is an arbitrary definition, it does evaluate the quantity of nutrients displaced by the root as it grows through the soil and which increases the concentration normal to the root over that present initially.

Root volumes for several species growing in soil are given in Table 4.1. For annual crops, root volume in the 0- to 20-cm layer of soil is usually less than 1% of the soil's volume. Hence, less than 1% of the available soil nutrients could be supplied by root interception.

Available nutrients, as defined in Chapter 1, implies a form or forms of nutrients that may be immediately absorbed by plant roots. In this book, the term refers to soluble nutrients plus those associated with the solid phase, other than those released from organic matter decomposition, that are in rapid (within 1 to 2 days) equilibrium with nutrients in the soil solu

TABLE 4.1 Volume of Roots in the Top 20 cm of Soil as a Percentage of Soil Volume

Plant species	Relative root volume (%)	Reference
Corn (*Zea mays* L.)	0.4	Mengel and Barber, 1974
Kentucky Bluegrass (*Poa pratensis* L.)	2.80	Dittmer, 1940
Winter rye (*Secale cereale* L.)	0.85	Dittmer, 1940
Oats (*Avena sativa* L.)	0.55	Dittmer, 1940
Soybean (*Glycine max* L.)	0.91	Dittmer, 1940
Soybean	0.4	Barber (unpublished)

tion. The significance of the quantity that reaches the root as a result of root interception depends on both the amount in the soil and the plant's requirement. A sample calculation is shown in Table 4.2 for a fertile north central U.S. Alfisol. The amounts shown, and subsequent calculation in this chapter of the supply mechanisms, will refer only to those nutrients whose concentration in solution is not affected by microbiological activity. The quantities are also the amounts in equilibrium soil systems before equilibrium is disturbed by root growth. These quantities can be measured in the laboratory.

Mass Flow

Mass flow is the movement of nutrients through the soil to the root in the convective flow of water caused by plant water absorption. The amount of nutrient movement by mass flow is related to the water used and nutrient concentration of that water. The movement from beyond the distance affected by diffusion and mass flow can be calculated per cm^2 of root surface by multiplying the rate of water entry, usually from 2 to 0.5×10^{-6} $cm^3/cm^2 \cdot s$ (cm/s), times the nutrient concentration in the equilibrium soil solution. Comparing the rate of supply with the rate of influx indicates the significance of mass flow. Alternatively, mass flow can be calculated by multiplying water use per plant by the concentration in the soil solution. The value can then be compared to the known needs of the plant. Multiplying use per plant by the number of plants per hectare gives values on a per hectare basis. The significance of mass flow varies widely with crop, climate, and moisture conditions, since each of these affects water use. The concentration of nutrients in soil solution also varies widely among soils. Average water use for producing an annual crop, such as corn, is about 2.5 to 3.0 million liters of water per hectare; Table 4.3 gives ranges of values for concentrations of

TABLE 4.2 Amounts of Available Nutrients[a] in an Average Fertile Silt Loam Soil Where Field Response of Maize to Added Nitrogen, Phosphorus, or Potassium is Unlikely[b]

Nutrient	C_{si} (kg/ha)	C_{si} (mg/L)	C_{li} (mg/L)	b
Nitrogen	200	120	300	0.4
Phosphorus	100	60	0.3	200
Potassium	400	240	7	34
Calcium	6000	3600	60	60
Magnesium	1500	900	40	23
Sulfur	100	60	26	2.3

[a]N, P, and K values adjusted from those shown in the edition.
[b]Soil bulk density of 1.2, and volumetric water content of 0.3 assumed, and soil depth for nutrient calculation of 20 cm.

nutrients in soil solutions. Values for a north central U.S. Alfisol were previously given in Table 4.2. Generally, the greater the soil weathering, the lower the concentration of nutrients in the soil solution.

The relative importance of each supply mechanism in Table 4.2 for producing corn on the Alfisol is shown in Table 4.4. In making these calculations, it was assumed that the roots intercepted 1% of the total available nutrients in the soil. Mass flow supply was calculated by multiplying the concentrations in Table 4.2 by 2.5 million L to get the supply per hectare. The mean values for total nutrient content of a hectare of corn are from Barber and Olsen (1968). Root interception accounted for only a fraction of the supply for all nutrients except calcium, where the quantity in the soil was very large compared to the calcium required for corn growth. Mass flow could supply the crop's nutrient demand for all nutrients except nitrogen, phosphorus, and potassium. Any nutrient requirements not supplied by root interception and mass flow were assumed supplied by diffusion. Most of the phosphorus and potassium, and part of the nitrogen, were presumably supplied by diffusion.

Diffusion

When root interception and mass flow do not supply the root with sufficient quantities of a particular nutrient, continued uptake reduces the concentration of available nutrients in the soil at the root surface. This in turn causes a concentration gradient perpendicular to the root surface, with nutrients subsequently diffusing along the gradient toward the root surface.

Nutrient diffusion occurs because of the thermal motion of molecules, sometimes termed Brownian movement. When a concentration gradient exists, there is greater movement from the zone of higher concentration to the zone of lower concentration. Hence, net movement occurs toward the

TABLE 4.3   Range of Concentrations of Major Nutrients in the Soil
            Solution

Nutrient	Amount in solution (μmol/L)
NO_3^-	100–50,000
NH_4^+	100–2,000
$H_2PO_4^-$ and HPO_4^{2-}	1–50
K^+	100–4,000
Ca^{2+}	100–5,000
Mg^{2+}	100–5,000
SO_4^{2-}	100–10,000

Source: Barber (1974a) (updated).

TABLE 4.4 Relative Significance of Root Interception, Mass Flow, and Diffusion in Supplying Corn with Its Nutrient Requirements from a Fertile Alfisol Silt Loam (kg/ha)

Nutrient	Amount needed for 9500 kg grain/hecatre	Approximate amount supplied by		
		Root interception	Mass flow	Diffusion
Nitrogen	190	2	150	38
Phosphorus	40	1	2	37
Potassium	195	4	35	156
Calcium	40	60	150	0
Magnesium	45	15	100	0
Sulfur	22	1	65	0

zone of lower concentration until concentration has been equalized. Because roots absorb nutrients, equilibrium does not occur, and nutrients continue to diffuse to the root along the concentration gradient. The distance the concentration gradient extends from the root depends on the rate of diffusion. The distance for diffusive-nutrient movement through the soil to the root is usually in the range of 0.1 to 15 mm. Hence, only soil nutrients within this soil zone contribute to diffusive nutrient supply to the root.

Diffusion Coefficient

The rate of diffusion is expressed by the diffusion coefficient, measured in cm^2/s. Fick's first law of diffusion, Equation 4.1, describes the steady state diffusive flux J in µmol:

$$J = -DA \frac{dC}{dx} \tag{4.1}$$

where D is the diffusion coefficient, A is the area for diffusion, and dC/dx is the concentration gradient in $\mu mol/cm^3 \cdot cm$. Equation 4.1 applies to steady-state diffusion; the minus sign indicates that movement is from higher to lower concentration. This equation is used primarily to calculate D since J and dc/dx can be measured experimentally.

Fick's second law, Equation 4.2, applies to transient-state diffusion. It is more applicable to plant root–soil situations.

$$\frac{dC}{dt} = D \frac{d^2C}{dx^2} \tag{4.2}$$

Hence, dC/dt is the rate of change in concentration with time at a fixed linear distance. At a given value of t, the concentration gradient can be determined from Equation 4.2.

In plant roots, nutrients are diffusing radially toward a cylindrical sink; Fick's second law with radial coordinates becomes

$$\frac{dC}{dt} = \frac{1}{r} \frac{d}{dr} \left(rD \frac{dC}{dr} \right) \tag{4.3}$$

where r is the radial distance from the center of the cylinder.

Size of Diffusion Coefficients in Soils

Table 4.5 reports values for the diffusion of ions in solution, soils, and within minerals themselves (solid-state diffusion). Diffusion is either salt diffusion, where a cation and an anion diffuse together, or counter diffusion, where one cation diffuses in one direction and a counter cation diffuses in the oppo-

TABLE 4.5   Diffusion Coefficients for Diffusion of Ions in Solution, in Soils, and Minerals

Ion	Medium	$D_1(cm^2/s)$	Reference
K^+	Water at 25°C	1.98×10^{-5}	Parsons (1959)
$H_2PO_4^-$	Water at 25°C	0.89×10^{-5}	Edwards and Huffman (1959)
NO_3^-	Water at 25°C	1.9×10^{-5}	Parsons (1959)
Ca^{2+}	Water at 25°C	0.78×10^{-5}	Parsons (1959)
Mg^{2+}	Water at 25°C	0.70×10^{-5}	Parsons (1959)
		D_e	
K^+	Soil	$10^{-7}-10^{-8}$	Barber (1974b)
$H_2PO_4^-$	Soil	$10^{-8}-10^{-11}$	Barber (1974b)
NO_3^-	Soil	$10^{-6}-10^{-7}$	Barber (1974b)
K^+	Mineral	10^{-23}	Talibudeen et al. (1978)

site direction in order to maintain charge balance. In counter diffusion, ions may even be labeled isotopically so that the counter ion is the same as the diffusing ion except for the isotopic label. Diffusion coefficients measured in this manner are called self-diffusion coefficients. Self-diffusion coefficients are frequently measured in soils; however, they are not truly representative of the diffusion that occurs in the soil next to the plant root. In this situation, ions such as hydrogen released from the root may commonly serve as counter ions during cation diffusion. The rate of diffusion of an ion also depends on the diffusion coefficient of both the ion and the counter ion.

Fick's law was developed for diffusion through a uniform medium, such as water or air. Soil is a variable medium, however, especially when viewed on a microscale. Viewed on a macroscale, diffusion of ions through soil is influenced by the proportion of the soil volume that is occupied by water, the tortuosity of the diffusion path, and chemical and physical effects of the solid phase on ion movement. Such values of the diffusion coefficient are called effective diffusion coefficients D_e, because ion movement is at the rate that would occur if the soil were a uniform medium and the diffusion coefficient of the nutrient were the size of the effective diffusion coefficient. Diffusion from a treated block of soil into an untreated block of soil gives a concentration gradient with distance affected by time that follows Equation 4.4. This relation was developed from Fick's second law

$$\frac{C}{C_o} = 1/2[1 - erf(x/2(D_e t)^{1/2})] \tag{4.4}$$

Here C/C_0 is the concentration at a point in space and time as a fraction of initial concentration C_0, *erf* is an error function made up of many integrals of the diffusion equation, and x is the distance in cm. Values of D_e for soils may also be obtained by measuring diffusion from the soil to an exchange resin membrane and using the relation

$$D_e = M_t^2 \pi / 4 C_0^2 t \qquad (4.5)$$

where M_t is the total amount diffusing to 1 cm^2 of exchange resin paper in time t, and C_0 is the initial uniform concentration of the ion in the soil. Equation 4.5 assumes C in the soil at the exchanger surface remains equal to zero, which is essentially true for small values of t.

Diffusion of ions in soil has usually been calculated on the basis of diffusion through the total soil volume, since this determines the flux per m^2 of root surface. Since ion diffusion is predominately through soil water and since soil water constitutes only a fraction of the soil's volume, the value for D will be reduced on a volume basis according to the proportion of the soil that is water. In addition, the liquid path in the soil is tortuous, so the value of D is further reduced. When nonadsorbed ions, such as nitrate, are considered, volumetric moisture content and tortuosity constitute the main factors that reduce D below the value in true solution. Some additional restrictions on ion diffusion may occur because of effects of the soil on water viscosity and the attraction of charged ions to soil surfaces. When adsorbed ions, such as exchangeable cations, and adsorbed phosphate are considered, D is reduced even further because of successive equilibration between ions in the adsorbed and solution phases as diffusion progresses.

Nye and Tinker (1977) have used Equation 4.6 to calculate effective diffusion coefficients of ions such as K$^+$ from readily measured soil parameters

$$D_e = \frac{D_l \theta f_l dC_l}{dC_s} \qquad (4.6)$$

Here, D_e is the effective diffusion coefficient in soil, D_l is the value of D in water, θ is the volumetric soil water content, f_l is the tortuosity or impedance factor that can be measured using nonadsorbed ions, and dC_l/dC_s is the reciprocal of the soil buffer power for the ion in question.

The average linear distance of diffusive movement of an ion with time is $(2Dt)^{1/2}$. An ion in water with a D_l of 1×10^{-5} cm^2/s would thus move 1.3 cm in 1 day. An ion in the soil such as K$^+$, with a D_e of 1×10^{-7} cm^2/s would move only 0.13 cm and an ion like H$_2$PO$_4^-$ with a D_e of 1×10^{-10} cm^2/s would move only 0.004 cm. These distances are important in determining the amounts of soil nutrients that may be close enough to reach the root diffusion during the growth of the plant.

Factors Influencing Mass Flow Flux

The rate of ion movement to the root by mass flow depends on the rate of water uptake, which in turn is affected by plant species, climate, and the soil's water level. Water influx occurs only during the day, when sunlight energy is available to evaporate water from the leaves. Values for water influx by corn roots of 2×10^{-6} cm³/cm² · s have been measured (Claassen and Barber, 1976). At this rate of water uptake, the linear rate of water movement at the root surface would average 0.17 cm/day in pure water, or 0.85 cm/day in soil of 20% (v/v) water content. Because of radial geometry, the water flux, and, therefore, ion flux by mass flow would decrease with distance from the root surface. With a root of radius 0.02 cm, water flux at 0.1 cm from the root surface would be one-fifth that at the root surface, or 0.03 cm/day in solution culture or 0.17 cm per day in soil with 20% (v/v) water content.

Autoradiographic Evidence for Mass Flow and Diffusion

When diffusion is the main source of supply for a particular nutrient, the nutrient will be depleted at the root surface and a concentration gradient extending perpendicular from the root will develop. Walker and Barber (1962) developed a procedure for using autoradiographs to verify the presence of ion accumulation or depletion around the root. Studies were made using ^{86}Rb as the radioactive isotope.

Corn was grown in soil that was uniformly labeled by mixing ^{86}Rb-labeled rubidium with the soil. The distribution of ^{86}Rb in the surface of a soil face on the side of a rectangular container was used to evaluate the presence of diffusive flux to plant roots. The exposed soil face was covered with Mylar film and supported with Plexiglas. In order to make an autoradiograph, X-ray film was placed between the Plexiglas and the Mylar film in a dark room and exposed long enough for the radiation to show the distribution of ^{86}Rb in the soil. Figure 4.2 shows an autoradiograph of ^{86}Rb distribution (*on the right*) and a photograph of the root distribution in the soil on the face of the box (*on the left*). The lighter areas near the roots indicate depletion of ^{86}Rb from the soil due to uptake by the corn root. The darker areas in the roots and root tips indicate concentration of ^{86}Rb in these areas. There was a concentration gradient extending perpendicularly to the root. Since corn roots extend about 3 cm/day, these depletion zones around the root had developed within a day. In this soil, the diffusion coefficient for Rb was approximately 5×10^{-8} cm²/s, as calculated from the ^{86}Rb concentration gradient around the root. One root was detached from the plant by removing a root section near the root base. The depleted zone around this detached root gradually disappeared as ^{86}Rb diffused into the depleted zone.

The properties of potassium and rubidium are relatively similar. Walker and Barber (1962) determined that the fractional uptake of potassium and rubidium by corn grown in soil used for autoradiographic experiments was

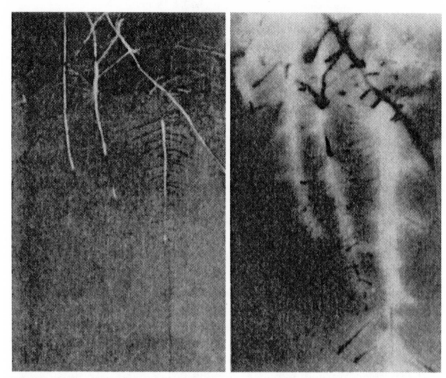

FIGURE   4.2   Autoradiograph (*right*) of corn roots shown in photograph (*left*) showing depletion (*lighter area*) of ^{86}Rb near root surface. Reprinted from Walker and Barber (1962) by permission of Martinus Nijhoff Publishers B.V.

similar over a series of treatments. Hence, rubidium was a satisfactory label for potassium. A subsequent experiment with ^{42}K-labeled potassium showed a similar area about the root. Therefore the autoradiograph in Figure 4.2 indicates changes that will also occur for potassium. Similar autoradiographic evidence for diffusive supply has been shown for phosphorus (Lewis and Quirk, 1967; Claassen et al., 1981), zinc (Wilkinson et al., 1968), and molybdenum (Lavy and Barber, 1964). Microtome sectioning of thin sections extending perpendicular to a plane of roots showed a depletion gradient perpendicular to the root (Bagshaw et. al., 1972; Claassen et al., 1986).

In Table 4.4, calculations indicated that mass flow would supply much more calcium and magnesium to the plant than was absorbed by the plant. This should cause an accumulation of these cations around the root and back diffusion into the soil. Using the same technique described in the preceding paragraphs, Barber and Ozanne (1970) obtained autoradiographs illustrating the accumulation of calcium around ryegrass (*Lolium rigidum*, Gaud) and subclover (*Trifolium subeterraneum* L.) roots. Pictures of the autoradiographs are shown in Figure 4.3. Ryegrass is low in calcium, so much more of this element was brought to the root by mass flow than was absorbed. The dark areas around the root are due to an accumulation of ^{45}Ca-labeled calcium. Because only solution calcium and not exchangeable calcium moves during water flow to the root, the experiment was conducted in a sandy soil to minimize the background level of exchangeable calcium present throughout the autoradiograph. Even though subclover absorbs more calcium than ryegrass, with the presence of calcium in subclover roots being observable from the autoradiograph, dark areas due to ^{45}Ca accumulation around the root can be seen for subclover as well as ryegrass. Lines in the picture are from segregation of soil particles when the sandy soil was poured into the container.

Calcium accumulation around the root may be either soluble calcium in solution or calcium sulfate precipitated at the root surface. Malzer and Barber (1975) demonstrated that calcium sulfate can precipitate near plant roots, growing in soil and also on plant roots growing in saturated calcium sulfate solution. The solubility of calcium sulfate in water at 25°C is 15.4 mmol/L. This concentration can readily be reached when mass flow moves calcium and sulfate to the root and water is absorbed by the root more rapidly than either of these elements.

FIGURE   4.3   Autoradiograph showing ^{45}Ca accumulation around ryegrass roots (*left*) and subclover roots (*right*), where mass flow has supplied more calcium than was absorbed.

Barber et al. (1963) also published an autoradiograph of ^{35}S accumulation about a corn root (Figure 4.4). The dense nature of the accumulation near the root rather than a pattern of back diffusion, as shown in Figure 4.3, suggests that precipitation of calcium sulfate was occurring. Barber (1974b) also published pictures of calcium carbonate accumulation around perennial roots (Figure 4.5). When perennials grow in high calcium, high pH sandy soil, the excess calcium coming to the root is precipitated as calcium carbonate. Over a long period of time, the layer of calcium carbonate accumulating around the root becomes large enough to be visible. Root respiration provides carbonate for reaction with calcium to form calcium carbonate.

This autoradiographic evidence and visual precipitation around plant roots verifies that ion concentrations at the root-soil interface can be altered by the balance between ion uptake and movement to the root by mass flow and diffusion. Accumulation or depletion occurred in accordance with changes predicted from mass flow and diffusion measurements.

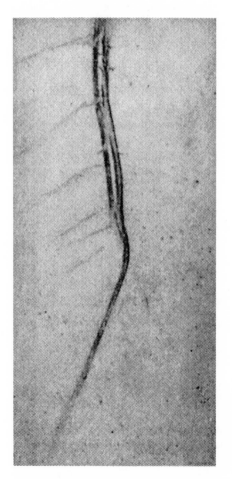

FIGURE 4.4 Accumulation of $^{35}SO_4^{2-}$ around two corn roots growing in soil. Reprinted with permission from Barber et al. (1963). Copyright 1963 American Chemical Society.

FIGURE   4.5   Calcium carbonate accumulation around a root of hickory (*Carya ovata*) growing in sandy soil (Barber, 1974a).

Factors Influencing Nutrient Supply by Diffusion

Nutrient Type and Concentration

When diffusion is a primary mechanism of nutrient supply to the root, the nutrient concentration at the root will always be lower than the nutrient concentration in the soil solution initially. This reduction in nutrient concentration will, in turn, reduce the rate of absorption. The lower the diffusion coefficient for a given nutrient, the greater will be the reduction in concentration at the root surface. Values of D_e for nitrate, phosphate, and potassium differ greatly. Values for a silt loam soil are NO_3^-, 2.5×10^{-6} cm²/s; $H_2PO_4^-$, 2.3×10^{-9} cm²/s; and K^+ 1.9 10^{-8} cm²/s. These differences in D_e result in different concentration gradients near plant roots. The differences are illustrated in Figure 4.6, where gradients predicted from the mechanistic model, to be described in Chapter 5, are shown. Values used in the model are common values for corn growing in a silt loam soil. The gradient was that predicted after 10 day's uptake by the root. Nitrate is not adsorbed by the soil; as indicated in

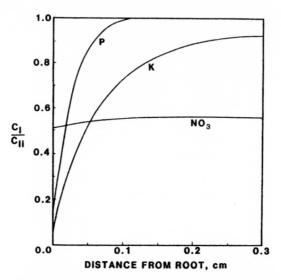

FIGURE   4.6   Calculated relative concentration (solution concentration, C_l, at 10 days as a proportion of the initial concentration, C_{li}) perpendicular to the root, where NO_3^-, $H_2PO_4^-$, and K^+ are calculated as if all supplied by diffusion.

Equation 4.6, it will have a large dC_l/dC_s and, hence, a large D_e. Phosphate, which is usually highly adsorbed by the soil, usually has a low value of dC_l/dC_s and, hence, a low D_e. The low D_e of phosphate gives a steep gradient of dC_l/dx perpendicular to the root. The value of C_l at the root will be much less than C_l initially, C_{li}. Only phosphate in soil near the root will be close enough to reach the root; hence, it is the only phosphate positionally available for uptake during the growth of the plant. On the other hand, the high D_e for nitrate allows much of the nitrate in the soil to reach the root. The concentration at the root surface after uptake and diffusion have progressed for some time will depend on both the uptake characteristics of the plant root and the soil characteristics that determine the rate of nutrient diffusion to the root.

Impedance Factor

As indicated in Equation 4.6, the rate of nutrient diffusion to the root also depends on the value of the tortuosity or impedance factor. Impedance factors are usually determined by measuring the rate of diffusion of a nonadsorbed ion, such as ^{36}Cl, through the soil or by measuring the diffusion rate for tritiated water. Values depend on soil texture, soil bulk density, and soil moisture. Warncke and Barber (1972) measured ^{36}Cl diffusion in six silt loam soils, each at three soil moisture levels and four bulk densities. There was no significant difference between soils, so only average results are shown in Figure 4.7. Soil bulk density had a larger effect on f_l at higher moisture lev-

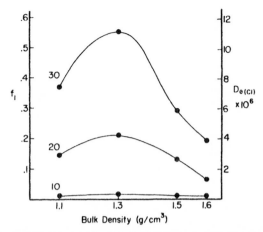

FIGURE   4.7   Average influence of soil bulk density and soil moisture level (w/w) on the rate of chloride diffusion and calculated tortuosity of the diffusion path for five silt loam soils. Values on curves are moisture levels. Reproduced from Warncke and Barber (1972) by permission of Soil Science Society of America.

els. Starting at a bulk density of 1.1 Mg/m³, f_1 increased as bulk density increased until a bulk density of 1.3 was reached. This is because more continuity of water occurs as soil air space is decreased. Increasing bulk density beyond 1.3 increased the tortuosity of the path once more, because the path around soil particles becomes more tortuous as soil particles are compressed closer together.

So and Nye (1989) investigated the effect of water level and soil bulk density on chloride diffusion in two soils and found that soil water had a large effect while soil bulk density had a small effect. Chloride diffusion can be used to measure f_1.

Barraclough and Tinker (1981) also measured impedance factors for selected soils; their values for the effect of θ on f_1 were similar to those of Warncke and Barber (1972). The f_1 values for clay and clay loam were much lower than for silt loam at similar θ values (0.1 versus 0.25). The relationship the authors obtained between θ and f_1 for silt loams and lighter textured soils is useful for calculating variation of D_e with f_1; it is shown in Equation 4.7:

$$f_1 = 1.6\,\theta - 0.172 \qquad (4.7)$$

Barraclough and Tinker investigated the effect of increasing bulk density on f_1, however, they found only a decrease. Possibly, they did not start at low enough values to give the initial increase observed by Warncke and Barber (1972).

The effect of bulk density on the soil's zinc diffusion coefficients is influenced by factors in addition to θ and f_1. Figure 4.8 gives the effect of bulk

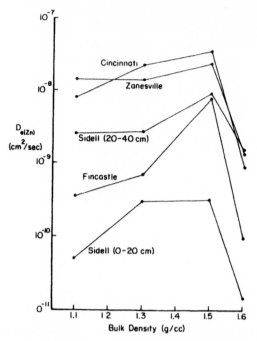

FIGURE   4.8   Influence of soil bulk density on Zn diffusion coefficients for five soils at 20% moisture (w/w). Reproduced from Warncke and Barber (1972) by permission of Soil Science Society of America.

density on Zn D_e values at 20% moisture for five silt loam soils that gave similar results for ^{36}Cl diffusion. Zinc diffusion reached a maximum at a soil bulk density of 1.5 rather than 1.3, which would indicate an interaction between f_l and dC_l/dC_s. Differences in D_e between soils reflect some wide differences with respect to dC_l/dC_s and initial levels of zinc in the soil. Values of θ would be 0.22, 0.26, 0.30, and 0.32 for bulk densities of 1.1, 1.3, 1.5, and 1.6.

Effect of Temperature on D_e

Diffusion of ions in water can be described by the Stokes–Einstein equation

$$D_l = \frac{k_B T}{6 \pi r_l \eta} \tag{4.8}$$

where k_B, T, r_l, and η are the Boltzmann constant, absolute temperature, ionic radius, and water viscosity, respectively. According to this equation, changes in the absolute temperature of 10 to 20°C will affect D_l directly only by 4 to 8%. However, temperature also causes a large change in water

TABLE 4.6   Effect of Temperature and Additions of Potassium to Raub Silt Loam on the C_l, dC_s/dC_l, and D_e Values for Potassium

Potassium added (μg/g)	°C	C_l (mmol/L)	dC_s/dC_l	$D_e \times 10^7$ (cm^2/s)
0	15	0.046	39	0.15
	29	0.089	23	0.39
50	15	0.174	12	0.50
	29	0.256	9.5	0.94
100	15	0.355	8.7	0.69
	29	0.516	3.3	2.7

Source: Reproduced from Ching and Barber (1979) by permission of the American Society of Agronomy.

viscosity: η is 1.139 at 15°C and 0.89 at 25°C. Decreasing temperature from 25°C to 15°C would thus give a D_l at 15°C only 0.78 of that at 25°C (Weast, 1982). Changes in temperature will also affect dC_l/dC_s. These changes will vary with ion and the nature of the adsorbant; some values are given in Table 4.6. Increasing temperature increased the concentration of potassium in solution, which caused a reduction in dC_l/dC_s and an increase in D_e.

Effect of Salt Concentration on D_e

Adding fertilizer salts or mineralizing organic matter to produce additional nitrates can increase the salt content of the soil solution; this may in turn influence the diffusion coefficient. Increasing salt concentration should displace some cations from the exchange sites, thereby increasing dC_l/dC_s. The magnitude of this effect will vary with the cation involved, its valence in comparison with the valence of complementary cations, the type of exchange material, and its degree of saturation with the cation. The relation between divalent and monovalent cations in solution will usually follow the relation given by the ratio law (Chapter 2) when salt content is increased. The relation usually occurring in soil is

$$\frac{a_{K_1}}{\left(a_{Ca_1} + a_{Mg_1}\right)^{1/2}} = \frac{a_{K_2}}{\left(a_{Ca_2} + a_{Mg_2}\right)^{1/2}} \qquad (4.9)$$

where subscripts 1 and 2 refer to activities before and after the soil's salt content has been changed.

The activity of divalent cations increases as the square as compared to the increase in activity of monovalents; hence, there will be a much greater increase in D_e for divalent cations as salt concentration is increased than for monovalent cations. Divalent-cation increase will be proportional to the increase in salt concentration. The increase in dC_l/dC_s for a monovalent cation, such as potassium will be relatively small.

Experimental Evidence for Nutrient-Supply Mechanisms

Mass Flow and Root Interception

To this point, we have calculated supply by diffusion by simply taking the difference between total plant uptake and calculated supply via mass flow and root interception. If we consider only situations where mass flow and root interception can supply plant demand, we can consider the relation between these two supply mechanisms and uptake. Al Abbas and Barber (1964) studied the relation between calcium uptake and calculated supply via mass flow and root interception for soybeans growing on six soils. The relation is shown in Figure 4.9. There is a close relation between these factors, but the supply averages four times the amount of uptake, which indicates that soybean roots require a high level of calcium at their surfaces before maximum calcium uptake occurs. In the same experiment, Al Abbas and Barber (1964) measured similar relations for magnesium (Figure 12.8). In this case, calculated supply agreed with uptake except where calculated soil supply was low; magnesium uptake then *exceeded* calculated supply. In this case, diffusion probably was supplying additional magnesium to the root.

Oliver and Barber (1966) attempted to evaluate the effect of root interception on calcium uptake by soybeans. They varied root interception by mixing soil with sand in order to reduce calcium supply and obtained a cor-

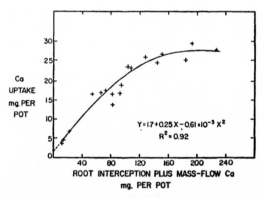

FIGURE   4.9   Relation between calculated calcium supply to roots via root interception and mass flow and calcium uptake for soybeans growing on six soils. Reprinted from Al Abbas and Barber (1964) by permission of Williams & Wilkens Co. copyright 1964.

relation having $r = 0.76$ between calculated root interception and calcium uptake minus that supplied by mass flow.

Diffusion

Calculations in Table 4.4 indicated that diffusion is the main mechanism for supplying phosphorus and potassium to roots growing in soil. If this is true, then any change in the soil that increases the rate of diffusive supply should increase uptake. Increasing soil water, or θ, was shown in Equation 4.6 to increase D_e, this was due to increases in both θ and f_1.

Cox and Barber (1992) have shown that as water content, θ, increases, the value of D_e increases according to $D_1 \theta f_1/b$ and this linearly increased predicted and observed phosphorus uptake (Figure 5.3). They used soils with θ of 0.13, 0.20, 0.26, and 0.40 at -33 kPa potential.

Place and Barber (1964) investigated the effect of θ on D_e values and subsequent ^{86}Rb uptake by corn roots. Raub silt loam was incubated with three levels of rubidium at five levels of soil water. ^{86}Rb was used to measure the rates of diffusion from a treated block of soil to an untreated block. The effect of rubidium concentration and percent soil water (w/w) on D_e is shown in Figure 4.10. D_e increased as water increased, with the degree of increase becoming greater the higher the level of exchangeable rubidium. All soils were leached after rubidium treatment, so that anion concentrations for all soils were similar.

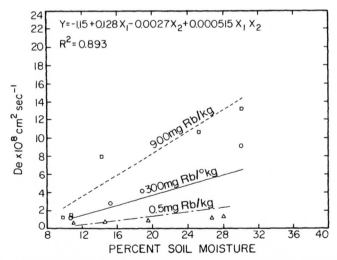

FIGURE   4.10   Relation of rubidium concentration and soil moisture level to D_e for self-diffusion of rubidium. Reproduced from Place and Barber (1964) by permission of Soil Science Society of America.

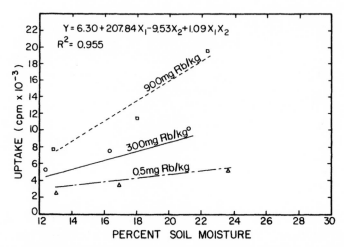

FIGURE   4.11   Relation of rubidium concentration and soil moisture level to rubidium uptake by corn roots. Reproduced from Place and Barber (1964) by permission of Soil Science Society of America.

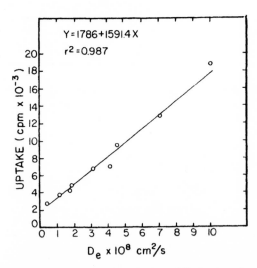

FIGURE   4.12   Relation between D_e for rubidium diffusion in soil as affected by soil moisture and rubidium concentration and rubidium uptake by corn roots. Reproduced from Place and Barber (1964) by permission of Soil Science Society of America.

Place and Barber (1964) also used this same soil to measure rubidium uptake by corn as affected by soil moisture level and rubidium concentration. Uptake was measured by exposing single roots to ^{86}Rb-labeled blocks of soil maintained at various rubidium and soil water levels; the effect of treatment on uptake is shown in Figure 4.11. The pattern was similar to that for diffusion; hence, correlating D_e with ^{86}Rb uptake gave the relation shown in Figure 4.12. Clearly, ^{86}Rb uptake was closely correlated with D_e and varied with soil water and exchangeable rubidium level. Rubidium and potassium behave similarly with regard to plant absorption.

In summary, calculations of the amounts of calcium and magnesium supplied by root interception and mass flow are in reasonable agreement with uptake. Autoradiographic evidence indicates an accumulation outside the root where calculations had predicted accumulation and depletion where calculations indicated that diffusion was supplying part of the nutrient to the root. Such evidence substantiates the usefulness of describing nutrient uptake by the mechanisms involved.

In Chapter 5 we will describe a mathematical mechanistic nutrient uptake model that can be used to predict nutrient uptake by the plant. The model describes the nutrient uptake process using information that has been discussed in the preceding chapters.

REFERENCES

Al Abbas, H., and S. A. Barber. 1964. The effect of root growth and mass flow on the availability of soil calcium and magnesium to soybeans in a greenhouse experiment. *Soil Sci.* **97**:103–107.

Albrecht, W. A., E. R. Graham, and H. R. Shepard. 1942. Surface relationships of roots and colloidal clay in plant nutrition. *Am. J. Bot.* **29**:210–213.

Bagshaw, R., L. V. Vaidyanathan, and P. H. Nye. 1972. The supply of nutrient ions by diffusion to plant roots in soil. VI. Effects of onion plant roots on pH and phosphate desorption characteristics in a sandy soil. *Plant Soil.* **37**:627–639.

Barber, S. A. 1974a. Nutrients in the soil and their flow to plant roots. In J. K. Marshall, Ed. *The Belowground Ecosystem: A Synthesis of Plant-Associated Processes.* Range Sci. Series No. 26, Colorado State University, Fort Collins. Pp. 161–168.

Barber, S. A. 1974b. The influence of the plant root on ion movement in soil. In E. W. Carson, Ed. *The Plant Root and Its Environment.* University Press of Virginia, Charlottesville. Pp. 525–564.

Barber, S. A., and R. A. Olson. 1968. Fertilizer use on corn. In L. B. Nelson, M. H. McVickar, R. D. Munson, L. F. Seatz, S. L. Tisdale, and W. C. White, Eds. *Changing Patterns in Fertilizer Usage.* Soil Science Society of America, Madison WI. Pp. 163–188.

Barber, S. A., and P. G. Ozanne. 1970. Autoradiographic evidence for the differential effect of four plant species in altering the calcium content of the rhizosphere soil. *Soil Sci. Soc. Am. Proc.* **34**:635–637.

Barber, S. A., J. M. Walker, and E. H. Vasey. 1963. Mechanisms for the movement of plant nutrients from the soil and fertilizer to the plant root. *J. Arg. Food Chem.* **11**:204–207.

Barraclough, P. B., and P. B. Tinker. 1981. The determination of ionic diffusion coefficients in field soils. I. Diffusion coefficients in sieved soils in relation to water content and bulk density. *J. Soil Sci.* **32**:225–236.

Ching, P. C., and S. A. Barber. 1979. Evaluation of temperature effects on potassium uptake by corn. *Agron. J.* **71**:1040–1044.

Claassen, N., and S. A. Barber. 1976. Simulation model for nutrient uptake from soil by a growing plant root system. *Agron. J.* **68**:961–964.

Claassen, N., L. Hendriks, and A. Jungk. 1981. Erfassung der Mineralstoffuerteilung im wurzelnchen Boden durch Autoradiographie. *Pflanzenernahrung Bodenkunde* **144**:306–316.

Claassen, N., K. M. Syring, and A. Jungk. 1986. Verification of a mathematical model by simulating potassium uptake from soil. *Plant Soil* **95**:209–220.

Cox, M. S., and S. A. Barber. 1992. Soil phosphorus levels for equal P uptake from four soils with different water contents at the same potential. *Plant Soil* **143**:93–98.

Dittmer, H. J. 1940. A quantitative study of the subterranean members of soybean. *Soil Conserv.* **6**:33–34.

Edwards. O. W., and E. O. Huffman. 1959. Diffusion of aqueous solutions of phosphoric acid at 25°. *J. Phys. Chem.* **63**:1830–1833.

Hanway, J. J. 1973. Experimental methods for correlating and calibrating soil tests. In L. M. Walsh and J. D. Beaton, Eds. *Soil Testing and Plant Analysis.* Soil Science Society of America, Madison, WI. Pp. 55–66.

Lavy, T. L., and S. A. Barber. 1964. Movement of molybdenum in the soil and its effect on availability to the plant. *Soil Sci. Soc. Am. Proc.* **28**:93–97.

Lewis, D. G., and J. P. Quirk. 1967. Phosphate diffusion in soil and uptake by plants. III. ^{31}P movement and uptake by plants as indicated by ^{32}P autoradiography. *Plant Soil* **26**:445–453.

Lund, Z. F., and H. O. Beals. 1965. A technique for making thin sections of soil with roots in place. *Soil Sci. Soc. Am. Proc.* **29**:633–634.

Malzer, G. L. and S. A. Barber. 1975. Precipitation of calcium and strontium sulfates around plant roots and its evaluation. *Soil Sci. Soc. Am Proc.* **39**:492–495.

Mengel, D. B., and S. A. Barber. 1974. Development and distribution of the corn root systems under field conditions. *Agron. J.* **66**:341–344.

Nye, P. H., and P. B. Tinker. 1977. *Solute Movement in the Soil-Root System.* Blackwell Scientific Publishers, Oxford, England.

Oliver, S., and S. A. Barber. 1966. An evaluation of the mechanisms governing the supply of Ca, Mg, K, and Na to soybean roots (*Glycine max*). *Soil Sci. Soc. Am. Proc.* **30**:82–86.

Parsons, R. 1959. *Handbook of Electrochemical Constants.* Academic Press, New York.

Place, G. A., and S. A. Barber. 1964. The effect of soil moisture and rubidium concentration on diffusion and uptake of rubidium-86. *Soil Sci. Soc. Am. Proc.* **28**:239–243.

So, H. B., and P. H. Nye. 1989. The effect of bulk density, water content and soil type on the diffusion of chloride in soil. *J. Soil Sci.* **40**:743–750.

Talibudeen, O., J. D. Beasley, P. Lane, and N. Rajendran. 1978. Assessment of soil potassium reserves available to plant roots. *J. Soil Sci.* **29**:207–218.

Walker, J. M., and S. A. Barber. 1962. Uptake of rubidium and potassium from soil by corn roots. *Plant Soil* **17**:243–259.

Warncke, D. D., and S. A. Barber. 1972. Diffusion of zinc in soils. II. The influence of soil bulk density and its interaction with soil moisture. *Soil Sci. Soc. Am. Proc.* **36**:42–46.

Weast, R. C. 1982. *Handbook of Chemistry and Physics.* The Chemical Rubber Co., Cleveland.

Wilkinson, H. F., J. F. Loneragan, and J. P. Quirk. 1968. The movement of zinc to plant roots. *Soil Sci. Soc. Am. Proc.* **32**:831–833.

CHAPTER 5

Modeling Nutrient Uptake by Plant Roots Growing in Soil

Developing an objective mechanistic model requires understanding the nutrient-uptake process well so that the equations used accurately describe the processes involved. In this chapter, I will outline the development of one such model (Barber and Cushman, 1981). A mechanistic model differs from a regression model, where coefficients are obtained statistically for unknown processes going on between "black boxes." In a mechanistic model, equations describing nutrient influx are combined with equations describing plant growth in order to describe nutrient uptake. Once a model has been developed and verified, it is useful for predicting the consequences of changing various soil and plant parameters with respect to nutrient uptake by the plant.

DEVELOPMENT OF THE MODEL

In developing the model, supply from the soil is assumed to be from a combination of mass flow and diffusion; supply by root interception will be ignored initially. This will have little effect on the supply of many nutrients, where root interception represents less than a few percentages of total supply. Uptake of nutrients by the plant root will be assumed to follow a Michaelis–Menten relationship between concentration and influx, as described in Chapter 3. In Chapter 4, I described radial diffusive flux to the root with Equation 4.3; however, I will now consider diffusion and mass flow acting simultaneously to supply nutrients to the root surface, which gives the equation

$$J_r = D_e \frac{\partial C_s}{\partial r} + v_0 C_1 \tag{5.1}$$

where J_r is flux to the root, D_e is the effective diffusion coefficient, r is the radial distance, C_s is the concentration of ions on the solid phase that readily equilibrates with C_1, the concentration of ions in the soil solution, and v_0 is the rate of water flux into the root.

For conservation of solute and because the area at r becomes smaller as r decreases

$$\frac{\partial 2\pi r J_r}{\partial r} = \frac{2\pi r \partial C_s}{\partial t} \tag{5.2}$$

Adding Equations 5.1 and 5.2, we obtain

$$\frac{\partial(rD_e \partial C_s / \partial r + rvC_1)}{\partial r} = \frac{r\partial C_s}{\partial t} \tag{5.3}$$

Using the relation $\partial C_s = \partial C_1 b$ (i.e., $b = \partial C_s / \partial C_1$) to convert C_s to C_1 and $r_0 v_0 = rv$ at r_0 we get

$$\frac{1}{r} \frac{\partial}{\partial r} \left(\frac{rD_e \partial C_1 b}{\partial r} + r_0 v_0 C_1 \right) = \frac{\partial C_1 b}{\partial t} \tag{5.4}$$

This can be reduced to

$$\frac{\partial C_1}{\partial t} = \frac{1}{r} \frac{\partial}{\partial r} \left(rD_e \frac{\partial C_1}{\partial r} + \frac{r_0 v_0 C_1}{b} \right) \tag{5.5}$$

where $r_0 =$ root radius.

When used with the appropriate boundary conditions, this continuity equation can be used to calculate the concentration gradient radially from the root as it changes with time. From this, we can calculate C_{lo}, the concentration in solution at the root surface, as it changes with time.

The development of the model using the continuity equation has the following assumptions:

1. The soil is homogeneous and isotropic.
2. Soil water conditions are maintained essentially constant near field capacity. No appreciable soil water gradient perpendicular to the root is assumed in the calculations of nutrient flux. The soil water gradient at this water level is usually relatively flat.
3. Nutrient uptake occurs only from nutrients in solution at the root surface.
4. Root exudates or microbial activity on the root surface do not influence nutrient flux.
5. Nutrients are moved to the root by a combination of mass flow and diffusion.
6. The relation between net influx and concentration can be described by Michaelis–Menten kinetics.
7. The roots are assumed to be smooth cylinders with no root hairs or mycorrhizae (except as otherwise noted).
8. D_e and b are assumed independent of concentration. (Because this is not true for some ions, values averaged over the concentration range of interest will be used in some cases.)
9. Influx characteristics are not changed by root age or plant age (except as noted).
10. Influx is independent of the rate of water absorption.

Some comments on these assumptions may be helpful. The first assumption guarantees the independence of location with respect to soil characteristics involved in determining nutrient flux. Where soil volumes differ, uptake can be calculated separately for each volume. The second assumption simplifies the nutrient transport mechanism, and the third assumption is necessary in order for assumption six to be used. Because there is limited evidence for effects from these factors, the fourth assumption is used. It assumes any effects are usually minimal. The fifth assumption describes what has been shown experimentally, and the sixth assumption selects a relationship between influx and concentration in solution. Other relations could be used, but this is the commonly used one. The seventh assumption is necessary to have radial symmetry. Root hairs may be taken into account by calculating flux to the root hair surface as well as the root cylinder. The eighth assumption is necessary to linearize the nutrient-transport equation.

For such nutrients as potassium, the assumption is approximately correct; for phosphorus, b and D_e vary with C_l. In this case, an average value for the C_l range between C_{li} and the mean value at r_0 can be used. This average value is assumed to be independent of C_l. The ninth assumption is used only to simplify calculations; it can be modified whenever time-dependent changes interest the investigator. The tenth assumption also simplifies calculations. At high values of C_l, there is information indicating that In is affected by v_0 for some nutrients. However, size of v_0 usually does not exert much effect on In. As more detailed information becomes available for some of the parameters assumed constant, variability in the parameters can be accounted for by incorporating functions describing their variations into the model.

Equation 5.5 requires an initial condition and two boundary conditions before it can be used to describe the concentration gradient perpendicular to the root. The initial condition is simply $C_{li} = C_{l0}$ at $t = 0$, which describes a uniform nutrient distribution in the vicinity of the root.

The inner boundary condition at the root surface, where $r = r_0$, can be developed by assuming that uptake follows Michaelis–Menten kinetics, so

$$J_r = \frac{I_{max}(C_l - C_{min})}{K_m + C_l - C_{min}}, \quad r = r_0, \ t > 0 \tag{5.6}$$

If we now substitute J_r from Equation 5.1 and use the relation $bC_l = C_s$, we get

$$Deb \frac{\partial C_l}{\partial r} + v_0 C_l = \frac{I_{max}(C_l - C_m)}{K_m = C_l - C_{min}}, \quad r = r_0, \ t > 0 \tag{5.7}$$

Equation 5.7 is the resultant expression for the inner boundary.

If we assume that roots do not compete for nutrients, the outer boundary, r_l will remain constant, where r_l is the half-distance between roots.

$$C_l = C_{li}, \quad r = r_l, \ t > 0 \tag{5.8}$$

If the concentration gradients extending from adjacent roots *do* overlap, then the outer boundary becomes $J_r = 0$ at $r = r_l, t > 0$.

The technique for solving this equation is described by Barber and Cushman (1981). Numerical solution including the Crank–Nicholson method to solve a finite-difference form of Equation 5.5 is involved.

Solving this equation gives the influx at the root surface with time. When diffusion supplies part of the nutrients to the root, concentration at the root decreases with time as uptake proceeds. Decreasing concentration at r_0 in turn decreases influx with time. Total uptake can then be obtained by summing influx over time; this would be the approach for a root that was not growing. In the usual situation for annual plants, the plant starts from seed, and new roots are continually being produced. Uptake by each new root starts at progressively later times during plant growth. Initial uptake by the plant roots can be expressed by

$$T = 2\pi r_0 L_0 \int_0^{t_m} J_r(r_0, S)dS \tag{5.9}$$

where T is total uptake at time t_m, L_0 is initial root length, and $J_r(r_0 S)$ is the influx at the root surface S.

If we incorporate root growth into this relation, we get

$$T = 2\pi r_0 L_0 \int_0^{t_m} J_r(r_0, S)dS$$

$$+ 2\pi r_0 \int_0^{t_m} \frac{df}{dt} \int_0^{t_{m-t}} J_r(r_0, S)dSdt \tag{5.10}$$

where df/dt is the rate of root growth. Equation 5.10 can then be solved to calculate nutrient uptake by plant roots growing in uniform soil systems (Barber and Cushman, 1981).

MODEL PARAMETERS

Parameters

The mathematical model has the following 11 parameters:

1. I_{max}, maximum influx at high concentrations, nmol/m² · s
2. K_m, nutrient concentration in solution $-C_{min}$ where In is one-half I_{max}, μmol/L
3. C_{min}, concentration in solution below which In ceases, μmol/L
4. L_0, initial root length, cm
5. k, rate of root growth, cm/s
6. r_0, mean root radius, cm
7. v_0, mean water influx, cm/s
8. r_1, half-distance between root axes, cm
9. D_e, effective-diffusion coefficient for the nutrient in the soil, cm²/s
10. b, buffer power of nutrient on the solid phase for nutrient in solution, dimensionless
11. C_{li}, initial concentration of the nutrient in the soil solution, μmol/L

Measuring the Model Parameters

The first three parameters, which describe In versus C_l, are measured in solution culture as outlined in Chapter 3. Parameters 4, 5, and 6 give the amount of root surface, its geometry, and its rate of change with time. They are determined by measuring root length and radii of roots growing in the soil investigated, after several periods of growth. A value for k can be calculated from root lengths at various values of t; depending on the growth period investigated, rate of root length increase can be described by either a linear, exponential, or sigmoid expression. Parameter 7, v_0, can be calculated from change in root surface area with time and total water use. Total water use is obtained by measuring water additions to weighed pots. Evaporative losses are estimated by subtracting water loss over the same periods from pots without growing plants. The half-distance between root axes is calculated from root density L_v in the soil. The approximate relation is $r_1 = 1/(\pi L_v)^{1/2}$. The last three parameters, which determine the soil's nutrient supply, are measured on soil in the laboratory using the same temperature, moisture, bulk density, and aeration conditions that will occur in the verification experiment. Values of C_{li} are obtained by measuring the concentration in displaced soil solution. Values for b are obtained from desorption curves of ΔC_s versus ΔC_l, and D_e is measured or estimated according to Equation 4.6.

VERIFICATION OF THE MODEL

Early modeling of nutrient fluxes around the root was done by Bouldin (1961), Passioura (1963), Nye (1966), Olsen and Kemper (1968), and Nye and Marriott (1969). These models, using varied assumptions, provided a theoretical description of the nutrient–concentration gradient perpendicular to the root. Verification of the model was difficult, however, because of the inability to measure the nutrient gradient accurately. Calculated gradients have been shown to be reasonably close to gradients shown by autoradiographs (Walker and Barber, 1962; Bhat and Nye, 1973).

Direct analytical measurements of the gradient were attempted by Farr et al. (1969), Bagshaw et al. (1972), and Claassen et al. (1986) where a plane of roots was grown across a block of soil. After a period of uptake, soil thin sections were taken to determine nutrient distribution perpendicular to the plane. These experiments indicate a gradient that was consistent with the size of the D_e value for the nutrient–soil combination. Claassen et al. (1986) used this type of experiment to verify their uptake model. They grew rape in soil at three potassium levels and measured the potassium gradient perpendicular to the root. They compared the observed values with the gradient calculated by the uptake model and obtained the close agreement shown in Figure 5.1.

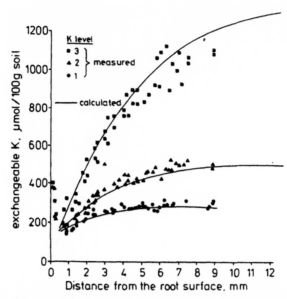

FIGURE   5.1   Measured (points) and calculated potassium concentration using the uptake model (lines) in the vicinity of rape roots (4 days old) at three different potassium levels. Reproduced from Claassen et al. (1986) with permission of Martinus Nijhoff Publishers, Dortrecht, The Netherlands.

With development of the model to predict nutrient uptake, verification can be accomplished by comparing observed uptake and predicted uptake. Brewster et al. (1976) and Claassen and Barber (1976) combined the theoretical model for the single root with an expression for the rate of root growth and predicted nutrient uptake over a period of growth.

An improvement in the model made by Barber and Cushman (1981) also accounted for competition between adjacent roots, which commonly occurs for potassium uptake. Itoh and Barber (1983b) accounted for the effect of uptake by root hairs that may occur for calculations of phosphorus uptake.

Pot Experiments

The models have been validated with a series of pot experiments using a range of plant species and soils. In the first experiment of this type Claassen and Barber (1976) grew corn for 17 days on four soils, one of which had five levels of potassium. Predicted potassium uptake was calculated using the Claassen and Barber model; observed uptake was from plant analysis. While a correlation of $r^2 = 0.87$ was obtained, the regression coefficient was 1.57, indicating more calculated uptake than that observed. A pot experiment of phosphorus uptake from six soils varying widely in properties was conducted by Schenk and Barber (1979b) using the Claassen and Barber model. An r^2 of 0.87 was obtained while the regression coefficient was 1.09. There was no root-to-root competition, so predicted uptake agreed more closely with observed uptake from plant analysis. There apparently was not appreciable root-to-root competition since phosphorus has a much smaller D_e than potassium. In both experiments the soil C_{li}, C_{si}, and θ, volumetric water, was measured for each soil and treatment and C_{li}, b, and D_e values were calculated. In subsequent experiments the Barber–Cushman model is used so that root-to-root competition would be included in the calculations. Figure 5.2 shows a comparison of predicted potassium uptake with observed potassium uptake in an experiment by Li and Barber (1991) using corn and three legume species grown in Chalmers silt loam (fine-silty, mixed, mesic Typic Haplaquolls) limed to pH levels of 5.72, 6.30, 7.22, and 8.30. The r^2 of 0.99 and regression coefficient of 0.98 indicated close agreement. More detail on this experiment is given in Chapter 21.

Cox (1991) compared predicted vs. observed phosphorus uptake (Figure 5.3) in an experiment where four soils varying in volumetric water levels at 33 kPa tension were used. The soils were Raub silty clay loam subsoil (fine-silty, mixed, mesic Aquic Argiudoll), Blount silt loam (fine, illitic, mesic Aeric Ochraqualf), Zanesville silt loam (fine-silty, mixed, mesic Typic Fragiudalf), and a Chelsea fine sand (mixed, mesic, Alfic Udipsamment). The soil volumetric water contents at 33 kPa tension were 0.40, 0.26, 0.20,

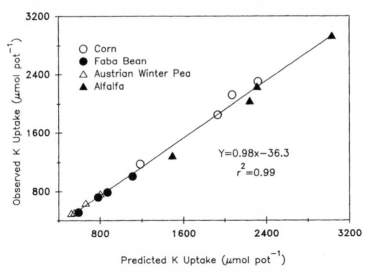

FIGURE 5.2 Relation between predicted potassium uptake using the Barber–Cushman model and observed uptake for corn, alfalfa, faba bean, and Austrian winter pea, grown at four soil pH levels. Reproduced from Li and Barber (1991) by permission of Marcel Dekkar Inc.

and 0.13 for Raub, Bount, Zanesville, and Chelsea soils, respectively. The soils were fertilized with 0, 200, and 400 Mg P/kg of soil using reagent grade monocalcium phosphate and then equilibrated for 40 days at 33 kPa soil water tension.

Corn was grown in 2.5 kg of soil (oven-dry basis) in 3-L pots. Five-day-old corn plants with four roots trimmed to 4 cm were planted in each pot and grown in a controlled climate facility for 13 days. Shoots and roots were then harvested and analyzed for phosphorus to calculate observed phosphorus uptake.

The relation between predicted phosphorus uptake and observed phosphorus uptake for corn grown for 12 days in six diverse soils is shown in Figure 5.3. The relation had an r value of 0.995 and a regression coefficient of 0.99. This experiment showed that the Barber–Cushman uptake model closely predicted phosphorus uptake by corn in this experiment, hence the model included all the significant parameters affecting phosphorus uptake and soil type had no influence other than the effect on the measured parameters.

Table 5.1 gives the regression and correlation coefficients for a number of experiments where predictions were made using the Barber–Cushman model for phosphorus and potassium. In an experiment by Itoh and Barber (1983b) where root hairs also had an effect on uptake, additional measurements were needed as described in Chapter 7.

Since the nutrient uptake predicted by the model may vary with nutrient, plant species, soil type, and environmental conditions, additional validation experiments were conducted using a wide range of conditions. The relation between predicted and observed uptake for these experiments given in terms of

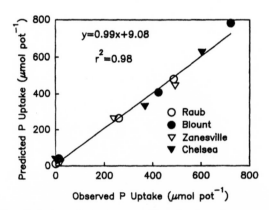

FIGURE   5.3   Relation between predicted phosphorus uptake using the Barber–Cushman model and observed phosphorus uptake by corn from four soils each at three levels of phosphorus. Reproduced from Cox (1991).

TABLE 5.1 Correlation and Regression Coefficients for the Relation between Predicted and Observed Nutrient Uptake Obtained in a Series of Pot Experiments Conducted to Validate the Model

Nutrient	Plant species	Soil	Treatment	Correlation coefficient	Regression coefficient	Source
Phosphorus	Six species	Raub silt loam	One P level	0.89[a]	0.98	Itoh and Barber (1983a) Figure 7.8
Potassium	Corn	Four Indiana soils	One K level	0.96	1.11	Shaw et al. (1983) Figure 10.9
Potassium	Soybeans	Raub silt loam	2 K × 2 bulk density	0.90	1.19	Silberbush et al. (1983)
Potassium	Soybeans	Raub silt loam	2 P × 2 K levels	0.97	1.09	Silberbush et al. (1983)
Phosphorus	Soybeans	Raub and Chalmers	3 cultivars × 2 soils	0.96	1.08	Silberbush and Barber (1983b) Figure 10.6
Potassium	Soybeans	Chalmers, Raub	3 cultivars × 2 soils	0.95	0.95	Silberbush and Barber (1983b)
Phosphorus	Sorghum Carrot	Six Malaysian		0.96 0.93	0.79 0.84	Bidin (1982)
Phosphorus	Corn	Five Columbian Two Indiana	2 rates P	0.99	0.99	Blanco (1989)
Phosphorus	Corn	Chalmers	5 pH levels	0.99	0.93	Chen and Barber (1990)
Potassium	Corn	Chalmers	5 pH levels	0.99	0.94	Chen and Barber (1990)

[a] Root hairs included in the calculation.

119

regression and correlation coefficients is shown in Table 5.1. When the experiment is discussed elsewhere in this book, the appropriate figure number is listed as the reference, otherwise the publication is cited. The correlation coefficients of all experiments indicated agreement between observed and predicted nutrient uptake. The regression coefficients were close to 1.0 for most experiments, which indicates the predicted uptake agreed with the observed uptake.

Field Experiments

Schenk and Barber (1980) grew three corn hybrids in the field and compared phosphorus uptake predicted from the Claassen-Barber model to observed phosphorus uptake. Root length and diameter measurements were made six days after planting for initial root length and at 47, 54, and 68 days. Uptake kinetics were adjusted for plant age according to the relation found by Jungk and Barber (1975; see Figure 9.5). Corn plants were harvested at each date listed to determine observed phosphorus uptake. Predicted phosphorus uptake was correlated with observed uptake with an r value of 0.91 and a regression equation of $y = 0.67x + 0.98$ where x is observed uptake and y is predicted phosphorus uptake. There was a high correlation between predicted and observed values, but predicted uptake averaged only 67% of observed uptake. The lower values may have been due to low C_{li} values, or to the effect of root hairs or mycorrhiza in increasing phosphorus uptake. Because of different soil parameters for each, phosphorus uptake for roots growing in the subsoil was calculated separately from uptake by roots growing in top soil.

Silberbush and Barber (1984) evaluated the model for predicting potassium uptake by five soybean cultivars growing on Chalmers silt loam, the Barber–Cushman model was used to predict uptake. Uptake was predicted separately for each soil layer, because the soil's supply parameters varied by layer. Predicted potassium uptake is compared with observed uptake in Figure 5.4; experiments conducted in two additional years gave similar results. Prediction of phosphorus uptake was also evaluated. On a soil with C_{li} of 52 μmol/L, predicted uptake agreed with observed uptake; however, on a soil with a C_{li} of 7.8 μmol/L, predicted uptake was less than observed uptake. This was probably due to the presence of root hairs or mycorrhizae or both, since these two factors affect phosphorus uptake but usually not potassium uptake. Measuring and predicting potassium uptake as well as phosphorus uptake on the same experiment is a method for determining if low predicted phosphorus uptake is due to faulty root measurements or to root hairs and/or mycorrhizae.

Field results obtained by Schenk and Barber (1980) for corn and by Silberbush and Barber (1984) for soybeans indicate that the prediction model can be used in the field as well as in pot experiments. The results for both pot experiments and field experiments indicate that prediction models described

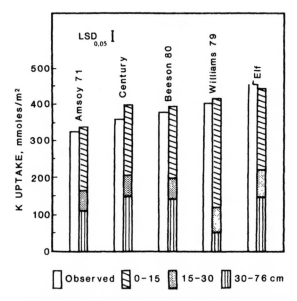

FIGURE   5.4   Relation between observed and predicted potassium uptake by five soybean cultivars growing in the field on Chalmers silt loam using the Barber–Cushman model. Predicted potassium uptake is shown for each soil depth. Reproduced from Silberbush and Barber (1984) by permission of American Society of Agronomy.

nutrient uptake satisfactorily in most cases. This means that for the soil-plant systems used, nutrient uptake could usually be described by assuming that the plant root did not exude anything to solubilize nutrients and that microorganisms in the rhizosphere did not have a significant effect on the uptake process. Simple uptake of nutrients from the soil solution by the root or root hairs was sufficient to describe uptake and predict amounts absorbed by the plant. This does not mean that the root behaves as a simple sink for nutrients in all cases. The effect of root exudates on nutrient uptake is discussed in Chapter 6, and in Chapter 7, the effect of mycorrhizae, root hairs, and soil microorganisms is discussed in relation to conditions where these properties appear to have an effect.

The results of the verification experiments indicate that the model parameters used here accurately describe uptake. In those cases where root hairs make significant contributions to uptake (usually for phosphorus) their measurement and inclusion result in agreement between predicted and observed uptake. Where mycorrhizae contribute to uptake, an additional relation will need to be included; this, however, usually only occurs at low soil phosphorus levels. Where root exudates (particularly H^+) contribute to uptake, this factor will also have to be recognized.

Since the model describes the uptake process rather accurately, it is useful to get an estimate of the relative effect on nutrient uptake of each parameter in the model by changing the value of each parameter separately while holding other parameters constant. This is called a sensitivity analysis.

The effect of changing the level of one parameter may be influenced significantly by the levels chosen for other parameters. If the nutrient supply to the root is so great that influx approaches I_{max}, changing the level of a parameter to increase nutrient supply to the root will have little effect on uptake, since the root is already absorbing at its maximum rate. If, on the other hand, soil parameters are greatly limiting nutrient supply to the root, changing parameters affecting influx into the root, such as I_{max}, will have little effect on predicted uptake. For the discussion on sensitivity analysis, effects of changing parameters are described under the assumption that the plant is getting insufficient quantities of the nutrient in question but is not greatly deficient. Hence, increases in supply by the soil or in rate of absorption by the root will result in increased uptake.

SENSITIVITY ANALYSIS

When a mathematical model accurately describes nutrient uptake, it is useful to conduct a sensitivity analysis to determine relative effects on predicted uptake when each parameter is changed independently. Silberbush and Barber (1983a) conducted such a sensitivity analysis using data from an experiment that investigated potassium uptake by Williams soybeans grown in pots of Raub silt loam in a growth chamber. The initial values used for each parameter are given in Table 5.2. Root growth was assumed to follow a linear relationship with time for the growth period from 10 to 20 days after planting. Four plants were grown in 2.5-L pots of soil having a bulk density of 1.25 Mg/m^3. Volumetric moisture content was maintained near 28% by daily addition of water. Sensitivity analysis was conducted by systematically changing individual values from 0.5 to 2.0 times the initial value, while all other parameters remained at the initial values shown in Table 5.2. For this analysis, the model was used to predict potassium uptake for a period of 10 days.

Results of the sensitivity analysis are shown in Figure 5.5. Predicted potassium uptake was most sensitive to k, root length, since r_1 was held constant this predicts a pattern of uptake where soil volume would be increasing at the same rate as root length. The second most sensitive parameter was r_0. Increasing r_0 increases the root's surface area for uptake, and as long as r_1 is sufficiently large so that little competition occurs, predicted potassium uptake increases. This increased uptake is curvilinear because of changes in radial geometry at the root surface as r_0 increases.

The soil supply parameters C_{li}, b, and D_e were the next most sensitive parameters, which suggests that with a constant root surface area, soil supply of potassium was restricting potassium uptake more than the kinetics of potassium absorption by the root. Changing I_{max}, K_m, and C_{min} had smaller

TABLE 5.2   Soil and Plant Parameters for the Cushman Mathematical Model, Their Symbols and Values for Potassium Uptake by Williams Soybeans Growing on Raub Silt Loam

Symbol	Parameter	Initial value
D_e	Effective diffusion coefficient in bulk soil	$3.47 \times 10^{-8} \text{cm}^2/\text{s}$
b	Soil-buffer power	24.0
C_{li}	Initial concentration in soil solution	250 μmol/L
v_0	Water influx to root	$5.0 \times 10^{-7} \text{cm/s}$
r_0	Root radius	0.015 cm
r_1	Half-distance between roots	0.2 cm
I_{max}	Maximal influx rate	70.5 nmol/m$^2 \cdot$s
C_{min}	Minimal concentration, where $In = 0$	1.4 μmol/L
Km	Concentration $-C_{min}$ when $In^a = \frac{1}{2} I_{max}$	10.3 μmol/L
L_0	Initial root length	250 cm
k	Root growth rate	0.03 cm/s

Source: Silberbush and Barber (1983a).
[a] In = net influx at r_0.

effects on predicted potassium uptake, and changing v_0 had little effect, indicating that most of the potassium was reaching the root by diffusion. Root competition became a factor only as r_1 was reduced below its initial value.

The degree of difference between parameters will vary with relative size of the parameters used. While it is interesting to evaluate each parameter separately, there are groups of parameters that interact. Hence, they are more realistically evaluated according to their interdependence.

Root Morphology Parameters

Root morphology and root density in the soil are described by k, r_0, and r_1. When only k varies, soil volume varies as well and predicted potassium uptake is linearly related to changes in k. A more realistic evaluation would be where soil volume is held constant so that r_1 decreases as k increases; we can then evaluate varying root length in terms of either r_1 or k. Figure 5.6 shows the relation between k and predicted potassium uptake; curve represents the case where all other parameters are held at their initial levels as in Figure 5.5. The effect of k in this case is linear because with a constant r_1 soil volume increases with the increase in k.

Curve kr_1 in Figure 5.6 represents the case where the sum of soil and root volumes remain constant. As k increases, r_1 decreases; this has no effect on predicted uptake until r_1 becomes small enough that competition between roots begins. Then predicted potassium uptake as compared to where r_1 is held constant becomes progressively less.

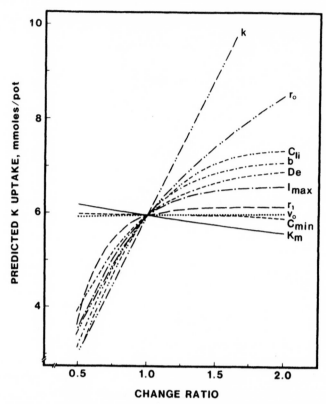

FIGURE   5.5   A sensitivity analysis of predicted potassium uptake using the Barber–Cushman model, showing the effect on predicted potassium uptake of varying each parameter while holding the remaining parameters constant at the level shown in Table 5.2. Reproduced from Silberbush and Barber (1983a) by permission of American Society of Agronomy.

Curve kr_1r_0 in Figure 5.6 represents the case where both root volume and soil volume are held constant. As k increases both r_1 and r_0 decrease, since the values $k2\pi r_0$ and $k2\pi r_1$ were held constant. We are thus simulating a system with increasingly finer roots while maintaining the same root volume growing in the same soil volume. Predicted potassium uptake was less than for curve kr_1 because of the reduction in root surface area as r_0 was reduced when k was increased. However, when k was reduced, predicted potassium uptake increased because r_0 increased.

Soil Parameters

When a soil is fertilized with potassium fertilizer, the three soil parameters are affected interdependently. First C_{li} increases, which may reduce the value

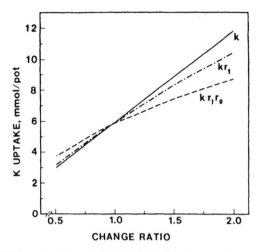

FIGURE   5.6   Effect on predicted potassium uptake of varying k, rate of root growth, where k, only k is varied; kr_1, r_1 is varied as k varies so that the sum of soil plus root volume remains constant; and kr_1r_0, as k is varied, both r_1 and r_0 are varied so that soil volume and root volume each remain constant (S. A. Barber, unpublished data).

of b since b equals $\Delta C_s/\Delta C_{li}$ and this may be curvilinear with C_{li}. The value of D_e is inversely related to the value of b so it will increase as b decreases. However, because of differences in the numbers and strengths of bonding sites for potassium, the relationships between C_{li}, b, and D_e will likely be different for each soil.

A sensitivity analysis of the effect of changing C_{li} on predicted potassium uptake is shown in Figure 5.7. Curve A is the same as that shown for C_{li} in Figure 5.5; all other parameters were held constant at their initial values. This represents the case of adding potassium where b is not affected by C_{li}, which is often the case for potassium. Curve B represents the case where b and D_e were varied with C_{li} according to the interdependence of these parameters; C_{si}, the diffusible potassium, was held constant. This is one extreme when different soils are being compared. The small difference between curves A and B indicates that C_{li} may be a useful parameter for determining available potassium. This is particularly true when competition between roots is negligible.

Interaction of Soil Parameters and Half-Distance between Roots

The significance of C_{li} as compared to b, D_e, and C_{si} ($C_{li} \times b$) is influenced greatly by r_1, which affects degree of root competition. As r_1 decreases and root competition for potassium increases, C_{si} becomes more important in determining uptake relative to C_{li}. A sensitivity analysis was conducted where all plant parameters were held constant except r_1, and soil parameters were varied according to the values shown in Table 5.3. Two values were

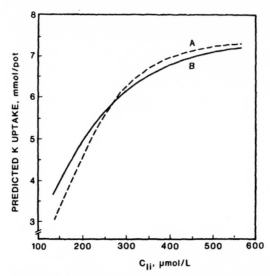

FIGURE 5.7 Effect on predicted potassium uptake of varying initial potassium concentration in solution C_{li} where A, all other parameters remained constant; and B, C_{li} time b remained constant with D_e varied according to the variation in b. Reproduced from Silberbush and Barber (1983a) by permission of American Society of Agronomy.

used for r_1, 0.09 and 0.45 cm, which represent root densities of 400 and 5 km/m³ (40 and 0.5 cm/cm³), respectively. For the soil system, C_{si} remained constant, when C_{li} decreased as more of the potassium was adsorbed on the solid phase; this caused an increase in b and a decrease in D_e. Predicted uptake was calculated with the Barber–Cushman model. When r_1 was 0.45 cm, there was 25 times more soil volume than when r_1 was 0.09 cm, since the rate of root growth was the same for all comparisons. Data in Table 5.3 show that changing C_{li} had little effect on predicted uptake when r_1 was small. When r_1 was large and roots did not compete, however, increasing C_{li} had a considerable effect on predicted uptake. In this case, the amount of nutrient in the soil did not limit the supply to the root.

The foregoing discussion applies to nutrients supplied to the root mainly by diffusion. With diffusion, an interaction between r_1 and D_e determines whether C_{li} or C_{si} is the more important in determining the supply of nutrient to the root. When mass flow supplies the majority of the nutrient to the root, the result is similar to having a very large D_e value and competition between roots occurs even at large r_1 values.

Soil Supply as Affected by Root Radius

Changing root radius has two effects. Since root surface equals $2\pi r_0 L$, increasing r_0 increases the surface area for absorption and increases uptake,

TABLE 5.3   Effect of Increasing Soil Root Density on Predicted Potassium Uptake as Influenced by Buffer Power of the Soil

Soil property	Soil		
	I	II	III
C_{li}, mmol/1000 cc	2	2	2
C_{li}, mmol/L	1	0.2	0.1
b	2	10	20
D_e, cm^2/s $\times 10^7$	5	1	0.5
Predicted uptake, mmol/pot			
with $r_1 = 0.09$ cm	23	21	20
with $r_1 = 0.45$ cm	269	121	71

Source: Reproduced from Barber (1981) with permission of the American Society of Agronomy and Soil Science Society of America.

as shown in Figure 5.5. The supply of nutrients to the root is influenced by r_0 because of radial geometry. When $D_e t/r^2_0 < 0.1$, the root cylinder can be considered a plane surface. In most cases, however, $D_e t/r^2_0 > 0.1$. For example, values in Table 5.2, using a period of 15 days, will give $D_e t/r^2_0$ of 200. When r_0 becomes small, the nutrient gradient perpendicular to the root is less and the concentration maintained at the root is higher. The effect of r_0 on the gradient perpendicular to the root for roots absorbing for 10 days is shown in Figure 5.8; data are for phosphorus uptake by soybeans. The smaller r_0 the more effective the nutrient absorption per cm^2 of root surface because of the greater supply of nutrient by the soil resulting from differences in radial geometry.

Root radius may also influence I_{max}. Peterson and Barber (1981) found that I_{max} increased as r_0 increased. This may be due to the greater area of plasma membrane within the cortex per cm^2 of root surface. Hence, in Figure 5.8, the rate of uptake by smaller roots could be somewhat less and the curves for small r_0 even flatter than those shown.

Water Influx

Increasing water influx v_0 causes a proportional increase in nutrient supply by mass flow. The overall effect on nutrient uptake will depend on the level of nutrients in solution at the root relative to the influx versus concentration curve. If the nutrient concentration is already high enough so that influx approaches or equals I_{max}, increasing v_0 will have little effect. If the nutrient concentration in solution is low, as is common for phosphate, then increasing v_0 also will have little effect; in this case, mass flow is supplying only a small fraction of the nutrient flux to the root. Barber (1974) reported that changing the amount of water transpired from 106 L/kg of shoot to 444 L/kg

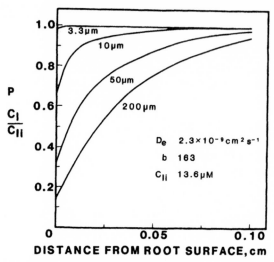

FIGURE   5.8   The effect of size of r_0 on the $C_l : C_{li}$ potassium gradient perpendicular to the root. Parameters used are shown in Table 5.2.

of shoot increased calcium uptake from 1.82 to 4.44 mmol/pot, because most of the calcium was being supplied to the root by mass flow. At the same time, however, increasing the water flux only increased manganese uptake from 10.9 to 14.5 mmol/pot. Mass flow supplied less than one-third of the manganese to the plant root in this experiment.

Interactions Involving Water Influx

As indicated above, the effect of v_0 depends both on C_{li} and on I_{max}. If $v_0 C_{li} = I_{max}$, mass flow will supply nutrients to the root at exactly the same rate that the root absorbs them. Hence, there will be no change in C_{li} at the root surface; this is represented by curve A in Figure 5.9. If $v_0 C_{li}$ is less than I_{max}, absorption by the root will reduce C_{li} at r_0 and curve B will result. If $v_0 C_{li}$ exceeds I_{max}, the ions in question will accumulate at the root surface, and curve C (where $v_0 C_{li}$ is 10 times I_{max}) will occur.

Root Nutrient-Influx Parameters

Maximum Influx

Increasing I_{max} has a greater effect on uptake the higher the level of nutrient concentration in solution C_{li}. If mass flow supplies nutrients to the root at a rate greater than I_{max}, then increasing I_{max} will increase uptake proportionately. Alternatively, when the level of a nutrient in the soil is low and diffusion is a major supply mechanism, then supply by the soil to the root will be the

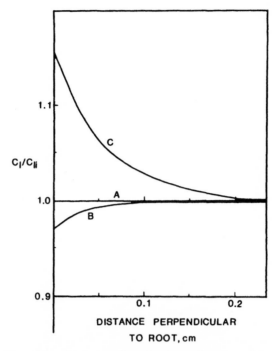

FIGURE   5.9   The effect of relative size of values of v_0, C_{li}, and I_{max} on distribution of ions normal to the root. The curves are shown where A, $v_0C_{li} = I_{max}$; B, $10v_0C_{li} = I_{max}$; C, $v_0C_l = 10\ I_{max}$. Other parameters were D_e, 6.3×10^{-8} cm²/s; b, 10; r_0, 0.01 cm; t, 4.32×10^5 s; v_0, 1×10^{-6} cm/s; I_{max}, 1 or 10 nmol/m² · s; K_m, 30 μmol/L; C_{min}, 2 μmol/L.

limiting factor in nutrient uptake. In this case, increasing I_{max} will have little effect on predicted uptake. These effects are illustrated in Figure 5.10 where the effect of changing I_{max} on predicted phosphorus uptake is related to C_{li} size. Relations similar to this can also be obtained for potassium. In Figure 5.10 increasing I_{max} had little effect on predicted phosphorus uptake when C_{li} was 0.015 μmol/ml but had a large effect when C_{li} was 0.20 μmol/ml. These sensitivity analyses can indicate the relative significance of conducting research on increasing I_{max}.

Michaelis–Menten Constant

Reducing the Michaelis–Menten constant K_m causes a greater nutrient uptake at the same solution concentration. In selecting plants that are efficient in absorbing nutrients from soil, those with absorption isotherms having a low K_m should be chosen. Changing K_m will have the most effect on uptake where nutrient levels in the soil solution are considerably below the level needed to give I_{max}. Some effects of changing K_m on the absorption of

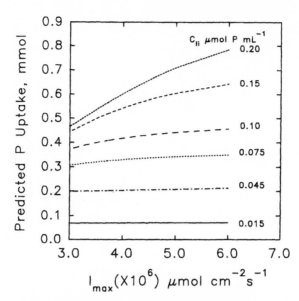

FIGURE   5.10   Influence of C_{li} value for phosphorus on effect of increasing I_{max} on predicted phosphorus uptake using the Barber–Cushman model. Other parameters were held constant.

nutrients are shown in Figure 5.5. The effect was much less than for most of the other parameters. Data in the literature indicate that it probably is difficult to find cultivars that vary greatly in K_m, hence, K_m is probably not a plant property that has a high potential for change leading to an efficiency gain.

Minimum Concentration

The minimum concentration C_{min} describes the level in solution at which nutrient uptake ceases. In soils where C_{li} is below C_{min} for a plant, the plant will not be able to absorb any of the particular nutrient. In fact, it may even return some of what was in the seed into the soil; this may occur with phosphorus, for example, in some phosphorus-deficient soils.

While lower C_{min} values will enable plants to grow at lower soil–nutrient concentrations, the effect of reducing C_{min} will be significant only where C_{li} is of the same order of magnitude. The effect is proportional to $(C_{li} - C_{min1})/(C_{li} - C_{min2})$.

There are many possible combinations of the 11 model parameters shown in Table 5.2. Sensitivity diagrams for values common for nitrogen are shown in Figure 8.3 and for phosphorus in Figure 9.7. In nature, some parameters vary more widely than others; there is a very wide range, for example, for the three soil variables. Either the Claassen–Barber or Barber–Cushman models can be used to investigate the effect on predicted uptake of conditions other than those used here for illustration.

REFERENCES

Bagshaw, R., L. V. Vaidyanathan, and P. H. Nye. 1972. The supply of nutrient ions by diffusion to plant roots in soil. VI. Effects of onion plant roots on pH and phosphate desorption characteristics in a sandy soil. *Plant Soil* **37**:627–239.

Barber, S. A. 1974. Influence of the plant root on ion movement in soil. In E. W. Carson, Ed. *The Plant Root and Its Environment.* University Press of Virginia, Charlottesville, Pp. 525–564.

Barber, S. A. 1981. Soil chemistry and the availability of plant nutrients. In R. H. Dowdy, J. A. Ryan, V. V. Volk, and D. E. Baker, Eds. *Chemistry in the Soil Environment.* American Society of Agronomy, Madison, WI. Pp. 1–22.

Barber, S. A., and J. H. Cushman. 1981. Nitrogen uptake model for agronomic crops. In I. K. Iskandar, Ed. *Modeling Waste Water Renovation-Land Treatment.* Wiley-Interscience, New York. Pp. 382–409.

Bhat, K. K. S., and P. H. Nye. 1973. Diffusion of phosphate to plant roots in soil. I. Quantitative autoradiography of the depletion zone. *Plant Soil* **38**:161–175.

Bidin, A. A. 1982. Phosphate in Malaysian Ultisols and Oxisols as evaluated by a mechanistic model. Ph.D. dissertation, Purdue Univ.

Blanco, J. O., 1989. Dynamics of phosphorus in five acid Columbian soils. Ph. D. dissertation, Purdue Univ.

Bouldin, D. R. 1961. Mathematical description of diffusion processes in the soil-plant system. *Soil Sci. Soc. Am. Proc.* **25**:476–480.

Brewster, J. L., K. K. S. Bhat, and P. H. Nye. 1976. The possibility of predicting solute uptake and plant growth response from independently measured soil and plant characteristics. V. The growth and phosphorus uptake of rape in soil at a range of phosphorus concentrations and a comparison of results with the predictions of a simulation model. *Plant Soil* **44**:295–328.

Chen, J.-H., and S. A. Barber, 1990. Soil pH and phosphorus and potassium uptake by maize evaluated with an uptake model. *Soil Sci. Soc. Am. J.* **54**:1032–1036.

Claassen, N., and S. A. Barber. 1976. Simulation model for nutrient uptake from soil by a growing plant root system. *Agron. J.* **68**:961–964.

Claassen, N., K. M. Syring, and A. Jungk. 1986. Verification of a mathematical model by simulating potassium uptake from soil. *Plant Soil* **95**:209–220.

Cox, M. S. 1991. Predicting soil phosphorus levels needed for equal uptake on soils with different water levels. MSc thesis, Purdue Univ.

Farr, E., L. V. Vaidyanathan, and P. H. Nye. 1969. Measurement of ionic concentration gradients in soil near roots. *Soil Sci.* **107**:385–391.

Itoh, S., and S. A. Barber. 1983a. Phosphorus uptake by six plant species as related to root hairs. *Agron. J.* **75**:457–461.

Itoh, S., and S. A. Barber. 1983b. A numerical solution of whole plant nutrient uptake for soil-root systems with root hairs. *Plant Soil* **70**:403–413.

Junhk, A., and S. A. Barber. 1975. Plant age and the phosphorus uptake characteristics of trimmed and untrimmed corn root systems. *Plant Soil* **42**:227–239.

Li, Y., and S. A. Barber. 1991. Calculating changes of legume rhizosphere soil pH and soil solution phosphorus from phosphorus uptake. *Comm. Soil Sci. Plant Anal.* **22**:955–973.

Nye, P. H. 1966. The effect of the nutrient intensity and buffering power of a soil, and the absorbing power, size, and root hairs of a root, on nutrient absorption by diffusion. *Plant Soil* **25**:81–105.

Nye, P. H., and F. H. C. Marriott. 1969. A theoretical study of the distribution of substances around roots resulting from simultaneous diffusion and mass flow. *Plant Soil* **30**:451–472.

Olsen, S. R.. and W. D. Kemper. 1968. Movement of nutrients to plant roots. *Adv. Agron.* **20**:91–151.

Passioura, J. B. 1963. A mathematical model for the uptake of ions from soil solution. *Plant Soil* **18**:221–238.

Peterson, W. R. and S. A. Barber. 1981. Soybean root morphology and potassium uptake. *Agron. J.* **73**:311–319.

Schenk, M. K., and S. A. Barber. 1979a. Root characteristics of corn genotypes as related to phosphorus uptake. *Agron. J.* **71**:921–924.

Schenk. M. K., and S. A. Barber. 1979b. Phosphate uptake by corn as affected by soil characteristics and root morphology. *Soil Sci. Soc. Am. J.* **43**:880–883.

Schenk, M. K., and S. A. Barber. 1980. Potassium and phosphorus uptake by corn genotypes grown in the field as influenced by root characteristics. *Plant Soil* **54**:65–76.

Shaw, J. K., R. K. Stivers, and S. A. Barber. 1983. Evaluation of differences in potassium availability in soils of the same exchangeable potassium level. *Commun. Soil Sci. Plant Anal.* **14**:1035–1049.

Silberbush, M., and S. A. Barber. 1983a. Sensitivity analysis of parameters used in simulating potassium uptake with a mechanistic mathematical model. *Agron. J.* **75**:851–854.

Silberbush, M., and S. A. Barber. 1983b. Prediction of phosphorus and potassium uptake by soybeans with a mechanistic mathematical model. *Soil Sci. Soc. Am. J.* **47**:262–265.

Silberbush, M., and S. A. Barber. 1984. Phosphorus and potassium uptake of field-grown soybeans predicted by a simulation model. *Soil Sci. Soc. Am. J.* **48**:592–596.

Silberbush, M., W. B. Hallmark, and S. A. Barber. 1983. Simulation of effects of soil bulk density and P addition on K uptake by soybeans. *Commun. Soil Sci. Plant Anal.* **14**:287–296.

Walker, J. M., and S. A. Barber. 1962. Uptake of rubidium and potassium from soil by corn roots. *Plant Soil* **17**:243–259.

CHAPTER **6**

Interaction of Plant Roots with the Soil and Environment

In Chapter 5, a mathematical model was developed that assumed that plant roots act as smooth cylindrical sinks for nutrients. The strength of the sink was then related to nutrient concentration in solution at the root surface. The model also assumes that roots do not interact with the soil to affect nutrient flux; nutrients are only taken from solution; and direct exchange between the solid phase and the plant root does not occur. The close relation between uptake predicted from this model and uptake observed during pot experiments with several plant species indicates that the model accurately describes the uptake process. However, evidence has also been obtained indicating, at least for some conditions, that plant roots *do* influence the rhizosphere soil and affect the rate of nutrient uptake. In addition, the soil's physical and chemical properties as well as light and temperature affect root growth; which in turn affects nutrient uptake. The influence of these factors is discussed in this chapter.

CHANGES IN THE RHIZOSPHERE

Changes in the rhizosphere that can affect nutrient flux to the root include changes in pH, salt concentration, concentration of complementary ions, microorganisms, and soil bulk density.

Changes in Rhizosphere pH

Plants growing in nutrient culture can increase or decrease the pH of the nutrient solution, depending on relative rates of absorption of anions and cations (Jackson and Adams, 1963; Kirkby, 1968). Cation uptake includes an exchange of cations across the plasma membrane for hydrogen or protons. Anion uptake represents an influx of nutrient anions and an efflux of carboxyl ions. If a plant takes up more cations than anions, the pH of the solution decreases; conversely, if the root takes up more anions than cations, the pH increases. A pH increase is usually accompanied by an increase in HCO_3^- level of the nutrient solution.

The form of nitrogen absorbed, NH_4^+ or NO_3^-, has a considerable effect on whether H^+ or HCO_3^- will be released by the roots. Barber (1974) determined the effect of nitrogen form on cation and anion uptake by ryegrass (*Lolium rigidum*). Data in Table 6.1 show that ryegrass absorbs large quantities of either nitrogen form as compared to the uptake of other cations. The ryegrass was grown in solution culture with either NH_4^+ or NO_3^- as the nitrogen source. Solution pH was adjusted to 6.5 daily by adding HCl or KOH. The amounts of HCl or KOH added agreed reasonably with measured differences in uptake of cations and anions calculated from plant analysis at the conclusion of the experiment.

Riley and Barber (1971) demonstrated that reductions in rhizosphere pH induced by supplying N as NH_4^+ rather than NO_3^- increased the uptake of phosphate by soybeans growing in Chalmers silt loam (Typic Argiaquoll). The soil was limed to three pH levels, and at each lime level, soybeans were supplied with NH_4^+ or NO_3^- nitrogen. A nitrification inhibitor was used to supress nitrification of NH_4^+. After three week's growth, plants were harvested. Roots were picked from the soil and loosely held soil shaken off. Strongly held soil remaining on the roots was within 0 to 2 mm of the root surface. The pH of this soil was measured by extracting plant root plus soil with deionized water for five minutes at a material to water ratio of 2:1. The water was removed by filtration and its pH determined. Similar measurements were made on nonrhizosphere soil. All treatments had similar plant growth, so phosphorus composition of the shoot reflected the soil's phosphorus availability. For this soil, decreasing soil pH increased the phosphorus concentration in the soil solution. The phosphorus concentration in the shoot was plotted against the pH of the bulk soil (Figure 6.1). The NH_4^+ and NO_3^- treatments gave separate linear relations between soil pH and phospho-

TABLE 6.1   Effect of Form of Nitrogen on the Balance between Cation and Anion Uptake by Ryegrass[a]

Ion	Uptake with nitrogen supplied [mmol (p^+) or (e^-)/pot	
	As NH_4^+	As NO_3^-
Ca^{2+}	0.20	0.46
Mg^{2+}	0.30	0.52
K^+	1.76	1.87
Na^+	0.03	0.03
NH_4^+	5.59	–
Total cations	7.88	2.88
SO_4^{2-}	0.97	0.33
Cl^-	0.45	0.17
$H_2PO_4^-$	0.37	0.34
NO_3^-	–	4.72
Total anions	1.79	5.56
HCl or KOH to adjust pH	6.70 KOH	1.60 HCl

Source: Barber (1974) with permission of University Press of Virginia.
[a] Values are means of three replicates.

rus concentration of the shoots. As had been described by Grunes (1959), the NH_4^+ treatment resulted in more phosphorus uptake than the NO_3^- treatment.

A plot of the pH of the rhizocylinder (plant root plus tightly adhering soil) versus phosphorus concentration of the shoot (Figure 6.2) showed that the NH_4^+ and NO_3^- treatments all fell along the same regression line. This suggests that soybean plants supplied with NH_4^+ reduced the pH at the root surface and thus increased phosphorus availability. Conversely, soybeans supplied with NO_3^- increased the pH at the soil-root interface and thus reduced phosphorus availability. Hence, if conditions are such that there is an imbalance between cation and anion uptake, the pH at the soil-root interface will change. When changes in soil pH cause a change in soil nutrient availability, there will be an effect of the plant root on soil nutrient availability.

In the same soybean experiment, plants were also analyzed for boron, since increases in soil pH frequently reduce boron availability to the plant (see Chapter 14). The relation between bulk soil pH and boron concentration of the soybean shoots is shown in Figure 6.3. The NH_4-treated soybeans had a higher boron content; there was also a reduction in the plant's boron content with increased soil pH. The relation between boron concentration in the shoot and pH of the rhizocylinder soil is shown in Figure 6.4. As with phosphorus, when the pH of rhizocylinder soil was correlated with boron con-

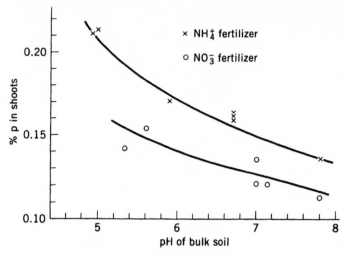

FIGURE 6.1   Relation between percent P in shoots of 3-week-old soybean seedlings fertilized with NH_4^+ or NO_3^- at different bulk-soil (nonrhizosphere) pH levels. 40 mgP/kg added to each soil. Reprinted from Riley and Barber (1971) by permission of Soil Science Society of America.

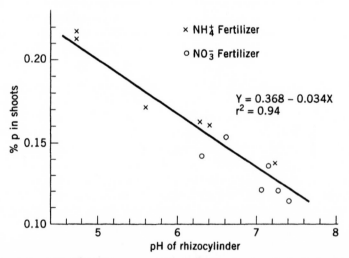

FIGURE 6.2   Relation between percent P in shoots of 3-week-old soybean seedlings fertilized with NH_4^+ or NO_3^- and rhizocylinder (rhizosphere soil and root) pH levels. 40 mgP/kg added to each soil. Reprinted from Riley and Barber (1971) by permission of Soil Science Society of America.

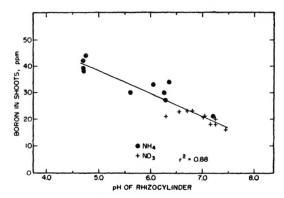

FIGURE   6.3   Relation between boron content of shoots of 3-week-old soybean seedlings fertilized with NH_4^+ and NO_3^- and pH of bulk soil (Barber, 1971).

centration in the shoot, both NH_4^+ and NO_3^- treatments gave values that fell along the same regression line. The NH_4^+ treatment caused an efflux of hydrogen from the root, which reduced soil pH at the root–soil interface and caused an increase in boron bioavailability. The effect of NO_3^- was in the opposite direction.

Smiley (1974) also found that the rhizosphere pH near wheat roots was lower than the initial soil pH where NH_4^+ was supplied and higher where NO_3^- was supplied. Differences of up to 2.2 units were recorded with NH_4^+ versus NO_3^- sources for wheat (*Triticum aestivum* L.) plants grown in pots. Differences of up to 1.2 units were noted for wheat grown in the field; these differences occurred among wheat varieties and plant genera.

In nature, both NH_4^+ and NO_3^- may be present; in many soils, however, most of the NH_4^+ is oxidized to NO_3^-. As a result, the balance between NH_4^+

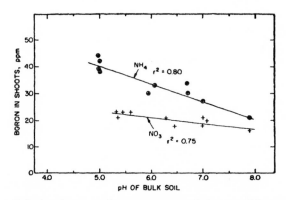

FIGURE   6.4   Relation between boron content of shoots of 3-week-old soybean seedlings fertilized with NH_4^+ and NO_3^- and pH of the rhizocylinder (Barber, 1971).

and NO_3^- uptake may often be such that no appreciable pH change occurs at the soil–root interface. In such cases, the root appears to have no effect on the soil other than acting as a sink for nutrients. With many plant species, the amount of NO_3^- in solution must be several times larger than the NH_4^+ level for cation and anion uptake from the solution or soil to be equal.

Nye (1981) calculated the theoretical pH distribution around roots resulting from efflux of H^+ or HCO_3^- as affected by pH buffering capacity, soil moisture content, initial soil pH, and PCO_2 of the soil. The effect of efflux of HCO_3^- or H^+ on a soil's pH depends on the initial pH of the soil. The closer the pH is to the soil's pH where the soil's acidity-diffusion coefficient is lowest, the greater the change in pH next to the root surface, since the radial distance that is affected is less. This pH is 5.3 at $PCO_2 = 0.01$ atm. Increasing the pH buffer capacity reduces the resultant pH change. A decrease in moisture level, on the other hand, increases the pH change, since it reduces the radial distance affected. The pH change is mainly in soil within 3 mm of the root surface.

Plant species vary in their relative degree of cation and anion uptake. Keltjens (1982) grew nine plant species in nutrient solution with NO_3^- as the nitrogen form. In all cases, anion uptake exceeded cation uptake and caused an HCO_3^- efflux. However, efflux per gram of plant varied from 0.40 mol(e^-) for bean to 1.45 mol(e^-)/kg dry matter for maize. Grinsted et al. (1982) found that rape seedlings reduced the pH of rhizosphere soil when plants became phosphorus deficient. This occurred even though the sole source of nitrogen was NO_3^-. Cunningham (1964) collected analyses of 62 common plant species grown in soil. He found that they contained on the average 2.5 mol(p^+) of cations and 3.6 mol(e^-) of anions per kg of plant; hence, on the average, they would release HCO_3^-. Nitrogen was calculated as an anion.

Marschner and Romheld (1983) developed a colorimetric procedure for rapidly measuring changes in pH along roots. Li and Barber (1991) found that roots of alfalfa (*Medicago sativa* L.), faba bean (*Vicia faba* L.), and Austrian winter pea (*Lathyrus hirsutus*) grown in Chalmers silt loam at pH levels of 5.72, 6.30, 7.22, and 8.30 reduced rhizosphere soil pH by 0.39 to 0.77 units and increased phosphorus availability by 20.8 to 241% (see chapter 21 for more detail). Israel and Jackson (1978) found that soybeans grown in a culture solution free of inorganic nitrogen, so that they received their nitrogen from rhizobium, released 1.08 mol H/kg of total plant dry weight produced. Haynes (1990), in an extensive review, examined the role of maintenance of cation–anion balance for active ion uptake in regulating rhizosphere pH. He concluded the generation of acidity or alkalinity in the rhizosphere is directly related to the cation–anion balance of ion uptake.

Increasing Salt Concentration at the Root–Soil Interface

When mass flow supplies more nutrients to the root–soil interface than the root can absorb, the concentration at the root surface increases, and the resultant concentration gradient decreases *away* from the root. The pattern of the salt-concentration gradient around the root was shown in Figure 5.9. Barber and Ozanne (1970) used autoradiographs to show calcium accumulation around plant roots (see Figure 4.3). Riley and Barber (1969) measured salt accumulation around soybean roots growing in Chalmers silt loam. Soybeans were grown for 2 and 3 weeks and then harvested. Salt accumulation was measured by extracting with water the salts from rhizosphere and nonrhizosphere soil. In order to vary salt accumulation, plants were grown at both low and high humidities (to vary transpiration rate) and also in soils with low- and high-soluble salt contents. Results after two and three weeks are shown in Figure 6.5. With all treatments and at both plant ages, salts accumulated at the root surface. The amount of salt moving to the root by mass flow as compared with plant uptake varied from 1.2 for the low-salt, low-transpiration treatment to 10 for the high-salt, high-transpiration treatment. Accumulation in the rhizocylinder tended to reach a maximum value that was 5 to 15 times the salt content in the nonrhizosphere soil.

Increasing the salt content increases the concentration of all soluble cations present in solution at the soil–root interface. When the concentration of the nutrients is higher than that required for maximum uptake, additional accumulation will have little effect on uptake. However, cations in solution are in equilibrium with exchangeable cations on the soil particles. Increasing

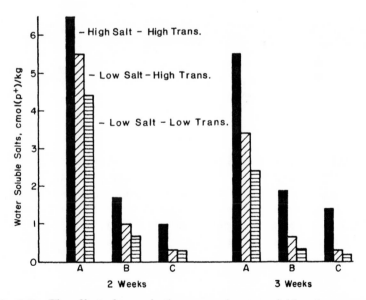

FIGURE   6.5   The effect of transpiration rate and water-soluble salt content of the nonrhizosphere soil *C* on the water-soluble salt content of the rhizosphere soil *B* and the rhizocylinder *A*. Reproduced from Riley and Barber (1970) by permission of Soil Science Society of America.

salt concentration will alter the relative distribution of cations in the soil solution because of Donnan equilibrium effects, as discussed in Chapter 2. The ratio law governing valence effects gives the relation shown in Equation 6.1.

$$\frac{K_1 + Na_1}{(Ca_1 + Mg_1)^{1/2}} = \frac{K_2 + Na_2}{(Ca_2 + Mg_2)^{1/2}} \tag{6.1}$$

If the solution in state 2 is more concentrated than in state 1, the concentration of divalent cations will have to increase as the square of the concentration of the monovalent cations in order to preserve the ratio. Hence, while salt accumulation will greatly affect the ratio Ca:K, it will not affect the concentration of K greatly. Since K is more likely to be deficient, salt accumulation may have little effect on nutrient uptake. The effect of changes in rhizosphere salt concentration on ion uptake will mainly be due to the influence of changes in the concentration of divalent cations on the uptake of monovalent cations. In solution-culture studies, Clark (1978) observed that the level of solution calcium had little effect on potassium uptake.

Microorganisms in the Rhizosphere and Nutrient Uptake

The soil at the root interface has been shown to have a higher population of microorganisms than the nonrhizosphere soil (Rovira and Davey, 1974). Since all nutrients absorbed by the root pass through this zone, increased levels of microorganisms could affect nutrient supply to the root. The $R:S$ ratio, where R is the amount of microorganisms in the rhizosphere and S is the number in the nonrhizosphere soil, has been used to characterize increased microbial activity near the root. Rovira and Davey (1974) report that $R:S$ values of 5 to 20 are common for bacteria, while fungi have values of 11 to 22. A major reason for higher microbial populations in the rhizosphere may be organic exudates from roots, debris from root cap cells, and sluffing off root hair and exodermal cells.

Foster (1981) investigated root surfaces of Bahiagrass, *Paspalum notatum* L., and wheat, *Triticum aestivum* L., growing in the field. He found that the root was initially bounded externally with a thin cuticle, but this cuticle was soon ruptured by mechanical action of the soil and action of microorganisms. This allowed mucilage to escape from the primary wall layer forming mucigel; hence, the mucigel came from a part of the outer root cell wall. The effect of rhizosphere microorganisms on nutrient uptake is discussed in Chapter 7.

Organic Exudates

Plant roots also exude organic compounds into the soil, which may affect nutrient flux into the root. Studies of exudates from roots grown aseptically in solution culture have shown a wide range of exuded compounds.

However, the total carbon released was usually less than 2% of the root carbon (Rovira and Davey, 1974), and only water-soluble materials were measured in these studies. Measured amounts of exudate from roots growing in soil tend to be much larger, due to contact with the soil medium, which may rupture root membranes and cause release of both water-soluble and insoluble exudates (Barber and Martin, 1976).

Labeling the carbohydrates of the plant with ^{14}C, by supplying ^{14}C-labeled CO_2 has made it possible to use procedures that measure both the quantity of materials released and the effect of microorganisms on the process. In one such study, plants were grown with their shoots in air containing ^{14}C-labeled carbon dioxide and their roots sealed into pots of soil maintained under sterile or nonsterile conditions. As the soil's atmosphere was changed, its carbon dioxide levels were determined. Both wheat and barley were grown under sterile and nonsterile soil conditions. The quantities of ^{14}C present in shoots, roots, and soil and the respired CO_2 expressed as a total of that recovered is shown in Table 6.2. Under sterile conditions respired CO_2 represents primarily respiration by the roots. In the case of nonsterile soil, it also represents release of carbon from compounds exuded from the root by the action of attendant microorganisms. In the sterile system, root respiration amounted to 1% of the total. The value 7.7% for nonsterile soil then represents 1.0% from root respiration plus 6.7% from exudate used by soil microorganisms. The total exudate in the nonsterile soil would appear to be 15.0% (6.7 + 7.7 + 0.6) of the total ^{14}C fixed, whereas in the sterile soil it represented only 6.3% of the total. Values for barley were similar.

Data of Barber and Martin indicate that growth of roots in the soil stimulates exudation and also that the presence of soil microorganisms stimulates exudation. Similar results were obtained from the experiments of Sauerbeck and Johnson (1976). Exudation of organic materials by plant roots increases microbial activity in the rhizosphere.

Changing Concentration of Complementary Ions

The influx of some ions is affected by concentrations of complementary ions present in the soil. The relative concentration of the complementary ion will

TABLE 6.2   Distribution of Photosynthetically Fixed Carbon (%) for Wheat Plants Grown under Sterile and Nonsterile Conditions for 3 Weeks

| Soil condition | Roots | Shoots | Soil | | |
			Water soluble	Insoluble	Respired
Sterile	24.5	68.2	0.0	5.5	1.0
Nonsterile	32.1	51.9	0.6	7.7	7.7

Source: Barber and Martin (1976). Reproduced with permission of New Phytologist Trust.

usually be different at the root surface than in the equilibrium soil because of the balance between nutrient flux to the root and nutrient influx into the root for each ion. Calcium may accumulate at the root, for example, while potassium is being depleted. Uptake of such ions as magnesium may be affected by the solution level of potassium (Maas and Ogata, 1971; see Chapter 12). Calcium influx may be affected by the level of potassium or magnesium (see Chapter 11), and zinc influx may be affected by the concentration of phosphate (see Chapter 19).

When we consider the effects of potassium on calcium uptake, we must know the levels of each of these ions at the root surface during uptake. Because potassium is supplied by diffusion, the level in solution at the root surface may be only one-tenth the initial level in the soil solution. Since calcium is largely supplied by mass flow, the level at the root may be three or more times that in solution initially. Hence, due to the action of the root, the Ca:K ratio at the root may be as much as 30 times the initial ratio in the soil solution. The extent of the effect on nutrient uptake will depend on whether the concentration of the complementary ion has an influence on the influx of the ion in question.

Changing Soil Bulk Density

After plant roots have grown into soil pores that are larger than the root tip, the root can expand and force soil particles together. Hence, the bulk density of the soil next to the root may be considerably greater than the initial soil value. The extent of the effect will depend on the type of root, the degree to which it increases in size, and the physical nature of the soil. Figure 4.1 illustrates the effect of the root on soil bulk density and particle distribution near a cotton root. Increasing soil bulk density will increase the concentration of available ions, C_{si}, per unit of soil volume. This will, in turn, create a steeper gradient for C_s to diffuse to the root, where diffusion is the supply mechanism. However, the concentration in the soil solution should remain the same.

SOURCE OF NUTRIENTS ABSORBED

The first edition of this book discussed the use of ratios of two similar cations to determine whether C_{li} or C_{si} was the immediate source of ions being adsorbed by the root. The strontium–calcium pair (Bole and Barber, 1971) and the potassium–rubidium pair (Baligar and Barber, 1978) were used since in solution culture uptake had the same ratio as was present in solution. In soil systems where most of these ion pairs reached the root by diffusion, the uptake ratio was the same as found in C_{si} while where mass flow supplied the nutrients to the root the ratio of uptake was the same as that in C_{li}. Table 6.3 gives an example of these results for potassium–rubidium. In both the experiments where diffusion supplied most of the ions to the root the ratio was that of C_{si}. In both cases the strontium or rubidium was added to the soil and incubated in the soil at field capacity water level for several weeks before grow-

TABLE 6.3   Comparison of Means of Potassium to Rubidium Ratios of Plant Uptake with Potassium to Rubidium Ratios in Soil Solution and on Exchange Sites[a]

Soil treatment	K/Rb		
	Uptake	Displaced solution	Exchangeable
Zanesville	5.0a	13.2b	4.8a
Zanesville + K	10.3a	20.8b	8.3a
Chalmers	13.5a	30.0b	10.2a
Chalmers + K	29.5a	70.8b	33.2a
Raub	3.1a	10.7b	3.1a
Raub + K	7.1a	23.8b	7.8a
Toronto	3.6a	17.5b	4.0a
Toronto + K	7.3a	25.3b	8.5a

Source: Reproduced from Baligar and Barber (1978) by permission of Soil Science Society of America.
[a]For each soil treatment, means followed by the same letter are not significantly different.

ing corn for 10 to 16 days. One is tempted to conclude that these results indicate that the roots are absorbing directly from the exchangeable ions and not from the solution at the root surface. This conclusion may not be correct since in a more recent evaluation of D_e, values for each of the ions were calculated. If we assume $D_e = D_l \theta f_l / b$ the only factors differing between the two ions are D_l and b. Values of D_l are not likely to vary greatly because of the similarity of each of the ions in the ion pairs. Values of b, however, will vary to the degree that C_{si}/C_{li} varies and the K/Rb ratio diffusing to the root will be closer to the C_{si} ratio than the C_{li} ratio, hence giving the results obtained. This effect may be accentuated as the ion level in the soil decreases, as occurs when a concentration gradient develops.

Effects of Root Respiration

In early research on soil nutrient availability, it was postulated that the HCO_3^- produced by root respiration dissolved nutrients from the soil. Nye and Tinker (1977) calculate that the amount of HCO_3^- formed during release of CO_2 by respiration could make only a negligible contribution to increasing the release of nutrients from the solid phase to the solution phase.

ENVIRONMENTAL EFFECTS ON ROOT GROWTH AND PHYSIOLOGY

This section discusses the effect of the environment on root parameters that control supply of nutrients to the shoot. These parameters include rate of root growth, root radius, and net nutrient influx as affected by the concentration of nutrients in solution outside the root. The environmental effects discussed are physical restraints on root growth that affect both rate of growth and morphology of the roots, temperature, light intensity, soil moisture tension, nutrient levels and distribution, soil aeration, and elemental toxicities.

Many investigations have evaluated effects of environmental factors on root growth rate or root growth rate relative to shoot growth rate. Fewer have investigated the effect on nutrient supply to the shoot.

Physical Restraint

Root growth is affected when there is a physical restraint; Figure 6.6 from Goss and Reid (1981) shows that even small pressure increments reduced the rate of root elongation of seminal barley roots. In this study, pressure was

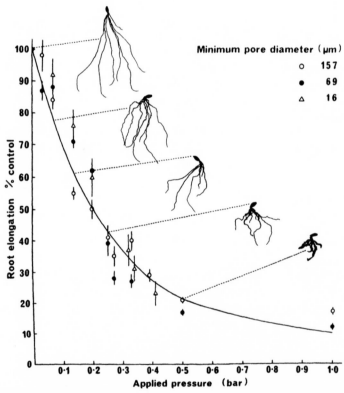

FIGURE 6.6   Relation between pressure applied to seminal barley (*Hordeum vulgare*) roots and root elongation. Barley was grown in beds of glass ballotini supplied with nutrient solution (Goss and Reid, 1981).

applied to a flexible container filled with glass ballotini. Ample nutrition was provided with aerated nutrient solution. Pore sizes varied from 15 to 150 μm. A constant pressure was maintained throughout growth of the plant. Normally, roots grown in solution grow with little resistance to elongation; roots grown in ballotini decreased in length and became thicker with increasing pressure. Roots growing in soil penetrate pore spaces and push soil particles aside as they grow; hence, they must exert pressure. Growth in pure sand can be restricted even more than growth in soil, because sand particles do not slip past one another as readily as do soil, silt, and clay particles. Hence, greater pressures are needed for root growth. Resistance of the soil to root penetration is called soil strength (Barley, 1962; Taylor and Gardner, 1963). As soil strength increases, rate of root elongation decreases and root diameter increases. Results obtained by Peterson and Barber (1981) illustrate the effect of soil strength on the soybean's root morphology. The authors studied two extreme cases by comparing roots grown in solution with roots grown in sand. They observed, as did Russell (1977), that the volume of the cortical cells did not change with resistance to growth but their shape did. Figure 6.7 shows cross sections of soybean roots grown in the two systems. The radius of the stele was not affected, but the radius of the cortex was increased greatly. The number of cortical cells in the cross section remained constant. Barley (1976) observed the same effect on cortical cross section of *Pisum sativum* L. roots.

Peterson and Barber (1981) also investigated the effect of change in root morphology on potassium influx per m^2 of root surface. Increasing root diameter by growing roots in sand produced roots that had a higher

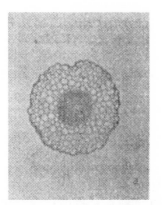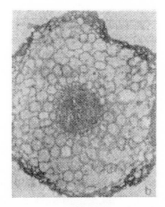

FIGURE   6.7   (*a*) Cross section of a solution-grown soybean (*Glycine max*) root. (*b*) Cross section of a sand-grown soybean root. Reproduced from Peterson and Barber (1981) by permission of American Society of Agronomy.

I_{max} for potassium uptake. The root radii for solution- and sand-grown roots were 0.17 and 0.22 mm, respectively. The corresponding I_{max} values were 46 and 71 nmol/m^2 · s, with the larger diameter root showing a greater influx. This may be due to the greater area of plasmalemma within the root for each m^2 of root surface, since influx was increased more than the root surface.

A layer of high-strength soil at the bottom of the tillage zone, sometimes called a tillage pan, can restrict root penetration into the subsoil and reduce the ability of the plant to absorb water and nutrients. A tillage pan is often one of the most obvious effects of physical restraint on root penetration. Soil strength also varies with soil moisture level; drying the soil and increasing soil water tension increases soil strength. Hence, soil moisture under unsaturated conditions has an effect on root penetration. In otherwise uniform and aerated soil, roots tend to proliferate in the wetter soil zones, because mechanical resistance is low rather than because the soil's water potential is low (Greacan and Oh, 1972). Soil bulk density can be used as a measure of the resistance to root penetration; as soil bulk density increases, the proportion of large pores decreases, which, in turn, inhibits root penetration. A bulk density greater than 1.80 Mg/m^3 usually restricts penetration of virtually all roots.

Hallmark and Barber (1981) grew soybeans in soil at two soil bulk densities, 1.25 and 1.45 Mg/m^3, and two levels of added potassium. The increased bulk density reduced root growth, and at the low level of potassium, this was associated with reduced potassium concentrations in the plant. At the higher level of potassium, however, there was no reduction in the plant's potassium concentration. Adding potassium helped reduce the detrimental effect of soil bulk density: There was a correlation ($r = 0.75$) between root surface area per pot and potassium concentration in the plant for plants grown at the low potassium concentration.

Silberbush et al. (1983) used data from Hallmark and Barber (1981) to determine the sensitivity of potassium uptake to changes in soil bulk density. They obtained the information shown in Figure 6.8, which shows potassium uptake decreasing with increased soil bulk density, while predicted influx increased to a maximum in this situation at a bulk density of 1.38 Mg/m^3. The effect of increasing soil bulk density on decreasing root growth was much greater than the effect on influx.

Temperature

Soil temperature affects root growth relative to shoot growth. Walker (1969) grew corn with constant shoot temperature of 25°C, while varying root temperatures in 1°C increments from 15° to 35°C. Shoot/root dry-weight ratios increased with increasing temperature from 15°C, reaching a peak at 29°C; for 23-day-old seedlings, the ratio varied from 2.4 to 5.66. Hence, when seedlings are grown in cold soil, additional root growth relative to shoot

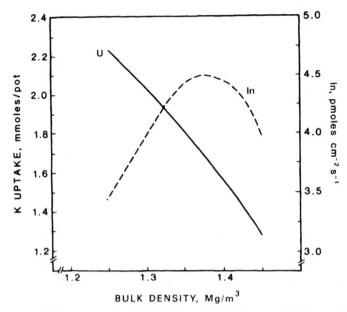

FIGURE 6.8   Effect of soil bulk density on simulated potassium uptake by Williams soybeans grown for 20 days on Raub silt loam and calculated potassium influx to roots when roots were 20 days old. Reprinted from Silberbush et al. (1983) by courtesy of Marcel Dekker, Inc.

growth will be beneficial in providing nutrients to the shoot. Since influx decreases with temperature, more root surface will be needed. The effect of temperature will vary with plant species; for species adapted to cooler climates than corn, the maximum shoot/root ratio will probably occur at a lower temperature.

Influx measurements by Mackay and Barber (1984) gave I_{max} values of 7.5 and 14 μmol/m² · s for corn at 18°C (25 days) and 25°C (18 days). Different ages gave seedlings of same stage of growth for measurement. Respective K_m values were 4.3 and 6.2 μmol/L, C_{min} was 0.3 μmol/L for both temperatures. Ching and Barber (1979) found values for potassium influx of 5.6 and 11.2 μmols/cm² · s and K_m of 14 and 28 μmol/L for 15° and 29°C of 16-day-old corn plants grown in solution culture. These experiments indicate an increase in influx kinetics as temperature increases.

Light

Light intensity affects root growth only indirectly, since it affects the quantity of photosynthate produced. This in turn affects root growth relative to shoot growth. Troughton (1980) reviewed existing research on effects of light on shoot/root ratio. He found that, in general, an increase in irradiance

TABLE 6.4   Linear Correlation between Root Weight and Total Weight per Plant under Different Light Intensities for Radish, Cucumber, and Bean Plants

Plant	Illumination (lux)	Dry weight (mg/plant)			Root (%)
		Root	Shoot	Total	
Radish	20,000	604	399	1003	60
Radish	15,000	501	466	967	48
Radish	10,000	240	408	468	38
Cucumber	20,000	244	950	1194	21
Cucumber	15,000	401	1148	1549	26
Cucumber	10,000	197	654	851	24
Bean	20,000	587	1615	2202	27
Bean	15,000	519	1009	2028	26
Bean	10,000	424	1566	1990	22

Source: Horvath et al., 1980. Reproduced with permission of Geobias International.

increased photosynthesis, which, in turn, results in an increase in root growth relative to shoot growth. Horvath et al. (1980) measured shoot and root growth in the field, where irradiance was varied by using degrees of shading. Reducing irradiance reduced root growth much more than shoot growth. Data in Table 6.4 from Horwath et al. (1980) for radish (*Raphanus sativus* L.), cucumber (*Cucumis sativus*), and bean (*Phaseolus vulgaris* L.) show a reduction in percent of radish and bean root's dry weight as illumination is reduced. Root and shoot weight of cucumber both decreased at similar rates. Silsbury (1971) studied the effect of light and temperature on the growth of ryegrass seedlings and found no effects of either on shoot/root ratio.

Soil–Water Tension

As water becomes more available to the plant, shoot growth usually exceeds root growth. Hence, plants that develop have larger shoot/root ratios, which can also mean that demand for nutrients per unit of root will increase. Desert plants usually have large root systems relative to the size of the shoot because of the difficulty in obtaining water. Presumably, only those plants with extensive root systems have survived such conditions.

Troughton (1980) reported results of relative growth rate (RGR) of shoots and roots under conditions where the osmotic pressure of the solution was varied to influence water availability. As the osmotic pressure was increased, RGR for shoots decreased more rapidly than RGR for roots. In an experiment with corn at an osmotic potential of −40 kPa, RGR for shoots and roots were 0.187 and 0.148, respectively. At −800 kPa, RGR was 0.019 for shoots and 0.066 for roots.

Effect of moisture is also influenced by the rate of root growth into the subsoil. In areas where droughts occur, root penetration into the subsoil can supply additional water to the plant. Depletion of water is most rapid from surface horizons because of higher root density in these areas. Plants having roots that can penetrate the subsoil at a more rapid rate than that at which moisture is depleted have an advantage over other plants. However, in terms of the plant's nutrition, the relative fertility of the subsoil is an important consideration. Frequently subsoils are less fertile than surface soils, since they have not been fertilized and have often been depleted by nutrient cycling to the surface during soil development.

Root growth in the upper 15 cm of soil is influenced by soil–water level in that layer. Kuchenbuch and Barber (1988) found for nine-yearly root measurements to 35 cm soil depth, corn root length density in the upper 15 cm of soil was significantly correlated ($r = 0.79$) with precipitation during the 3-week period prior to silking; measurements were made on a long-term field experiment where nutrients were maintained at adequate levels (see Figure 3.5).

Cox and Barber (1992) found, in a pot experiment with five soils varying in water level at the same water tension that at the same soil–water tension increased water content increased the phosphorus supply to the root surface. This increased phosphorus supply will influence root growth, the degree depending on the bioavailable phosphorus level of the soil (see Figure 5.3).

Nutrient Levels and Distribution

When nutrients are uniformly distributed throughout the topsoil, increasing the level of nitrogen or phosphorus generally increases the shoot/root ratio. Shoot growth may increase faster than root growth, or there may even be an actual decrease in the rate of root growth. Barber (1979) studied the effects of nitrogen, phosphorus, and potassium on corn root growth during field experiments. He found root density L_A was 6.6 km/m² with a high rate of nitrogen fertilization (225 kg/ha), while at a low nitrogen-fertilization rate (66 kg/ha) it was 9.6 km/m². In a different experiment employing several rates of phosphorus and potassium, root density for high levels of both was 5.0 km/m²; with low phosphorus and high potassium it was 7.2 km/m², and with low potassium and high phosphate, it was 5.3 km/m². Hence, adding nitrogen and phosphate resulted in decreased root growth, while adding potassium had no effect.

Applying fertilizer to only part of the root system affects distribution of roots between the fertilized and unfertilized soil; adding phosphate to part of the soil volume and not the remainder stimulates root growth in the volume fertilized with phosphate. Zhang and Barber (1992) found the effect of phosphorus placed in part of the soil on root length density in the fertilized volume increased as the ratio of C_{si} of the fertilized soil to C_{si} of the unfertilized soil increased (see equation 21.1 and Figure 21.5).

Anghinoni and Barber (1980) studied the effect of phosphate additions to varying volume proportions of the soil. They obtained the relation $y = x^{0.68}$

(Figure 2.1.), where y is the proportion of total root length in the phosphorus-treated soil and x is the proportion of the soil's volume treated with phosphorus. For example, when phosphorus was added to 20% of the soil's volume, there was 33% of the total root length in the phosphorus-fertilized volume and only 67% in the remaining 80% of the soil's volume. However, the total root length present may still be less than where no phosphate was added, providing sufficient phosphate were available for reasonable growth.

Aeration

Root growth and active nutrient uptake require energy. Photosynthate from the shoot is used for growth and is also respired by the roots. Root respiration uses oxygen and produces carbon dioxide, so gas exchange between air in the soil's pore space and the atmosphere is needed. Soil microorganisms also compete for oxygen, often using as much as the roots. The minimum oxygen level required in soil air may vary with plant species but must usually drop below 10% before root growth noticeably suffers.

Level of soil moisture also affects air exchange between the soil and atmosphere; diffusion of oxygen is 10^4 times faster through air than through water. Hence, there will be an interaction between moisture tension in the soil and soil aeration. In flooded soils, diffusion of oxygen into the soil's pore space is slow. However, plants such as rice (*Oriza sativa* L.) have cortical passages, aerenchyma, in their roots through which oxygen diffuses to the root. The roots receive their oxygen for respiration in this manner. Drew (1979) has shown that corn root structure is modified to provide arenchyma when roots develop under anoxic conditions. This may be an adaptive mechanism to aid in plant survival.

Level of carbon dioxide in the soil air also affects root growth rate (Geisler, 1980). Maximum root growth occurs at 1 to 2% by volume of CO_2. Plants tolerate CO_2 levels as high as 16%, though growth may be somewhat restricted (e.g., 80% of optimum) at such levels.

Toxic Elements

Levels of such elements as aluminum, manganese, and hydrogen may reduce root growth; the degree of reduction will vary with element and plant species. Each of these three elements may become toxic in acid soils, where there will also be lower levels of available calcium and magnesium than in neutral soils.

Aluminum

Aluminum is one of the most common toxic elements in acid soils. At a given soil pH, the level of exchangeable and solution aluminum will vary with the nature of the solid phase. Appreciable quantities of soluble aluminum are generally present at pH values below 5.2 on soils having amorphous aluminum compounds that can release aluminum into solution; Ultisols and Oxisols usually have large quantities of such materials. Aluminum solubility is reduced as pH increases; above pH 5.2, little soluble aluminum tends to be present (see Chapter 2).

Soluble aluminum in the soil reduces root penetration and increases the probability of drought. Soluble aluminum is particularly a problem in subsoils, because their acidity cannot readily be corrected by liming. Subsoils of Ultisols and Oxisols are frequently of low pH (often below 5.0) and have toxic aluminum levels (more than 0.2 cmol/kg).

Levels of aluminum required to inhibit root growth vary with species, cultivar, and soil. For many crops, levels of exchangeable aluminum above 2 mmol/kg will reduce root and shoot growth. Some species that grow well in acid soils, such as cranberry (*Vaccinium macrocarpon*), can tolerate high levels of exchangeable aluminum. Cotton (*Gossypium herbaceum* L.) and wheat (*Triticum vulgare* L.) are particularly susceptible to aluminum toxicity. Aluminum-injured roots are characteristically stubby and spatulate in appearance (Foy, 1974). Lateral root development is also inhibited: The laterals are thickened and lack fine branches.

Manganese

Manganese toxicity occurs less frequently than aluminum toxicity, since plants can tolerate considerably higher levels of soluble manganese. Manganese often requires a combination of both low pH and reducing conditions, situations that tend to be mutually exclusive. The pH almost invariably rises in flooded acid soils. Manganese toxicity usually affects the shoot more than the root (Foy, 1974).

Hydrogen

Plant roots can tolerate pH values in nutrient solution as low as pH 4.0. At this level, aluminum, manganese, and other toxic metals are brought into solution, so that hydrogen is not usually the main factor affecting root growth. Low pH may inhibit the nitrogen-fixing activity of rhizobia on plant roots, causing legumes to be nitrogen deficient.

Calcium Deficiency

Leached acid soils are usually very low in calcium, and this low level of calcium can also affect root growth. Pearson (1966) reported the effects of calcium on the extension of cotton tap roots. When calcium levels in the soil solution were below 0.75 mmol/L, root growth was restricted. The degree of restriction varied with soil, though data for soils and culture solution followed the same relation shown in Figure 6.9, however, when expressed as the ratio of calcium to total cations in solution. When the ratio was below 0.2, root growth was reduced. Under most soil situations, calcium is the

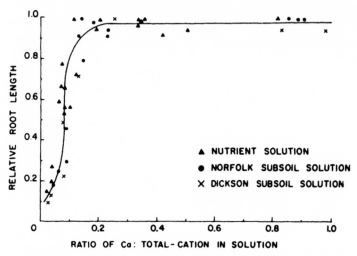

FIGURE   6.9   Effect of ratio of calcium to total cations in subsoil solutions, and in nutrient solution, on primary cotton (*Gossypium hirsutum*) root elongation. Reproduction from Pearson (1966) by permission of American Society of Agronomy.

dominant ion, so these data pertain only to systems extremely low in calcium and where potassium represents a large share of the cations. Higher levels of potassium depress calcium uptake. Calcium is not translocated in the phloem to the roots, so it must be absorbed by the root as it grows for calcium to be available for cell division. Hence, roots may not grow into calcium-deficient soils.

Herbicides

Herbicides used for weed control can also affect root growth, particularly if used in excess. The usual effect is stunted root growth, which reduces nutrient uptake. Effects vary with herbicide form. Investigations are being conducted to find growth regulators that increase root growth or change root morphology to increase nutrient uptake. Little specific information on useful growth regulators for such purposes was available at the time of writing.

SUMMARY

In some instances, modifications are necessary in the model describing nutrient flux into soil-grown roots described in Chapter 5. These modifications are due to the effect which the root may have on C_{li}, b, or D_e, thus increasing or decreasing nutrient supply by the soil. Changes in the rhizosphere may also alter nutrient-absorption characteristics of the roots, thereby affecting uptake.

The effect of the soil on root growth and morphology is of great importance in determining rate of nutrient supply to the plant. Rate of root growth and size of root radius are among the most sensitive parameters in determining nutrient uptake.

While many of the factors described in this chapter have an influence on nutrient uptake, it is fortunate that the size of their effect is minimal in soils where plant growth is not limited appreciably by soil conditions. Hence, the simpler model in Chapter 5 can be used in most situations.

REFERENCES

Anghinoni, I., and S. A. Barber. 1980. Predicting the most efficient phosphorus placement for corn. *Soil Sci. Am. J.* **44**:1016–1020.

Baligar, V. C., and S. A. Barber. 1978. Use of K/Rb ratio to characterize potassium uptake by plant roots growing in soil. *Soil Sci. Soc. Am. J.* **42**:575–579.

Barber, D. A., and J. K. Martin. 1976. The release of organic substances by cereal roots into soil. *New Phytol.* **76**:69–80.

Barber, S. A. 1971. The influence of the plant root system in the evaluation of soil fertility. *Proc. Int. Symp. Soil Fert. Evaln.*, New Delhi **1**:249–256.

Barber, S. A. 1974. Influence of the plant root on ion movement in soil. In E. W. Carson, Ed. *The Plant Root and Its Environment.* University Press of Virginia, Charlottesville. Pp. 525–564.

Barber, S. A. 1979. Growth requirements for nutrients in relation to demand at the root surface. In J. L. Harley, Ed. *The Soil-Root Interface.* Blackwell Scientific Publishers, Oxford, England. Pp. 5–20.

Barber, S. A., and P. G. Ozanne. 1970. Autoradiographic evidence for the differential effect of four plant species in altering the Ca content of the rhizosphere soil. *Soil Sci. Soc. Am. Proc.* **34**:635–637.

Barley, K. P. 1962. The effects of mechanical stress on the growth of roots. *J. Exp. Bot.* **13**:95–110.

Barley, K. P. 1976. Mechanical resistance of the soil in relation to the growth of roots and emerging shoots. *Agrochima* **20**:173–182.

Bole, J. B., and S. A. Barber. 1971. Differentiation of Sr-Ca supply mechanisms to roots growing in soil, clay, and exchange resin cultures. *Soil Sci. Soc. Am. Proc.* **35**:768–772.

Ching, P. C., and S. A. Barber. 1979. Evaluation of temperature effects on potassium uptake by corn. *Agron. J.* **71**:1040–1044.

Clark, R. B. 1978. Differential response of corn inbreds to calcium. *Commun. Soil Sci. Plant Anal.* **9**:729–744.

Cox, M. S., and S. A. Barber. 1992. Soil phosphorus levels needed for equal P uptake from four soils with different water contents at the same water potential. *Plant Soil* **143**:93–98.

Cunningham, R. K. 1964. Cation-anion relationships in crop nutrition. III. Relationships between the ratios of sum of cations: sum of the anions and nitrogen concentrations in several plant species. *J. Agric. Sci.* **63**:109–111.

Drew, M. C. 1979. Properties of roots which influence rates of absorption. In J. L. Harley and R. S. Russell, Eds. *The Soil-Root Interface.* Academic Press, New York. Pp. 21–28.

Foster, R. C. 1981. The ultrastructure and histochemistry of the rhizosphere. *New Phytol.* **89**:263–273.

Foy, C. D. 1974. Effects of aluminum on plant growth. In E. W. Carson, Ed. *The Plant Root and Its Environment.* University Press of Virginia, Charlottesville. Pp. 601–642.

Geisler, G. 1980. Morphogenetic factors in the root environment and their effects on root characteristics. In D. N. Sen, Ed. *Environment and Root Behaviour.* Geobios International, Jodhpur, India. Pp. 159–169.

Gerdemann, J. W. 1974. Mycorrhizae. In E. W. Carson, Ed. *The Plant Root and Its Environment.* University Press of Virginia, Charlottesville. Pp. 205–217.

Goss, M. J., and J. B. Reid. 1981. Interaction between crop roots and soil structure. In F. E. Shotten, Ed. *Aspects of Crop Growth M.A.F.F. Ref. Book.* Her Majesties Stationery Office, London. Pp. 34–48.

Greacan, E. L., and J. S. Oh. 1972. Physics of root growth. *Nature (New Biol.)* **235**:24–25.

Grinsted, M. J., M. J. Hedley, R. E. White, and P. H. Nye. 1982. Plant-induced changes in the rhizosphere of rape (*Brassica napus* var. Emerald) seedlings. I. pH change and the increase in P concentration in the soil solution. *New Phytol.* **91**:19–29.

Grunes, D. L. 1959. Effect of nitrogen on the availability of soil and fertilizer phosphorus to plants. *Adv. Agron.* **11**:369–396.

Haynes R. J. 1990. Active ion uptake and maintenance of cation-anion balance: A critical examination of their role in regulating rhyzosphere pH. *Plant Soil* **126**:247–264.

Hallmark, W. B., and S. A. Barber. 1981. Root growth and morphology, nutrient uptake, and nutrient status of soybeans as affected by soil K and bulk density. *Agron. J.* **73**:779–782.

Horvath, I., E. Mihalik, and E. Takacs. 1980. Effect of light on root production. In D. N. Sen, Ed. *Environment and Root Behavior.* Geobios International, Jodhpur, India. Pp. 231–255.

Israel, D. W., and W. A. Jackson. 1978. The influence of nitrogen nutrition on ion uptake and translocation by leguminous plants. In C. S. Andrew and E. J. Kamprath, Eds. *Mineral Nutrition of Legumes in Tropical and Subtropical Soils.* Commonwealth Scientific and Industrial Research Organization, Melbourne, Australia. Pp. 113–129.

Jackson, P. C., and H. R. Adams. 1963. Cation-anion balance during potassium and sodium absorption by barley roots. *J. Gen. Physiol.* **46**:369–386.

Keltjens, W. G. 1982. Nitrogen metabolism and K-recirculation in plants. In A. Scaife, Ed. *Plant Nutrition 82. Proc. Ninth Int. Plant Nutr. Colloq.* **1**:283–287.

Kirkby, E. A. 1968. Influence of ammonium and nitrate nutrition on the cation-anion balance and nitrogen and carbohydrate metabolism of white mustard plants grown in dilute nutrient solutions. *Soil Sci.* **105**:133–144.

Kuchenbuch, R. O., and S. A. Barber. 1988. Significance of temperature and precipitation for maize root distribution in the field. *Plant Soil* **106**:9–14.

Li, Y. and S. A. Barber. 1991. Calculating changes of legume rhizosphere pH and soil solution phosphorus from phosphorus uptake. *Commun. Soil Sci. Plant Anal.* **22**:955–973.

Mass, E. V., and G. Ogata. 1971. Absorption of magnesium and chloride by excised corn roots. *Plant Physiol.* **47**:357–360.

Mackay, A., and S. A. Barber. 1984. Soil temperature effects on root growth and phosphorus uptake by corn. *Soil Sci. Soc. Am. J.* **48**:818–823.

Marschner, H., and V. Romheld. 1983. In vivo measurement of root-induced pH changes at the soil-root interface. *Z. Pflanzenphysiol.* **111**:241–000.

Nye, P. H., and P. B. Tinker. 1977. *Solute Movement in the Soil-Root System.* Blackwell Scientific Publishers, Oxford, England.

Nye, P. H. 1982. Changes of pH across the rhizosphere induced by roots. *Plant Soil* **61**:7–26.

Pearson, R. W. 1966. Soil environment and root development. In W. H. Pierre, D. Kirkham, J. Pesek, and R. Shaw, Eds. *Plant Environment and Efficient Water Use.* American Society of Agronomy, Madison, WI. Pp. 95–126.

Peterson, W. R., and S. A. Barber. 1981. Soybean root morphology and K uptake. *Agron. J.* **73**:316–319.

Riley, D., and S. A. Barber. 1969. Bicarbonate accumulation and pH changes at the soybean (*Glycine max* L. Merr) root-soil interface. *Soil Sci. Soc. Am. Proc.* **33**:905–908.

Riley, D., and S. A. Barber. 1970. Salt accumulation at the soybean (*Glycine max* L. Merr) root-soil interface. *Soil Sci. Soc. Am. Proc.* **34**:154–155.

Riley, D., and S. A. Barber. 1971. Effect of ammonium and nitrate fertilization on phosphorus uptake as related to root-induced pH changes at the root-soil interface. *Soil Sci. Soc. Am. Proc.* **35**:301–306.

Rovira, A. D., and C. B. Davey. 1974. Biology of the rhizosphere. In E. W. Carson, Ed. *The Plant Root and Its Environment.* University Press of Virginia, Charlottesville. Pp. 153–204.

Russell, R. S. 1977. *Plant Root Systems: Their Function and Interaction with Soil.* McGraw-Hill, New York.

Sauerbeck, D. R., and B. G. Johnsen. 1976. Root formation and decomposition during plant growth. *Soil Organic Matter Studies.* Int. Symposium, Braunschweig, Germany.

Silberbush, M., W. B. Hallmark, and S. A. Barber. 1983. Simulation of effects of soil bulk density and P addition on K uptake by soybeans. *Commun. Soil Sci. Plant Anal.* **14**:287–296.

Silsbury, J. H. 1971. The effects of temperature and light on dry weight leaf area changes in seedling plants of *Lolium perenne* L. *Aust. J. Agric. Res.* **22**:177–187.

Smiley, R. W. 1974. Rhizosphere pH as influenced by plants, soils, and nitrogen fertilizers. *Soil Sc. Soc. Am. Proc.* **38**:795–799.

Taylor, H. M., and H. R. Gardner. 1963. Penetration of cotton seedling taproots as influenced by bulk density, moisture content, and strength of soil. *Soil Sci.* **96**:153–156.

Troughton, A. 1980. Environmental effects upon root-shoot relationships. In D. N. Sen, Ed. *Environment and Root Behaviour.* Geobios International, Jodhpur, India. Pp. 25–41.

Walker, J. M. 1969. One-degree increments in soil temperatures affect maize seedling behaviors. *Soil Sci. Soc. Am. Proc.* **33**:729–736.

Zhang, J., and S. A. Barber. 1992. Maize root distribution between phosphorus-fertilized and unfertilized soil. *Soil Sci. Soc. Am. J.* **56**:819–822.

CHAPTER **7**

Rhizosphere Microorganisms, Mycorrhizae, and Root Hairs

Up to this point, roots have been considered as smooth cylindrical absorbing organs. In this chapter, I consider free-living microorganisms, associative microorganisms, mycorrhizae, and root hairs, each of which affects nutrient influx into roots.

FREE-LIVING RHIZOSPHERE MICROORGANISMS

The rhizosphere is that zone of soil influenced by the root. It can have as much as 10 times the population of microorganisms in the bulk soil. These microorganisms may be free living, or they may form symbiotic relationships with the root, called mycorrhizae. In this section, I will discuss the relation of nonpathogenic free-living microorganisms to nutrient supply. Most investigations have concerned phosphorus, since supply of this element is usually believed to be susceptible to change. Microorganisms present in the rhizosphere include bacteria, fungi, and other organisms. The number of microorganisms in the rhizosphere is larger because of root exudates and material sloughed off root surfaces. These materials serve as a source of energy for the growth of microorganisms.

Rovira (1979) reports that recent studies on the effect of rhizosphere microorganisms have departed from plate counts of increases in numbers of microorganisms in the rhizosphere to a study of the colonization of root surfaces. Only part of the root is covered with microorganisms. Studies have also been made of "generation time," where the doubling time of the population on the root is measured. Reported values for doubling time range from 5 to over 100 hs.

In order to alter the rate of phosphorus flux to the root, the microorganisms would have to alter one of the following: (1) the size or morphology of the root surface; (2) the influx kinetics of the root; (3) the concentration of phosphorus in soil solution at the root surface; or (4) transport of phosphorus in organic form to the root for release and absorption. Little information is available on any of these effects. Experiments have been conducted where plants were grown in sterile environments and phosphorus uptake was compared with uptake under nonsterile conditions. Barber and Rovira (1975) and Barber et al. (1976) investigated these effects on phosphorus uptake and found the results varied according to microflora present, length of the experiment, and plant age. Nonsterile roots absorbed and transported more phosphorus during the first 30 min of each experiment, but the difference decreased with time until eventually there was no difference. Hence, the difference between sterile and nonsterile roots was regarded as of little consequence in a field situation.

If microorganisms changed the pH of the soil near the root, they could alter the rate of phosphorus supply, since in many soils C_{li} for phosphorus is pH-dependent. Competition among microorganisms for phosphorus could even *reduce* the amount of phosphorus supplied to the root. Tinker (1980) calculated, however, that the amount of phosphorus in microorganisms of the rhizosphere amounted to only 3% of one day's uptake of phosphorus by the plant; hence, the competitive effect should be negligible.

The fact that reasonable agreement is obtained between observed uptake by plants and uptake predicted by mechanistic simulation models that do not include effects of microorganisms suggest that microbial effects on nutrient uptake may be small.

ASSOCIATIVE RHIZOSPHERE MICROORGANISMS

Nitrogen fixation by grass–bacteria associations in tropical soils has been described by Dobereiner and Day (1975). While not a direct effect of the root on the soil, it *is* an effect whereby the root provides the substrate for the nitrogen-fixing activity of the bacteria. On the basis of 6- to 12-h experiments on C_4H_4 reduction. it was estimated that N_2-fixation rates as high as 1 kg/ha · day could be obtained. Bahia grass (*Paspalum notatum*, Flugge) is the principal grass with which N_2 fixation has been observed. Long-range values of 50 kg/ha in 3 months have been estimated; differences between cultivars varied over a six-fold range. Only the upper end of the range of N-fixation values has usually been mentioned; the lower end is about only one-tenth the highest value. Most of the bacterial activity was associated with the root and could not be removed by washing. Little activity occurred in the soil itself. The fixation of nitrogen by associative rhizosphere microorganisms has been mainly observed on tropical grasses and sugarcane (*Saccharum*). There is no definitive evidence that associative rhizosphere microorganisms supply nitrogen for corn (Dobereiner and Solomone, 1992). For plant species other than tropical grasses and sugar cane, practical and theoretical studies make it seem that nitrogen fixation is not very great (Wild, 1988, p. 533). Kennedy and Tchan (1992) reviewed the recent advances in biological nitrogen fixation and found that nonleguminous field crops other than sugar cane and tropical grasses sometimes benefit from association with diazotrops, however, the potential benefit from N_2 fixation is usually gained from spontaneous associations that rarely can be managed as part of agricultural practice.

MYCORRHIZAE

Mycorrhizae are soil fungi that form a symbiotic relation with the root of a host plant. The fungi supply nutrients, particularly phosphorus, and the plant supplies carbohydrate for the growth of the fungi. Mycorrhizae are non-pathogenic, present in most soils, and form symbiotic relations with most plant species. I am concerned primarily with the effect of mycorrhizae on improving the nutritional status of the host plant. Mycorrhizae are classified into five groups: ecto, endo, ericoid, arbutoid, and orchidacious mycorrhizae; the first two are of major importance and the only ones discussed here. Research on mycorrhizas was reviewed by Tinker (1982), with special emphasis on endomycorrhizae.

Ectomycorrhizae

Ectomycorrhizae form a compact sheath, or mantle, of hyphae over the root surface. Infected roots are short, stubby, and more branched than non-mycorrhizal roots. Ectomycorrhizae infect primarily *Pinacea*, *Betulacea*, and *Lagaceae* species. The hyphae from the sheath penetrate between epidermal and cortical cells to form a network of intercellular hyphae, referred to as the Hartig net. Strands of hyphae growing into the soil constitute the surface for absorbing nutrients. Exchange of materials between the host and the fungus occurs primarily in the Hartig net.

Abundant evidence (Harley, 1969) indicates that ectomycorrhizal infection increases nutrient, particularly phosphorus, uptake and improves plant growth. The principal mechanism for increasing nutrient uptake is the increase in absorbing surface area. Harley (1969) reports that hyphae function much as do root hairs in absorbing nutrients from the soil. Because of the small size of hyphae and the difficulty in detecting all of them, information is not available on the surface area of hyphae per unit m² of root surface. However, as indicated in Figure 7.1, the large number present would suggest that their surface area is much greater than the root's.

Ectomycorrhizal infection of the roots greatly increases nutrient absorption. Pine seedlings planted in Puerto Rico in a noninfected soil failed to grow even when provided with fertilizers. Inoculating the pines with mycorrhizal fungi resulted in rapid growth of most of the seedlings (Voggo, 1971).

Endomycorrhizae

Most plant species, cultivated and native, growing in soil are infected by endomycorrhizae. The principal group is nonseptate vesicular-arbuscular mycorrhiza; the genus is Endogene. Vesicular-arbuscular (VA) mycorrhizae are widely distributed geographically. They appear not to infect *Cruciferae* and *Chenopodeaceae* but do infect most other species.

These VA mycorrhizae indicate a distinctly different type of infection of the root from that shown by ectomycorrhizae. They do not affect the morphology of the root; the infection is illustrated in Figure 7.2. A fungal hypha, or spore, enters the epidermis, and hyphae spread between and into the root's cortical cells. Simultaneously, hyphae grow into the soil. Branched structures

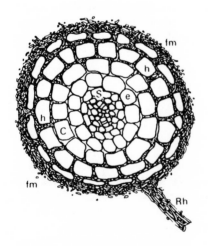

FIGURE 7.1 Diagrammatic representation of ectomycorrhiza. fm, fungal mantle; Rh, rhizomorph; S, stele; C, cortex; e, endodermis; h, host cell. Reproduced from Tinker (1980) by permission of American Society of Agronomy, Crop Science Society of America, and Soil Science Society of America.

called arbuscules and spherical vesicles are formed within the root. An extensive network of mycelium forms within the root cortex, and arbuscules form within the cortical cell. They form by hypha entering the cell and then branching dichotomously and repeatedly to form a dense cluster of fine hyphae, which occupy much of the cell's volume. The arbuscules appear to be transfer organs between fungus and root. Vesicles that form between the cortical cells are globular storage organs, about 50 μm in diameter.

Vesicular-arbuscular mycorrhizae have only recently (since 1970) received widespread attention. They did not draw much attention earlier,

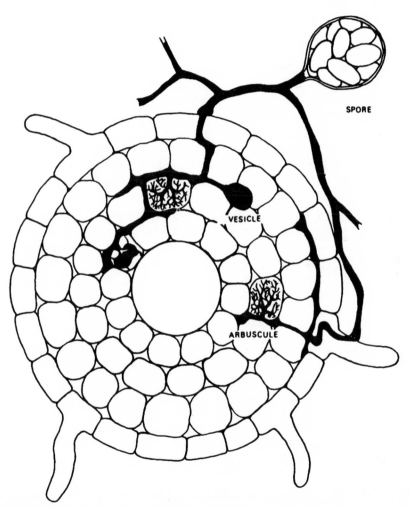

FIGURE   7.2   Diagrammatic representation of endomycorrhiza. Reproduced from Tinker (1980) by permission of American Society of Agronomy, Crop Science Society of America, and Soil Science Society of America.

because they did not affect the growth of the root and could not be cultured in the absence of the living root. The latter property makes it difficult to culture new species of mycorrhizae and introduce them into the soil.

Degree of Infection

The most significant morphology with respect to nutrient absorption by mycorrhizae is the length and radius of the hypha and the number of hyphae per m^2 of root surface. Degree of infection of roots varies widely with the nature of the host plant, nutritional status of the plant, soil, and climatic conditions where plants are growing. These factors have not been studied in detail; the usual measurement reported is the percentage of the root length infected. One method involves cutting roots into 1.5-cm-long segments and then determining the percentage of segments infected. Length of active mycellium in the soil is the important parameter in determining the effect of mycorrhiza in increasing P uptake. It is difficult to separate and measure mycellium that may have a length of 80 cm/cm of root. However, this value will certainly vary with host plant and soil conditions. Few measurements have been made of the radius of the hypha, which are usually less than half the radius of root hairs. A value of 1 to 3 μm would probably cover the range.

The incidence of infection does not appear to be influenced by soil pH or soil texture (Tinker, 1980). Usually, lower infection occurs with higher levels of available phosphorus in the soil.

Sufficient infection of the roots with versicular-arbuscular mycorrhiza (VAM) to increase phosphorus uptake by plant roots usually does not occur until 2 weeks after planting the crop. Because of this our experiments with rapidly growing plants in pots do not exhibit increased phosphorus due to the presence of VAM as the plants usually were harvested within 2 to 3 weeks of planting since longer plant growth resulted in root density in the soil much greater than that occurring in the field.

Contribution to Phosphorus Uptake

The main benefit from VA mycorrhizae that has been established is the increase in phosphorus uptake, particularly by infected plants growing in low-phosphorus soils. Some information suggests that zinc and copper uptake may also be increased. When mycorrhizal and nonmycorrhizal plants are compared, the degree of difference in growth decreases as levels of additional phosphorus are increased. The greatest benefit appears to be in soils where available phosphorus is low enough so that appreciable responses to added phosphorus fertilizer would occur. Yost and Fox (1979) found that the soil-solution level at which mycorrhizal infection ceased to increase yield was in the range of 0.1 to 1.6 mg P/L (3 to 51 μmol/L) depending on species. The authors grew seven plant species at 10 levels of soil phosphorus using both mycorrhizal (nonfumigated soil) and nonmycorrhizal (fumigated soil) treatments. At the lowest soil-solution phosphorus level, 0.003 mg/L (0.1

μmol/L) plant growth was 25 times greater on the nonfumigated soil. However, *Brassica chinensis*, a species that does not form mycorrhizal associations, grew better in the fumigated soil. This was probably because of concurrent control of soil-borne pathogens. These researchers did find a relationship between the degree of infection in the roots with mycorrhizae and an increase in phosphorus uptake by mycorrhizal plants over non-mycorrhizal plants. Mycorrhizae are very important for some plants: Cassava (*Mannihot esculenta* Crantz) is highly dependent on its mycorrhizae and would be an extremely phosphorus-demanding crop if it were not for the mycorrhizae present on its roots (Howeler et al., 1982).

Sanders and Tinker (1973) showed that VA mycorrhizal infection increased the uptake of phosphorus per unit length of onion root by a factor of 4. The increase in uptake presumably comes from absorption by the mycellial hyphae and transfer through the hypha. This transfer is presumably by plasma streaming to the cortical cells of the root.

Soil Disturbance and VA Mycorrhiza Infection

O'Halloran et al. (1986) observed phosphorus concentration and shoot weight of six-leaf maize plants grown in soil zero tilled since 1968 was greater than in conventionally tilled soil. Disturbing zero-tilled soil reduced the phosphorus concentration to that grown in tilled soil. Evans and Miller (1988) found higher phosphorus concentrations and higher VA mycorrhizal infection on maize plants grown on undisturbed soil. Fairchild and Miller (1988) found that the greater the number of previous maize crops grown on untilled soil the greater the effect. The effect did not occur when the soil had been irradiated or when nonmycorrhizal species were grown. The tillage effect occurred in potted soil as well as in field experiments. Similar results were obtained for *Trifoleum subteraneum* L. by Jasper et al. (1989) who found colonization of new roots occurred much more rapidly for plants grown in undisturbed soil. Vivekananda and Fixen (1991) also observed similar effects with corn grown in the field.

McGonigle et al. (1990) hypothesized that tilling or disturbing the soil disrupted an extraradical mycelium system that was capable of connecting to a newly developing root system and supplying the plant with extra nutrients, particularly phosphorus. When the soil has been disturbed before planting, a common practice, a new extraradical mycelium system would need to be established before VA mycorrhiza could supply the plant with extra phosphorus. Additional experiments gave evidence supporting this hypothesis.

Evidence obtained with ^{32}P-labeled soil phosphorus indicates that the source of phosphorus for mycorrhizae is the same as for the root, soil solution $H_2PO_4^-$. The $^{32}P:^{31}P$ ratio of phosphorus in plants infected with mycorrhizae was the same for plants without mycorrhizae (Sanders and Tinker, 1971; Hayman and Mosse, 1972). The degree to which mycorrhizae increase phosphorus uptake depends on the length of hyphae per cm of root, their distribution, and their phosphorus-influx characteristics. Hattingh et al. (1973) have shown that hyphae can extend several centimeters from the root and the hyphae on onion roots absorbed ^{32}P-labeled phosphate placed as much as 27 mm from the root. While hyphae serve a similar role as root hairs, their distribution around the root is different. Hyphae often grow parallel to the root, which makes them less effective than those growing perpendicularly to the root surface. Some researchers have considered that hyphae may be able to absorb phosphorus at lower solution phosphorus concentrations than roots. Roots usually absorb phosphorus only down to 0.2 μmol/L. It is not presently possible to study the kinetics of phosphorus influx of hyphae; we can only infer kinetics by comparing uptake kinetics for roots without mycorrhizae with those for roots containing mycorrhizae.

Simulation of phosphorus uptake by hyphae can be done by assuming that hyphae have phosphorus-uptake kinetics similar to roots. Calculations of predicted phosphorus uptake by hyphae are compared with predicted uptake by roots using the following assumptions: Soil parameters: D_e, 2.3×10^{-9} cm^2/s; C_{li}, 13.6 μmol/L; b, 136. Plant parameters: I_{max}, 7 nmol/m^2 · s; K_m, 5 μmol/L; C_{min}, 0.2 μmol/L; v_0, 5×10^{-7} cm/s; r_0, root, 0.15 mm; hypha, 0.005 mm; r_1, root, 2.8 mm; hypha, 0.55 mm; L_0, root, 10^3 cm; hypha, 3×10^4 cm (30 cm of hypha per cm of root), k value zero for both. Roots and hypha occupy the same soil volume and no growth was assumed for this uptake comparison. Predicted phosphorus uptake by the root was 17.5 μmol and uptake by the hypha was 37.1 μmol.

Since the hypha and root compete for uptake, the total uptake would be less than the sum of the two and would depend on their distribution. Uniform distribution and no influence of root uptake on hypha uptake and vice versa were assumed. The uptake model can be used to calculate predicted uptake with many different assumptions for values of the parameters involved. The results from the calculations here agree reasonably with commonly observed behavior. (A calculation error occurred in the first edition of this book.)

Phosphorus uptake by hyphae depends on the rate of diffusion of phosphorus to the root. Because of radial geometry and differences in r_0, the phosphorus gradient normal to the hypha is almost flat, as indicated in Figure 7.3, while for plant roots, a steeper gradient exists, and the phosphorus concentration is much lower at the root surface than at the hyphae surface. This factor would increase phosphorus uptake per m^2 of hyphae as compared to uptake per m^2 of root. Since the phosphorus gradient is less, hyphae would be able to absorb phosphorus from low-phosphorus soils more effectively than would the larger radius plant roots. Mycorrhiza would not have to be able to absorb at lower soil-solution phosphorus levels, since levels around mycorrhizal hypae would be higher than those around roots.

The soil phosphorus level above which mycorrhizae do not benefit plant growth must be one where roots plus root hairs can absorb sufficient phosphorus to maximize plant growth. This level varies according to soil phos-

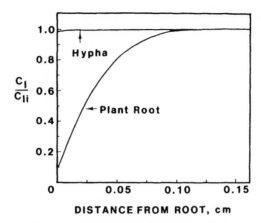

FIGURE   7.3   Phosphorus concentration gradients in soil normal to a, hypha with $r_0 = 0.005$ mm; and b, plant root with $r_0 = 0.15$ mm. Calculations from a simulation model. Values for other parameters are given in the text.

phorus level and the root system of each plant species. When plants with mycorrhizal-infected roots grow better on soil treated with rock phosphate than do plants with uninfected roots, this does not mean that mycorrhizae solubilize the phosphate. Rather, they may supply a much greater surface area to absorb the small amounts of slightly soluble phosphate that *does* come into solution. The ^{32}P studies suggest that solubilization is not the reason for better phosphorus utilization by mycorrhizal-infected roots.

Uptake of other nutrients supplied to the root by diffusion may also be increased whenever mycorrhizae provide more surface for adsorbing nutrients that diffuse only short distances. In Table 7.1, copper and zinc uptake were increased by mycorrhizal infection. Adding phosphate reduced uptake, which may have been due to a reduction in mycorrhizal activity on the mycorrhizal plants.

Li et al. (1991) used Plexiglas pots designed to permit spatial soil zones for root growth and for mycorrhiza hyphae growth. They grew white clover (*Trifolium repens* L.) and to separate soil zones they used a 30-μm mesh nylon net that permitted hyphae to pass but not roots. They found hyphae length density in the soil of 3 to 4 m/cm³ of soil. They attributed 53 to 62% of copper uptake to mycorrhizae hyphae.

Tinker et al. (1982) indicate that 10 to 12% of the carbon that would normally go into the shoot of the host goes to the root in plants infected with mycorrhizae. Here, it is metabolized by mycorrhizae and appears as additional CO_2. However, because of the effectiveness of mycorrhizae in supplying phosphorus to the host, the net benefit from added phosphorus, in most cases, far outweighs the loss due to diversion of photosynthate to the fungus. In some soils, however, soil fauna feed on mycorrhiza, thereby reducing the amount of fungus present and its effectiveness.

TABLE 7.1   Effect of Presence of Mycorrhiza and Level of Added Phosphate on Soybean Growth and Composition of 40-Day-Old Soybean Plants

	P added (mg P/kg soil)	Plant yield (g/pot)	Plant composition		
			P (g/kg)	Zn (mg/kg)	Cu (mg/kg)
Nonmycorrhizal	0	1.06d^a	0.70g	16.4d	5.5bc
	25	0.98d	0.78g	15.8d	6.2b
	75	2.21b	1.18f	14.2e	5.7bc
	200	2.81a	2.36c	17.6	5.2c
Mycorrhizal	0	1.72c	1.42e	56.5a	8.2a
	25	2.44ab	1.81d	35.7b	7.4a
	75	2.64a	2.71b	28.5d	6.0b
	200	2.67a	3.59a	28.4c	6.2b

Source: Reproduced from Lambert et al. (1979) by permission of the Soil Science Society of America.
[a]Means within columns followed by different letters are significantly different at 5% probability level.

Introducing Mycorrhizae

There is an opportunity in crop management for introducing new mycorrhizae endophytes that will give greater phosphorus uptake. Types most likely to be introduced would be those producing greater surface area of hypha per unit of infected roots. This increase may be due to both greater length and a greater radius. Investigators (as reported by Tinker, 1980) have found that there are endophytes that give greater phosphorus uptake than natural endophytes occurring in the field. Developing new endophytes for plants growing on soils low in phosphorus is one method of increasing phosphorus-use efficiency in low-phosphorus soils and especially where phosphorus fertilizer applications are low (Mosse, 1973).

ROOT HAIRS

Root hairs are tubular extensions of the root epidermal cells and occur as a result of lateral cell growth. Root hairs arise in the region of maturation just beyond the zone of elongation (Esau, 1977). The length of time during which they exist is indicated by the distance along the root over which hairs are present. Root hair growth can increase the area of the outer surface of the epidermal cell by 2 to 10 times. While most species have root hairs, a few have none or extremely small root hairs.

Morphology

Root hairs vary in length from 0.1 to 1.5 mm and in diameter, from 0.005 to 0.025 mm (Dittmer, 1949). Root hair length and diameter are relatively uniform within a species. Root hair density varies from 50 to 500 million per m^2 of root surface. Root hair density varys with environmental factors such as aeration, presence and numbers of soil microorganisms, soil moisture, soil nutrient status, and soil physical properties.

Sterile roots have been found to form shorter and less dense root hairs than nonsterile roots. Root hair growth is more irregular on nonsterile roots and denser where root hairs grow into air of high humidity, such as that found in soil pores. Probably more root hairs grow in pores, because there is no resistance to growth. Barley and Rovira (1970) showed that clay density could control root hair growth. Pea (*Pisum sativum*) grown into clay with a voids ratio of 1.0 had no root hairs, while those grown into clay with a voids ratio of 1.2 had root hairs. Hence, external pressure inhibits root hair growth.

Root hairs are more prevalent in soils at lower moisture content, probably because of the additional voids present into which root hairs can grow. There is some evidence that there are fewer root hairs on plants growing in soils with large amounts of available phosphorus. Some species, such as onion and carrot, produce none or extremely short, stubby root hairs.

Cormack (1962) observed that some plant species had alternating long and short cells in the same cell row of the epidermis, with root hairs only forming from the short cells. Dittmer (1949) found that almost all root hairs are straight and protrude perpendicularly to the root surface; they represent nonseptate protuberances of the epidermal cell.

The effect of solution nitrogen, phosphorus, and potassium concentration on root hair growth of tomato, rape (*Brassica oleracea* L.), and spinach (*Spinacea oleracea* L.) was investigated by Fohse and Jungk (1982). Using solutions varying from 2 to 1000 μmol/L for each nutrient individually, the authors observed that nitrogen and phosphorus had a marked effect on root hair length. Their results, shown in Figures 7.4 and 7.5, indicate a 2- to 10-fold increase in root hair length when nitrogen and phosphorus concentrations in solution were decreased to 2 and 10 μmol/L, respectively. Similar investigations with ammonium, potassium, magnesium and calcium showed little effect on root hair length or density. Root hair density on spinach was affected by solution level of nitrate or phosphorus. Density was three- to six-fold greater at 2 to 10 μmol/L as compared to 100 and 1000 μmol/L. No effects were observed on root hair densities for rape or tomato. Hence, density effects were both nutrient- and species-specific. Effects on root hair growth occurred when nitrate in phosphorus concentrations were 10 μmol/L or less. While this may not be a common level for nitrate in soil solution, it is a common value of C_{l0} for phosphorus. Hence, root-hair stimulation should occur when phosphorus levels in the soil are low.

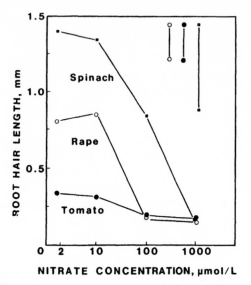

FIGURE   7.4   Effect of nitrate concentration of the nutrient solution on the root-hair length of tomato, spinach, and rape, Vertical bars represent significant differences, $p = 0.05$, by Dunnett's test (Foehse and Jungk, 1982).

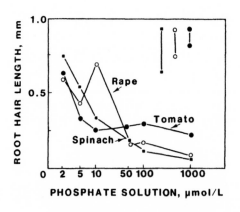

FIGURE   7.5   Effect of phosphate concentration of the nutrient solution on the root-hair length of tomato, spinach, and rape Vertical bars represent significant differences, $p = 0.05$, by Dunnett's test (Foehse and Jungk, 1982).

Mackay and Barber (1984) grew maize in solution culture at phosphorus concentrations of 1, 10, and 500 μmol/L and compared it with maize grown for 21 days in two soils where the soil solution phosphorus, C_{li}, was 9 (S_1) and 40 (S_2) μmol/L. There was no significant decrease in root hair length as solution phosphorus increased. However, root hair length was two to three times longer in soil culture than in solution culture. Root hair length averaged 0.40 mm in S_1 and 0.25 mm in S_2 as compared to an average of 0.14 mm in the solution culture. Root hair density was two to three times greater in soil than in solution. Hence root hair growth in soil is different from root hair growth in solution.

In a study of the effect of soil water level and soil P level on root hair growth, Mackay and Barber (1985) found a larger effect of soil water level than of soil P level. Maize was grown in pots of soil (Raub silt loam, Aquic Argiudoll) in a controlled climate chamber for 21 days with three volumetric soil water levels of 22% (M_0), 27% (M_1), and 32% (M_2) and five soil C_{li} P levels of 0.81, 12.1, 21.6, 48.7, and 203.3 μmol P L^{-1}. As in the previous experiment root hair growth was evaluated by measuring percentage of root length having root hairs, root hair density (number of root hairs per unit area) in the sections having root hairs, and average root hair length. As volumetric soil water level decreased from 32 to 22%, proportion of the root length with root hairs increased (Figure 7.6). Root hair density in the sections of the root with root hairs decreased as soil water level decreased and as soil P level increased. Root hair length was not affected significantly by soil P level, however, soil water level had a large effect on root hair length with the average overall phosphate levels being 0.40 mm at M_0, 0.33 mm at M_1, and 0.26 mm at M_2. As a result the root surface area, RSA, decreased with an increase in soil phosphorus level and an increase in soil water level (Figure 7.7). The kPa soil water tension for the three soil water levels were -173 (M_0), -33 (M_1), and -7.5 (M_2).

The pronounced effect of soil water content on root hair growth as compared to soil P level is supported by field results of Mackay and Barber (1985). At two locations, the phosphorus level (C_{si}) in the topsoil (0–20 cm) was 94 and 96 mg/kg, while the level in the subsoil was less than 1 mg/kg at both locations. The soil water level in the topsoil in the first site decreased from 22 to 15% (w/w) during the growing season. In the second site it decreased from 27.5 to 19.5% (w/w). However, in the subsoil of both soils the soil water level remained above 28% at both sites. Root hair growth in the topsoil increased during the growing season at both sites, while little change occurred in root hair density and length in the subsoil.

In a pot experiment, the effect of cyclic wetting and drying of three soils on root hair growth of maize showed that increasing soil water level from -175 to -7.5 kPa, resulting in a decline of root hair growth behind the root cap in all three soils. Drying the soil had the opposite effect on root hair growth. Root hairs produced in dryer soil persisted after the soil water level was increased. Since root hairs grow in air-filled pore space, these root hairs may be more effective in absorbing nutrients after the soil water level is increased.

Root Hairs and Phosphorus Uptake

Experiments have demonstrated that root hairs increase phosphorus uptake. Using systems varying only in void ratio of the clay, Barley and Rovira

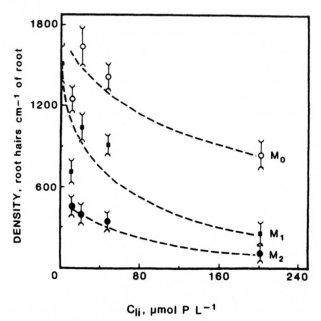

FIGURE   7.6   Effect of soil water and phosphorus level on root hairs per cm of corn root in sections having root hairs from 21-day-old plants grown at three soil water potentials of -175 (M), -33 (M_1) and -7.5 (M_2) kPa having soil water contents of 22, 27, and 32% (w/w), respectively, and five soil phosphate levels. Standard deviations are represented by vertical lines at each data point. Reproduced from Mackay and Barber (1985) by permission of Martinus Nijhoff Publisher. B.V.

(1970) found that systems with root hairs absorbed 78% more phosphorus than those without root hairs. They also presented evidence to show that this increase did not occur for roots growing in solution culture and that phosphorus-uptake rate for portions of the root without root hairs was similar to that where root hairs were present. This suggests that root hairs provide a greater surface for phosphorus diffusion; thus they increase the supply from the soil rather than the capacity of the root to absorb.

The pattern of phosphorus concentration in the soil normal to the root has been investigated by using ^{33}P and making autoradiographs of the soil and root systems. Brewster et al. (1976) measured depletion of phosphorus in the soil near rape (*Brassica napus* L.) roots, and Hendricks et al. (1981) measured depletion of phosphorus near corn and rape roots. Both found that depletions corresponded with the length of root hairs. Lewis and Quirk (1967) observed the same pattern for phosphorus uptake by wheat, where they labeled the soil phosphorus with ^{32}P and measured ^{32}P distribution with autoradiographs.

Temple-Smith and Menary (1977) used ^{32}P labeling of the soil and autoradiography to identify phosphorus depletion zones around cabbage and let-

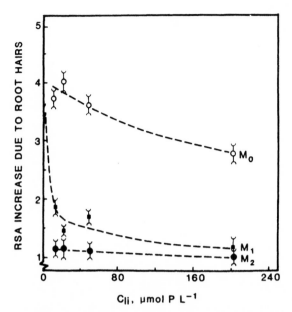

FIGURE   7.7   Increase in root surface area of 21-day-old corn grown at three soil water levels of M_0 (22%), M_1 (27%), and M_2 (32%) and five soil phosphate levels. Increase in root surface area calculated from data for root length, radius, and total root hair length and radius. Standard deviations are presented by vertical lines at each data point. Reproduced from Mackay and Barber (1985) by permission of Martinus Nijhoff Publisher. B.V.

tuce roots. They used two soils, one high and one low in phosphate sorption capacity. The soil with the higher sorption capacity produced a depletion zone that had a maximum diameter of 0.7 to 0.8 mm, while the soil with the low sorption capacity had a depletion zone of 3.0 to 3.5 mm; the latter included the effects of root hairs. Depletion was evident, because root hairs adsorbed much of the phosphorus in the root hair zone.

Itoh and Barber (1983a) studied the effect of root hairs on absorption of phosphorus by six plant species. The plant species used, root radius, root hair length and radius, and the number of root hairs per cm of root are shown in Table 7.2. There was a wide variation among these species in root hair morphology; onions had essentially no root hairs, while Russian thistle had many long root hairs. Photographs of the root hairs are shown in Figure 7.8.

Itoh and Barber used the Barber–Cushman mathematical model to predict phosphorus uptake by each of the six species grown to three ages. Comparison of predicted and observed phosphorus uptake is shown in Figure 7.9. Predicted uptake by Russian thistle and tomato was less than half of observed uptake. The dashed line indicates predicted phosphorus uptake

TABLE 7.2   Characteristics of Root and Roots Hairs of Six Plant Species

Plant	Root radius (mm)	Number of hairs per (cm)	Length (mm)	Radius (mm)	Root hair surface/ Root surface
Wheat (*Triticum aestivum* L.)	0.108	560	0.29	0.0057	0.7
Lettuce (*Lactuca sativa* L.)	0.124	1270	0.30	0.0048	1.6
R. thistle (*Salsola kali* L.)	0.056	890	0.60	0.0039	3.8
Tomato (*Lycopersicon esculentum*, M.)	0.107	1650	0.43	0.0043	2.5
Onion (*Allium cepa* L.)	0.225	1180	0.04	0.0110	0.2
Carrot (*Daucus carota* L.)	0.107	1810	0.04	0.0040	0.3

Source: Reproduced from Itoh and Barber (1983a) by permission of American Society of Agronomy.

equal to observed uptake. Tomato and Russian thistle had much longer root hairs than the other four species. These root hairs apparently increased phosphorus uptake. In order to evaluate this hypothesis, uptake by root hairs was calculated with the same model, using parameters for the root hairs. Overlapping zones of phosphorus supply to adjacent root hairs, and between root hairs and roots, was corrected for. Predicted uptake for roots plus root hairs compared with observed uptake is shown in Figure 7.10. Again, the dashed line indicates predicted uptake equal to observed uptake. When phosphorus uptake by root hairs was included, good agreement was obtained between observed and predicted uptake. These results show that under soil conditions where D_e for phosphorus was 2×10^{-9} cm^2/s, presence

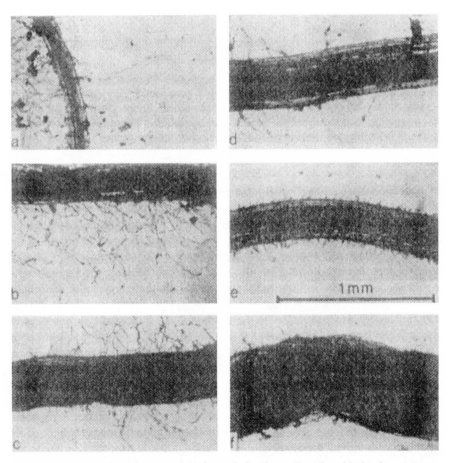

FIGURE   7.8   Microphotographs of root hairs for a, Russian thistle; b, tomato; c, lettuce; d, wheat; e, carrot; and f, onion. Plants were grown in soil with C_{li} of 15.5 μmol/L.

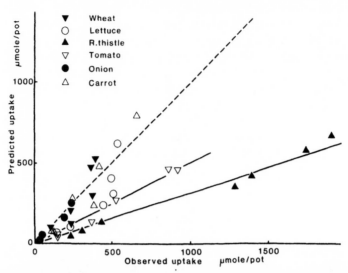

FIGURE   7.9   Comparison of observed P uptake by six plant species with predicted P uptake where root-hair effects were not considered. The dashed line corresponds to the situation where predicted uptake equals observed uptake. Solid lines
are regression lines for R. thistle (*lowest line*) and tomato. Reproduced from Itoh and
Barber (1983a) by permission of American Society of Agronomy.

of root hairs longer than 0.3 mm increased phosphorus uptake. The width of
the depletion zone around plant roots is influenced by $(D_e t)^{1/2}$, with the mean
linear diffusion distance of an ion being $(2D_e t)^{1/2}$. If root hairs are to have little influence, the concentration gradient should extend past the length of the
root hairs within a few days. Assuming a time period of 3 days and a D_e of 2
$\times 10^{-9}$ cm²/s, the mean diffusion distance will be 0.32 mm. This agrees with
experimental results, where there was little effect of root hairs on uptake
when root hairs were less than 0.3 mm in length. While root hairs did not
appreciably increase phosphorus uptake by lettuce and wheat in experiments
by Itoh and Barber (1983a) using Raub silt loam with a D_e of 2×10^{-9} cm²/s,
root hairs probably would have, if the researchers had used a soil with a D_e of
2×10^{-10} cm²/s because the mean diffusion distance would have been only 0.10
mm. Conversely if these species had been grown in a soil with a D_e for phosphorus of 2×10^{-8} cm²/s, there might not have been much effect from the root
hairs of any of the species, because the mean diffusion distance for three
days would have been 1.02 mm, which is greater than the mean length of the
root hairs on any of the species.

Itoh and Barber (1983b) developed a simulation program that modifies the
Cushman model in order to calculate additional nutrient uptake due to the
effect of root hairs. They ran a sensitivity analysis that showed effects of root

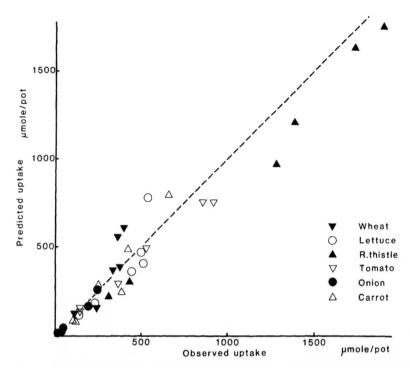

FIGURE   7.10   Comparison of observed P uptake by six plant species with predicted P uptake where root hairs P was considered. The dashed line is where predicted uptake equals observed uptake. The regression of these values gave $y = 0.98x + 18$, with $r = 0.89$. Reproduced from Itoh and Barber (1983a) by permission of American Society of Agronomy.

hair length, I_{max}, phosphorus uptake by root hairs, and root hair density on the absorption of phosphorus from Raub silt loam, where D_e was 2×10^{-9} cm^2/s; results are shown in Figure 7.11. Doubling I_{max} or the number of root hairs increased uptake asymptotically to a maximum. The percentage increase from doubling I_{max} was 10% and from doubling root hair number, it was 15%, however, doubling root hair length more than doubled uptake. The effect of a unit increase in root hair length increased with each increase in root hair length because of the radial nature of nutrient flux to the root. In addition the higher rate of phosphorus supply to the root would maintain a higher C_{lo} value at the root surface.

Figure 7.11 shows that root hair length is a prime factor in increasing nutrient uptake. Root hair length may often be the reason for differences among species in ability to absorb phosphorus. Since phosphorus commonly has small D_e values, it is the nutrient most often influenced by root hair morphology. On soils low in available phosphorus, D_e is usually less

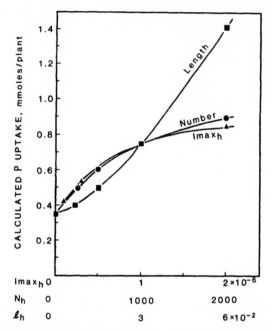

FIGURE   7.11   Effect on calculated P uptake of varying I_{max_h}, I_{max} for root hairs; N_h, number of root hairs/cm of root; and l_h, root-hair P length. Figure reprinted from Itoh and Barber (1983b), by permission of Martinus Nijhoff Publishers B.V.

than 1×10^{-9} cm²/s. Hence, the presence of root hairs often may double the rate of phosphorus uptake. On soils with higher available phosphorus levels, D_e may be between 5×10^{-9} and 5×10^{-8} cm²/s; in this case, root hairs should have a minimal influence on phosphorus uptake.

When phosphorus levels were lower, Bhat and Nye (1973) observed that root hairs were longer. This may be a plant-adaptive mechanism to aid in absorbing more phosphorus from low-phosphorus soils.

Many root hairs grow into voids in the soil. These root hairs probably have little effect on phosphorus uptake, because they are surrounded by air. The greatest effect of root hairs should be on soils with high soil moisture levels, where a higher proportion of root hairs is surrounded by soil from which phosphorus can diffuse to the root.

REFERENCES

Barber, D. A., and A. D. Rovira. 1975. Rhizosphere microorganisms and the absorption of phosphate by plants. *Ann. Rep.* ARC Letcombe Laboratory, England, 1974. Pp. 27–28.

Barber, D. A., G. D. Bowen, and A. D. Rovira. 1976. Effects of microorganisms on absorption and distribution of phosphate in barley. *Aust. J. Plant Physiol.* **3**:801–808.

Barley, K. P., and A. D. Rovira. 1970. The influence of root hairs on the uptake of phosphate. *Commun. Soil Sci. Plant Anal.* **1**:287–292.

Bhat, K. K. S., and P. H. Nye. 1973. Diffusion of phosphate to plant roots in soil. I. Quantitative autoradiography of the depletion zone. *Plant Soil* **38**:161–175.

Brewster, J. L., K. K. S. Bhat, and P. H. Nye. 1976. The possibility of predicting solute uptake and plant growth response from independently measured soil and plant characteristics: V. The growth and phosphorus uptake of rape in soil at a range of phosphorus concentrations and a comparison of results with the predictions of a simulation model. *Plant Soil* **44**:295–328.

Cormack, R. G. H. 1962. Development of root hairs in Angiosperms. II. *Bot. Rev.* **28**:446–464.

Dittmer, H. J. 1949. Root hair variations in plant species. *Am. J. Bot.* **36**:152–155.

Dobereiner, J., and J. M. Day. 1975. Nitrogen fixation in the rhizosphere of tropical grasses. In W. D. P. Stewart, Ed. *Nitrogen Fixation by Free Living Microorganisms.* Cambridge University Press, New York. Pp. 39–56.

Dobereiner, J., and G. Salomone. 1992. Biological dinitrogen fixation in maize. In *Abstract from the International Symposium on Environmental Stress: Maize in Perspective.* EMBRAPA, Belo Horizonte, Minas Gerias Brazil, March 8–13, 1992, A.T. Machado. President, EMBRAPA. Sete Lagoas, Brazil.

Esau, K. 1977. *Anatomy of Seed Plants*, 2d ed. John Wiley, New York.

Evans, D. G., and M. H. Miller. 1988. Vesicular-arbuscular mycorrhizas and the soil disturbance-induced reduction of nutrient absorption by maize. I. Causal relations. *New Phytol.* **110**:67–74.

Fairchild, G. L., and M. H. Miller. 1988. Vesicular-arbuscular mycorrhizas and the soil disturbance-induced reduction of nutrient absorption in maize. II. Development of the effect. *New Phytol.* **110**:75–84.

Fohse, D., and A. Jungk. 1982. Root hairs in relation to the mineral nutrition of plants. Personal Communication.

Harley, J. L. 1969. *The Biology of Mycorrhiza*, 2nd ed. Leonard Hill, London.

Hattingh, M. J., L. E. Gray, and J. W. Gerdemann. 1973. Uptake and translocation of ^{32}P-labeled phosphate to onion roots by endomycorrhizal fungi. *Soil Sci.* **116**:383–387.

Hayman, D. S., and B. Mosse. 1972. Plant growth responses to vesicular-arbuscular mycorrhiza. III. Increased uptake of labile P from soil. *New Phytol.* **71**:41–47.

Hendriks, L., N. Claassen, and A. Jungk. 1981. Phosphate depletion at the soil-root interface and the phosphate uptake of maize and rape. *Zeit. Pflanzenernahrung Bodenkunde* **144**:486–499.

Howeler, R. H., C. J. Asher, and D. G. Edwards. 1982. Establishment of an effective endomycorrhizal association on cassava in flowing solution culture and its effects on phosphorus nutrition. *New Phytol.* **90**:229–238.

Itoh, S., and S. A. Barber. 1983a. Phosphorus uptake by six plant species as related to root hairs. *Agron. J.* **75**:457–461.

Itoh, S., and S. A. Barber. 1983b. A numerical solution of whole plant nutrient uptake for soil-root systems with root hairs. *Plant Soil* **70**:403–413.

Jasper, D. A., L. K. Abbott, and A. D. Robson. 1989. Hyphae of a vesicular-arbuscular mycorrhizal fungas maintains infectivity in dry soil, except when soil is disturbed. *New Phytol.* **112**:101–107.

Kennedy, I. R., and Y. T. Tchan. 1992. Biological nitrogen fixation in non-leguminous crops: Recent advances. *Plant Soil* **141**:93–118.

Lambert, D. H., D. E. Baker, and H. Cole, Jr. 1979. The role of mycorrhizae in the interactions of phosphorus with zinc, copper, and other elements. *Soil Sci. Soc. Am. J.* **43**:976–980.

Lewis, D. G., and J. P. Quirk. 1967. Phosphate diffusion in soil and uptake by plants. III. [31]P movement and uptake by plants as indicated by ^{32}P autoradiography. *Plant Soil* **26**:445–453.

Li, X.-L., H. Marschner, and E. George. 1991. Acquisition of phosphorus and copper by VA-mycorrhizal hyphae and root to shoot transport in white clover. *Plant Soil* **136**:49–57.

Mackay, A. D., and S. A. Barber. 1984. Comparison of root and root hair growth in solution and soil culture. *J. Plant Nutr.* **7**:1745–1757.

Mackay, A. D., and S. A. Barber. 1985. Effect of soil moisture and phosphate level on root hair growth of corn roots. *Plant Soil* **86**:321–331.

Mackay, A. D., and S. A. Barber. 1987. Effect of cyclic wetting and drying of a soil on root hair growth of maize roots. *Plant Soil* **104**:291–293.

McGonigle, T. P., D. G. Evans, and M. H. Miller. 1990. Effect of degree of soil disturbance and mycorrhizal colonization and phosphorus absorption by maize in growth chamber and field experiments. *New Phytol.* **116**:629–636.

Mosse, B. 1973. Advances in the study of vesicular-arbuscular mycorrhiza. *Ann. Rev. Phytopathol.* **11**:171–196.

O'Halloren, I. P., M. H. Miller, and G. Arnold. 1986. Absorption of P by corn (*Zea mays* L.) as influenced by soil disturbance. *Can. J. Soil Sci.* **66**:287–302.

Rovira, A. D. 1979. Biology of the soil-root interface. In J. L. Harley and R. S. Russell, Eds. *The Soil-Root Interface.* Academic Press, New York. Pp. 145–160.

Sanders, F. E., and P. B. Tinker. 1971. Mechanism of absorption of phosphate from soil by *Endogone* mycorrhizas. *Nature (London)* **233**:278–279.

Sanders, F. E., and P. B. Tinker. 1973. Phosphate flow into mycorrhizal roots. *Pestic. Sci.* **4**:385–395.

Temple-Smith, M. G., and R. C. Menary. 1977. Movement of ^{32}P to roots of cabbage and lettuce grown in two soil types. *Commun. Soil Sci. Plant Anal.* **8**:67–79.

Tinker, P. B. 1980. Role of rhizosphere microorganisms in phosphorus uptake by plants. In F. E. Khasawneh, E. C. Sample, and E. J. Kamprath, Eds. *The Role of Phosphorus in Agriculture.* American Society of Agronomy, Madison, WI. Pp. 617–654.

Tinker, P. B. 1982. Mycorrhizas: The present position. *Trans. Twelfth Int. Cong. Soil Sci.*, New Delhi, India 150–166.

Tinker, P. B., and F. E. Sanders. 1975. Rhizosphere microorganisms and plant nutrition. *Soil Sci.* **119**:363–368.

Tinker, P. B., D. P. Stribley, and R. C. Snellgrove. 1982. The relationship between phosphorus concentration and growth in plants infected with vesicular-arbuscular mycorrhizal fungi. In A. Scaife, Ed. *Plant Nutrition 1982. Proc. Ninth Int. Plant Nutr. Colloq.*, Coventry, England **2**:670–675.

Vivekananda, M. and P. E. Fixen. 1991. Cropping systems effects on mycorrhizal colonization, early growth, and phosphorus uptake by corn. *Soil Sci. Soc. Am. J.* **55**:136–140.

Voggo, J. A. 1971. Field inoculations with mycorrhizal fungi. In E. Hacskaylo, Ed. *Mycorrhizae. Proc. First North American Conference on Mycorrhizae* 187–196.

Wild, A. (Ed.). 1988. *Russell's Soil Conditions and Plant Growth*, 11 ed. Longman Scientific and Technical copublished by John Wiley, New York.

Yost, R. S., and R. L. Fox. 1979. Contribution of mycorrhizae to P nutrition of crops growing on an Oxisol. *Agron. J.* **71**:903–908.

CHAPTER 8

Nitrogen

Soil nitrogen is primarily in the organic fraction of the soil. The atmosphere above the soil contains 79% nitrogen, but this nitrogen can only be used by leguminous plants, which have symbiotic nitrogen-fixing microorganisms, such as *Rhizobium,* in nodules on their roots. The nitrogen in the mineral fraction includes ammonium fixed in the clay minerals, exchangeable ammonium on cation-exchange sites and in solution, and nitrate in soil solution. The amount of mineral nitrogen is generally small compared with that in the organic fraction. Nitrogen in the organic matter came initially from the atmosphere via plants and microorganisms that have since decomposed and left resistent and semiresistant organic compounds in the soil during development. The bulk of soil nitrogen is present in the upper soil horizon, where the bulk of organic matter is located. Organic matter has an average of 5% nitrogen (w/w), the plow layer of cultivated soils usually contains from 0.02 to 0.4% nitrogen (w/w). For a more detailed discussion of nitrogen in agricultural soils see Stevenson (1982, 1986).

FORMS OF NITROGEN IN THE SOIL

Nitrogen in the soil falls into five categories: (1) Nitrogen in organic matter; (2) mineral nitrogen in the soil solution and on exchange sites; (3) nitrogen in plant residues in the soil; (4) ammonium fixed in clay minerals; and (5) gaseous nitrogen in the soil's atmosphere. Interchange between various forms is primarily via microbiological activity.

Nitrogen in Organic Matter

Nitrogen in organic matter occurs largely as NH_2 groups. Organic compounds containing nitrogen are mainly amino acids and hexosamines. Identifying the soil's organic matter compounds usually requires soil treatment to separate organic matter from the mineral fraction. This treatment often breaks down proteins into their constituent amino acids. Amino acids and hexosamines are common organic forms of nitrogen. Because of the great diversity in organic compounds that can be present in the soil and changes occurring during extraction, not all such compounds are known; 20 to 50% of the nitrogen may be in the amino acid form; while 5 to 10% occurs as hexosamines. This leaves about half the nitrogen compounds unidentified.

Organic nitrogen may be divided into a readily mineralizable fraction (Stanford and Smith, 1972) and a stable fraction. The readily mineralizable fraction is usually less than one-third of total organic soil nitrogen.

Inorganic Nitrogen

Inorganic nitrogen in the soil occurs mainly as NH_4^+ and NO_3^-; in some soils with high pH, small amounts of NO_2^- may occur. However, NO_3^- is generally the principal form, since nitrification by microorganisms converts NH_4^+ and NO_2^- to NO_3^-. Inorganic nitrogen usually constitutes less than 5% of total soil nitrogen in soils with appreciable organic matter contents, since most of the nitrogen is organically bound. In soil, there is a dynamic process of nitrogen release from the organic fraction through microbiological decomposition of organic matter; this process is called ammonification. At the same time, inorganic nitrogen will be transformed into organic nitrogen by microbial decomposition of carbonaceous plant materials. Because of microbiological activity, levels of ammonium and nitrate in the soil do not remain constant. When inorganic nitrogen is added to the soil, some is transformed to organically bound forms, which may later be released to inorganic forms once more.

When ammonium is added to the soil or released from organic matter by mineralization, it is usually nitrified rapidly to nitrate. The ratio of NH_4^+ to NO_3^- found in the soil depends on the presence of satisfactory conditions for nitrification; nitrification is inhibited by low soil pH and anaerobic conditions.

Where nitrification readily occurs, most of the mineral nitrogen occurs as nitrate. Reisenauer (1964) summarized nitrate concentrations in soil solutions from 879 soil samples. The distribution of values according to concentration is shown in Table 8.1: 61% of the values were between 50 and 150 mg/L. Current values may be higher due to greater use of nitrogen fertilizer in recent years. Reisenauer did not summarize corresponding NH_4^+ values. Ammonium is similar in size to potassium and is held on soil exchange sites with a similar strength. More of the ammonium (up to 10 times) will be present as an exchangeable cation than is found in solution; hence, solution concentrations are usually low. Buffering power for this distribution has not been investigated to any extent; presumably, buffer power will be similar to that observed for potassium on the same soil.

Plant Residue Nitrogen

Many crops are harvested for grain, with their stalks, leaves, and roots remaining in the field. Since these residues contain nitrogen, this nitrogen is present in the residues until the residue has been decomposed by soil microorganisms. Part of the nitrogen in the residue will then be released into the soil as ammonium, and part of it will go on to form new soil organic matter.

Two approaches can be used to measure the release of nitrogen from crop residues. The most common approach is to measure inorganic nitrogen levels in the soil periodically for several months after incorporating residues. The results are then compared with inorganic nitrogen values for soils treated similarly but without incorporated residues. This is generally done in randomized replicated plots. The greater amount of inorganic nitrogen where residues

TABLE 8.1   Summary of Literature Values for the NO_3^- Nitrogen in Displaced Soil Solution

Nitrate-nitrogen concentration (mg/L)	Fraction of samples (%)
0–25	4.9
26–50	14.3
51–100	28.8
101–150	32.2
151–200	10.5
201–300	2.7
301–400	4.9
401–500	1.0
>501	0.8

Source: Reisenauer (1964) by permission of Federation of American Societies of Experimental Biology.

were present is assumed to represent nitrogen released from residues. The second approach is to conduct experiments where residues are produced and the nitrogen released during the succeeding year is then measured by uptake of nitrogen by a crop or crop yield. Again, randomized replicated plot experiments are used with control plots where no residues were added. Bartholomew (1965) reviewed results from experiments where the first approach was used. When plant residues having a carbon to nitrogen (C:N) ratio greater than 20 were added to soil, nitrate and ammonium levels in the soil decreased as microorganisms used up carbon from the residues. Hence, nitrogen was immobilized. If the C:N ratio was less than 20, on the other hand, nitrogen was released as microorganisms decomposed the residues. Rates of decomposition of plant residues are influenced by temperature, moisture, and soil nutrient levels Stevenson (1986). When cereal straws and other low-nitrogen residues were added to soil, nitrogen was immobilized during decomposition of plant residues and transformed into relatively stable organic nitrogen forms. When all of the residue was decomposed, decomposition of the microbes released inorganic nitrogen into the soil.

Present theory is that new organic matter is formed from microbial tissue that becomes relatively resistant to decomposition. Nitrogen will be released to the soil during early stages of decomposition, where C:N ratios of the residues are below 20. Nitrogen that is immobilized in soil organic matter will then be released at the rate of decomposition for soil organic matter. In the field, this rate is usually between 1 and 3% per year when calculations are based on levels of total soil organic matter.

The second approach measures nitrogen supplied to a succeeding crop. Barber (unpublished) has measured nitrogen released from residues of corn grown with several levels of nitrogen and from soybeans (Table 8.2). Corn grown without nitrogen fertilization was used to measure nitrogen release. The amount of nitrogen supplied from the nitrogen fertilization of the previous crop, or soybeans, was measured by comparing corn yield and leaf nitrogen composition resulting from treatments given to the preceding crop. With nitrogen fertilization of continuous corn, 20 to 30% of the nitrogen applied the first year was effective the second year. Much of this nitrogen was assumed to have been taken up by the first crop and returned to the soil via corn residues. With soybeans, all of the additional nitrogen supplied for the following corn crop was assumed to come from soybean residues. The amount from a 3500 kg/ha soybean crop (grain yield) averaged 60 kg N/ha. The residue was not analyzed. Assuming average nitrogen contents, however, it would appear that 50 to 75% of the nitrogen in either corn or soybean residues was available for the succeeding crop.

Ammonium Fixed in Soil Minerals

Clay minerals will "fix" ammonium just as they do potassium; both ions are similar in dehydrated size. Nommik and Vahtras (1982) reviewed the litera-

TABLE 8.2   Effect of Residual Nitrogen From Corn and Residual Nitrogen from Soybeans on Corn Yield (6-Year Average)

Treatment		Corn yield (kg/ha)	Yield increase due to N (kg/ha)	Residual as % of response in year N applied
Previous year	Year of corn yield			
Corn 0N[a]	Corn 0N	2585		
Corn 67N	Corn 0N	3085	500	20
Corn 135N	Corn 0N	3405	820	18
Corn 201N	Corn 0N	4095	1510	29
Soybeans	Corn 0N	5030	2365	–
Corn 0N	Corn 0N	2665	–	–
Corn 0N	Corn 67N	5190	2525	–
Corn 0N	Corn 135N	7110	4445	–
Corn 0N	Corn 201N	7820	5155	–

Source: Barber (unpublished data).
[a]Rate of nitrogen application kg/ha.

ture on ammonium fixation in soils, and concluded that fixation occurred mainly in vermiculite and illite clays, where potassium removal left spaces for ammonium fixation. Fixed ammonium is defined as ammonium that cannot be readily exchanged by other cations. Two aspects of ammonium fixation are of interest; one is the amount of native fixed ammonium in the soil, and the second is the amount of ammonium fixation that occurs when ammonium fertilizers are applied.

Stevenson (1959) found 3.5 to 7.5% of the nitrogen in surface soils in the central United States occurred as fixed ammonium. The relative proportion of total nitrogen as fixed ammonium increased with depth because of the decrease in the soil organic matter content. Amounts of fixed ammonium equivalent to 5 to 15 mmol(p^+)/kg have been reported in central U.S. and European soils. Fixed ammonium has also been found in rocks, so that some of the fixed ammonium in soil clays may have persisted since the minerals were synthesized. Very little of the native fixed ammonium becomes available to nitrifying microorganisms or plants. This nitrogen is included in total soil nitrogen determinations if HF treatments are used but not in the usual oxidative methods of soil nitrogen analysis.

Soils have been found to fix amounts of added ammonium ranging from 5 to 15 mmol(p^+)/kg of soil; the information on availability of fixed ammonium to plants and microorganisms is incomplete. However, if applied ammonium fertilizer becomes fixed, it would appear that much of it becomes available, since the ammonium form of nitrogen is usually as effective as the nitrate form under situations where leaching and denitrification losses are minimal. In nitrification experiments, recently fixed ammonium has been found to be less available to nitrifying organisms than exchangeable ammonium. It is not certain whether this is simply a matter of rate of availability or whether some of the fixed ammonium is actually unavailable.

In addition to ammonium fixation, ammonia (NH_3) can be fixed by organic matter. Furthermore, the ammonia-organic matter association is resistant to decomposition. The extent of this reaction and its significant in removing nitrogen from the pool available to plants lacks definition, however; it is most significant in soils having high organic matter levels.

Gaseous Forms

Denitrification of nitrate produces nitrous oxide, N_2O, and dinitrogen, N_2. These gases are in the soil pore space and can be lost to the atmosphere. Nitrogen in the soil atmosphere is used by symbiotic root microbes, such as *Rhizobium,* and also by nonsymbiotic microorganisms, such as *Azotobacter* and *Clostridium,* that fix nitrogen in the soil.

NITROGEN-UPTAKE KINETICS

The nitrogen requirement of most nonleguminous plants is large, so the nitrogen flux at the root surface is larger than that for most other ions.

Ammonium Versus Nitrate

Plant roots may take up either NH_4^+ or NO_3^- from the soil, since one or both may be present, depending on soil conditions. When NH_4^+ is the only nitrogen source, cation uptake exceeds anion uptake, and H^+ is released from the root to balance the charge. When NO_3^- is the only nitrogen source, anion uptake usually exceeds cation uptake and OH^- or HCO_3^- is released to balance the charge. Barber (1974) measured anion and cation uptakes by ryegrass when NH_4^+ or NO_3^- was the only source of nitrogen. The data are shown in Table 6.1.

Some plants prefer NH_4^+, while others prefer NO_3^-. Barber (unpublished data, 1979) studied nitrogen uptake characteristics of tall fescue (*Festuca arundinacia* Schreb) and reed canary grass *(Phalaris arundinacia* L.). When both NH_4^+ and NO_3^- were present in equal amounts, these plants absorbed NH_4^+ 1.3 to 2.2 times faster than NO_3^-. Warncke and Barber (1973) investigated relative rates of NH_4^+ and NO_3^- uptake by corn using solution-culture experiments in a controlled-climate chamber. They found no significant difference in relative rates of absorption for NH_4^+ and NO_3^-. The researchers used five NH_4^+/NO_3^- ratios, ranging from 8.40 to 0.17. Solution pH was monitored and controlled near pH 5.8. The NH_4^+/NO_3^- ratio of uptake varied with solution concentration of nitrogen: As nitrogen solution concentration was increased, the ratio of uptake moved toward 1.0. These data are shown in

TABLE 8.3   Influence of NH_4^+/NO_3^- Ratio and Nitrogen Concentration of the
Nutrient Solution on the Ratio of NH_4^+ to NO_3^- Absorbed during Corn
Seedling Growth from 13 to 18 Days

Mean N concentration in solution (μmol/L)	Solution NH_4^+/NO_3^- ratio				
	8.40	2.46	1.05	0.49	0.17
	(NH_4^+/NO_3^- absorbed)				
16	6.82	2.88	1.11	0.56	0.17
67	9.40	2.78	0.97	0.50	0.15
303	4.91	2.38	1.45	0.70	0.28
1,507	3.71	2.05	1.48	0.68	0.33
6,015	3.50	2.09	1.04	0.69	0.31
L.S.D.					
0.05	4.33	0.67	0.42	0.16	1.14

Source: Reproduced from Warncke and Barber (1973) by permission of American Society of Agronomy.

Table 8.3. Uptake was measured over a 5-day period, starting with 13-day-old corn seedlings. Solution concentrations were maintained by frequent additions.

Temperature and solution pH also influence relative uptake rates of NH_4^+ and NO_3^-. Ammonium absorption increases as pH increases, whereas NO_3^- uptake tends to decrease with increasing pH (Van den Honert and Hooymans, 1955; Fried et al., 1965; Jungk, 1970). When both NH_4^+ and NO_3^- are present, absorption of NH_4^+ is greater than NO_3^- at 8°C with NH_4^+ influx reaching a maximum at a media temperature of 25°C. However, NO_3^- absorption also increased with temperature, becoming greater than NH_4^+ at 23°C and further increasing up to 35°C (Lycklama, 1963).

When both NH_4^+ and NO_3^- are present in the same nutrient media, NH_4^+ has been reported to inhibit NO_3^- uptake by some plant species (Fried et al., 1965; Minotti et al., 1969a). Minotti et al. (1969b) reported that NH_4^+ may affect the NO_3^- absorption process rather than the reduction of NO_3^-. Nitrate is generally assumed not to influence the uptake of NH_4^+, although Lycklama (1963) did observe a slight decrease in NH_4^+ absorption rate by ryegrass as NO_3^- concentration in solution increased. We might expect NO_3^- and NH_4^+ to mutually influence one another's uptake and metabolism, because both are sources of N and both are metabolized through the same pathways after NO_3^- has been reduced to NH_4^+.

Reisenauer (1978) found that wheat seedlings were larger when both NH_4^+ and NO_3^- were present than when only NO_3^- was present. Warncke and Barber (1973) also found that the presence of some NH_4^+ yielded larger corn plants than when only NO_3^- was present. Plants' organic-acid concentrations are greater when NH_4^+ is the nitrogen source than when NO_3^- is the source; the change in organic acids may have some effect on plant growth.

I_{max} Values

Nitrogen absorption is assumed to follow Michaelis-Menten kinetics and have a maximal influx value. Edwards and Barber (1976b) obtained an I_{max} value for 18-day-old corn seedlings of 30 nmol/m $\cdot$ s. Bhat et al. (1979) obtained values for 8-day-old rape (*Brassica napus* L.) on the order of 500 nmol/m^2 $\cdot$ s. However, these values decreased to less than one-tenth for 21-day-old plants. Warncke and Barber (1974) obtained values of 120 nmol/m^2 $\cdot$ s for 7- to 13-day-old corn and similar values for forage sorghum (*Sorghum bicolor* L.). Gregory et al. (1979) measured nitrogen inflow by wheat (*Triticum aestivum* L.) roots growing in the field in England of 1.0 nmol/m $\cdot$ s. Teo et al. (1992) obtained an average value of 475 nmol/m^2 $\cdot$ s for I_{max} of three rice (*Oryza sativa* L.) varieties. Values for K_m and C_{min} were 62 and 3 μmol/L, respectively.

Edwards and Barber (1976a) used 18-day-old corn seedlings to measure the effect of reducing the amount of roots per gram of top on nitrogen influx in roots supplied with nitrogen. They reduced root/shoot ratio by using split-root experiments or trimming the roots. Nitrogen demand by the shoot per unit of root was thus increased. The effect of these treatments on I_{max} for nitrogen and on nitrogen uptake per plant was shown in Table 3.3. With trimmed roots, I_{max} increased slightly as root trimming increased, but the increase was not enough to compensate for the reduced amount of roots; hence, uptake per plant was less. With split-root systems, decreasing the proportion of roots exposed to nitrogen increased I_{max} more than for the trimmed root system; however, the increase in rate was not enough to compensate for reduced root length supplied with nitrogen. Thus, nitrogen uptake per plant was also reduced with this treatment. The, reason for the difference between trimmed-root and split-root treatments was probably due to the length of time the treatment was imposed on a given corn plant. Split-root treatments were imposed from the time that 6-day-old plants were placed in nutrient solution. The percent nitrogen in the shoot decreased as fewer roots were supplied with nitrogen; however, there was little difference between treatments with respect to root nitrogen concentration. It is probable that the lower nitrogen concentration of the shoot caused I_{max} to increase by a feedback mechanism between shoot and root. Barber (unpublished data) investigated the effect of root trimming on nitrogen uptake by fescue and reed canary grass. In both species, trimming the roots did not increase I_{max} for the remaining roots. Hence, it appears that increases in I_{max} due to additional shoot demand per unit of root must develop gradually if at all.

The increase in I_{max} resulting from split-root treatments was related to the reduction in nitrogen concentration in the shoot and not the root. This relation is similar to that for potassium, as shown in Figure 3.7.

Factors Influencing Nitrogen Influx

Effect of pH

Van den Honert and Hooymans (1955) measured the effect of solution pH on nitrate uptake by maize; uptake was measured for 10 to 60 min. They found that nitrate influx decreased by one-third as pH was increased from pH 5.0 to 7.8. This decrease could not be compensated for by increasing nitrate concentration. Lycklama (1963), using perennial ryegrass, found that nitrate uptake by this species reached a maximum at pH 6.2. In reviewing data from the literature, Van den Honert and Hooymans (1955) found that the effect of pH on nitrate uptake was highly variable.

Effect of Temperature

As with most nutrients, starting at low temperatures, then increasing temperature, increases nitrogen uptake rate; which proceeds to a maximum and decreases at still higher temperatures. Van den Honert and Hooymans (1955) found that nitrate uptake by maize increased with increasing temperature over the temperature range 5° to 30°C. Lycklama (1963), on the other hand, found that the maximum rate of nitrate uptake by ryegrass occurred between 20° and 25°C. Barber (unpublished data) found that the maximum value for I_{max} occurred at 30°C for corn and at 25°C for fescue and reed canary grass (see Figure 3.9).

Effect of Plant Age on I_{max}

Edwards and Barber (1976b) investigated the effect of corn plant age on I_{max} for nitrogen uptake by corn; five ratios of NH_4^+/NO_3^- were used. Since there appeared to be no difference in ratio for I_{max} values for nitrogen ($NH_4^+ + NO_3^-$) uptake, values for the different ratios were averaged together. Seven ages of corn were investigated, with the corn germinated at varied dates so that nitrogen influx for all ages could be determined on the same day. Older plants were grown in the greenhouse. The I_{max} for nitrogen uptake as affected by plant age is shown in Table 8.4. As plants became older, I_{max} decreased; the average value at 58 days was only 0.044 the I_{max} at 18 days. This decrease in influx with age is similar to that found for other nutrients.

Barber (unpublished data) studied I_{max} for nitrogen uptake by tall fescue and reed canary grass. He obtained values ranging from 50 to 145 nmol/m² · s. There was variation within this range due to plant age, ion form, and plant history.

Values of K_m for Nitrogen Uptake

The value of K_m determines the slope of the relation between net influx and nitrogen concentration. Edwards and Barber (1976b) found values ranging from 12 to 20 µmol/L for nitrogen uptake by corn; K_m did not vary with plant

TABLE 8.4   Inflluence of Corn Plant Age on Total Plant Weight, Root Length per Plant, and Maximum Nitrogen-Uptake Rate

Plant age (days)	Total dry matter (g/plant)	Root length (m/plant)	I_{max} (nmol/m$^2 \cdot$s)
15	0.38	19	8.66a[a]
18	0.59	26	10.12a
21	1.10	39	9.18a
24	1.57	63	8.74a
33	5.27	123	3.25b
46	21.80	479	0.68b
58	61.46	971	0.45b

Source: Reproduced from Edwards and Barber (1976*b*) by permission of the American Society of Agronomy.
[a]Means within column followed by different letters are significantly different at 5-% probability level.

age. Barber obtained values varying from 15 to 25 µmol/L for fescue and reed canary grass. Mugwira et al. (1980) studied the nitrate uptake of triticale (*Triticasecale,* Wittmack), wheat, and rye (*Secale* sp.) at solution nitrate concentrations varying from 0.1 to 10 mmol/L. They found that mean nitrogen-uptake rate increased between 0.1 and 1.0 mmol/L, but not above this concentration. Data indicated that K_m should be much less than 100 µmol/L. Hassan and van Hai (1976) measured ammonium-uptake kinetics by citrus roots; obtaining a K_m value for phase 1 uptake of 59 µmol/L for 60-day-old citrus; by 180 days, the K_m value had increased to 95 µmol/L. Bhat et al. (1979) measured K_m for nitrate uptake by rape grown in the growth chamber. They obtained a value of 34.4 µmol/L, which was similar to the value of 33 µmol/L found for ryegrass by Lycklama (1963) and to the values of 21 to 30 µmol/L obtained by van den Honert and Hooymans (1955) for maize.

Many of the values reported for K_m have been in the vicinity of 15 to 30 µmol/L. These values are relatively low, considering the unbuffered behavior of nitrate in the soil. However, in comparison to values for potassium (15 to 20 µmol/L) and phosphorus (2 to 5 µmol/L) they are somewhat higher. The values are also low compared to nitrate concentrations in the soil solution, which commonly range between 3 and 15 µmol/L. Because K_m values are low relative to C_{li} and *b* is low, nitrogen influx will probably be close to I_{max} until most of the nitrate in the soil has been depleted.

Minimum Concentration

The concentration below which nitrogen influx ceases varies with species. Warncke and Barber (1974) investigated four plant species and obtained C_{min}

values of 1.7, 2.7, 2.4, and 1.4 μmol/L for forage sorghum, grain sorghum, soybeans, and bromegrass (*Bromus inermus*), respectively. Edwards and Barber (1976b) obtained values varying between 3 and 9 μmol/L for corn varying in age from 15 to 58 days. Plant age did not affect the value obtained. Barber (unpublished data) obtained values of 6 to 8 μmol/L for fescue and reed canary grass. Olsen (1950) studied nitrate uptake by rye and found a C_{min} value of 0.3 μmol/L. This is an order of magnitude lower than values observed for other crops. The values of C_{min} indicate that plants can effectively deplete nitrogen to low soil solution levels. This is important for reducing nitrate contamination of groundwater, rivers, and lakes, since the nitrate remaining in the soil after cropping is susceptible to being leached from the soil.

NITROGEN SUPPLY FROM THE SOIL

Soil microorganisms are continually breaking down soil organic matter for a source of energy, which causes mineralization of nitrogen in the organic matter. Such mineralization is a source of nitrogen for many crops.

The amount of nitrate and ammonium in the soil that can be transported to the root depends on (1) net release from organic matter mineralization, (2) nitrification of ammonium to nitrate, (3) denitrification of nitrate, (4) loss of nitrate from the soil by leaching, (5) loss of NH_4^+ by volatilization, (6) nitrogen uptake by the crop, (7) nitrogen additions in fertilizers, (8) nitrogen additions by rainfall, (9) nitrogen additions by free-fixing microorganisms, and (10) nitrogen fixation into organic forms by microorganisms. The number of these processed make predicting nitrate supply to the plant difficult. Under many situations, however, quantities involved in some of these processes are small. A diagram of the processes involved is shown in Figure 8.1. Each of the processes is discussed briefly in turn.

Mineralization

Nitrogen supply due to mineralization of soil organic matter has been studied by Stanford and Smith (1972). They measured the mineralization potential of 39 widely differing soils by incubating each soil at 35°C for 30 weeks leaching inorganic nitrogen from the soil at intervals during the incubation period, and measuring it. With most soils, cumulative net nitrogen that was mineralized was linearly related to the square root of time. It was assumed that rate of mineralization was proportional to the quantity of nitrogen in the mineralizable substrate $dN/dt = -k_N$, where N is the amount of potentially mineralizable soil nitrogen. Values for the mineralization potential varied from 20 to 300 mg/kg of air-dry soil. This potential was 5 to 40% of total nitrogen in the soil, with an average of 18% for the 39 soils studied. The mineralization rate constant k_N did not vary significantly

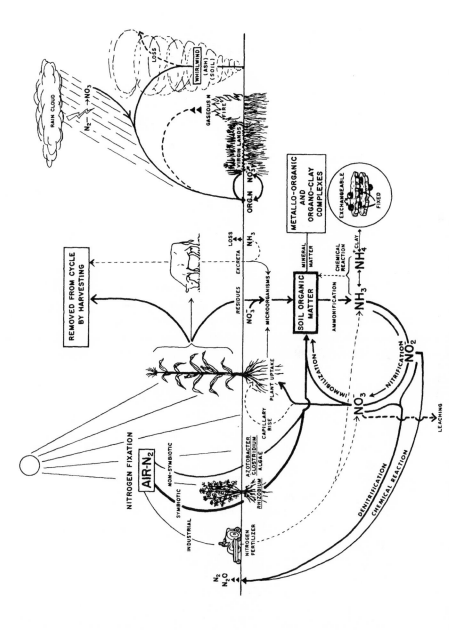

FIGURE 8.1 Nitrogen cycle in the soil. Reproduced from Stevenson (1965) by permission of American Society of Agronomy.

among most of the soils; it averaged 0.054 ± 0.009/week; the time to mineralize one-half of the potential was 12.8 ± 2.2 weeks. Integrating the expression $dN/dt = -k_N$ gives an equation describing the mineralization process

$$\log(N_0 - N_t) = \log N_0 - \left(\frac{k_N t}{2.303}\right) \qquad (8.1)$$

where N_0 is the soil nitrogen mineralization potential and N_t is the cumulative amount of nitrogen mineralized during time t.

Stanford and Smith (1972) used a laboratory procedure to determine mineralization. Its application to the field was evaluated by Smith et al. (1977); who measured inorganic nitrogen production on fallow plots in soil contained in plastic bags buried in the field. Their results indicated that the soil's nitrogen mineralization potential in the field was in agreement with values calculated from laboratory measurements. These experiments indicate that only part of the soil nitrogen is in the pool from which mineralization occurs. The remainder is assumed not to be available for mineralization; it resides in the relatively stable soil organic matter.

Smith et al. (1980) evaluated the Stanford and Smith procedure and found it to give low values because of soluble organic materials that were leached during successive leaching treatments used. This would eliminate measurement of nitrogen mineralized from soluble organic compounds, because they would be leached from the soil during the removal of inorganic nitrogen. For some soils, there was as much nitrogen in the soluble organic fraction as there was mineralized. Hence, N_0 could be as little as one-half the "true" value.

A second procedure for estimating mineralization of nitrogen from organic matter is to measure the rate of decomposition of organic matter, assuming that all organic matter can be decomposed equally. Barber (1979) measured the reduction in soil organic matter content in plots that were fallowed for six years; the rate was 2.4% per year. In a similar study in Iowa, Larson et al. (1972) measured organic matter decrease during 11 years of fallowing, obtaining a value of 1.9% per year. Hence, we might realistically expect about 2% of the nitrogen in the soil to be mineralized each year. Total soil nitrogen varies from 0.02 to 0.4% (w/w). If we assume a plow layer 20 cm thick and a bulk density of 1.3, this amounts to an annual release value varying from 10 to 200 kg N/ha. Many soils are in the total nitrogen range 0.05 to 0.1% N, thus they may release 25 to 50 kg N/ha/year.

The release of nitrogen during soil organic matter mineralization is one part of a cycle. If soil organic matter level is to be maintained, an equal amount of nitrogen must be transformed into new organic matter each year through crop residues and roots. Decomposition of crop residues usually results in formation of new soil organic matter, so that the level in the soil is maintained.

Nitrification

Nitrification is the oxidation of ammonium to nitrate, with nitrite as an intermediate step. Nitrification is an autotropic phenomena, largely due to *Nitrosomonas* and *Nitrobacter* microorganisms. The rate of nitrification is influenced by nutrient supply, temperature, pH, aeration, moisture, organic matter, and the presence of inhibitors.

McLaren (1976) has described the process mathematically as follows:

$$NH_4^+ \xrightarrow[\underset{P}{4}]{1} NO_2^- \underset{3}{\overset{2}{\rightleftharpoons}} NO_3^- \tag{8.2}$$

where the numbers refer to the reaction rate constants and P is the sum of the N_2O, NO, and N_2 generated. Oxidations (Reactions 1 and 2) are assumed to be first-order reactions, while reductions (Reactions 3 and 4) within anaerobic crumb microsites are assumed to follow zero-order reactions.

$$\frac{d(NH_4^+)}{dt} = -k_1(NH_4^+) \tag{8.3}$$

$$d(NO_2^-)dt = k_1(NH_4^-) - k_2(NO_2^-) + k_3 - k_4 \tag{8.4}$$

$$\frac{d(NO_3^-)}{dt} = k_2(NO_2^-) - k_3 \tag{8.5}$$

$$\frac{dP}{dt} = k_4 \tag{8.6}$$

Reasonable values for reaction rate constants are $k_1 = 0.02/h$, $k_2 = 0.04/h$, $k_3 = 0.015$ mg/kg N $\cdot$ h, and $k_4 = 0.01$ mg/kg N $\cdot$ h. Since the value of k_2 is larger than k_1, this means that nitrite usually does not accumulate. Values for k_3 and k_4 depend greatly on oxygen levels present within the soil.

Oxidation of NH_4^+ to NO_2^- is accomplished by the microorganism *Nitrosomonas*, with the conversion from NO_2^- to NO_3^- carried out by *Nitrobacter*. Since these microorganisms obtain most of their energy from the process itself, the rate of reaction is equal to the microbial growth rate in the soil (McLaren, 1969).

After addition of ammonium to soil, nitrification begins. Nitrate is formed and ammonium disappears, so the level of nitrate in the soil increases in hyperbolic fashion over time. The time for nitrification depends on such factors as populations of nitrifiers, aeration, soil pH, and temperature. However, under conditions of near neutral pH and at 25°C temperature with adequate aeration, ammonium may be nitrified at a rate of 10 to 20 kg/ha/day.

Nitrification also causes acidification of the soil, since the reaction is

$$NH_4^+ + 2O_2 \rightarrow NO_3^- + H_2O + 2H^+ \qquad (8.7)$$

Prolonged use of ammonium fertilizer results in acidification of the soil so liming is required to compensate for the acidity generated.

Denitrification

Nitrogen is lost from the soil to the atmosphere by denitrification, an anaerobic process that follows the reaction

$$NO_3^- \rightarrow N_2O \rightarrow N_2 \qquad (8.8)$$

It is not certain that N_2O is an intermediary in producting N_2, though the sequence in Equation 8.8 *does* explain several reactions observed in soils. Denitrification is a biological process involving anaerobic bacteria that are able to use nitrate in place of oxygen as a hydrogen acceptor. Under aerobic conditions, oxidation of a carbohydrate, such as glucose, can be described as follows:

$$C_6H_{12}O_6 + 6O_2 \rightarrow 6CO_2 + 6H_2O \qquad (8.9)$$

However, under anaerobic conditions in the presence of nitrate, some bacteria are capable of nitrate respiration, as follows:

$$C_6H_{12}O_6 + 4NO_3^- \rightarrow 6CO_2 + 6H_2O + 2N_2 \qquad (8.10)$$

Production of N_2O results in less H_2O and additional H^+, production of N_2O versus N_2 varies with aeration and pH. With increased pH and decreasing aeration, the proportion of N_2 to N_2O increases. The same is also true with increasing time.

Denitrification is a zero-order reaction, so its rate is relatively independent of the level of NO_3^- present in the medium. The rate of denitrification is influenced by pH, aeration, temperature, and level of decomposable organic matter present. The effect of each is discussed separately.

Effect of pH

There is a linear relation between pH and rate of denitrification. Denitrification tends to be low in acidic soils and increase with increasing soil pH. Rates of 30 mg N/kg/day may occur at pH 8.0 (Focht, 1974). A relation between pH and denitrification from Focht (1974) is shown in Figure 8.2.

Effect of Aeration

Absence of oxygen is required for NO_3^- to be used as an electron acceptor in place of oxygen. Increasing soil water content reduces aeration and hence

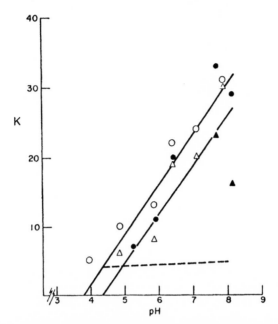

FIGURE   8.2   Effect of pH on denitrification rates (adapted from Focht, 1974). Rates of nitrate reduction (MgN/kg/day) are shown by the upper line (*circles*), rates of nitrous oxide reduction are shown by the lower line (*triangles*). Net production of N_2O_3 shown by the dashed line. Data from Nommik (1956), *open circles*; data from Wyler and Delwicke (1954), *closed circles*. Reproduced from Focht (1974) by permission of the Williams & Wilkins Co.

leads to denitrification. There is a negative linear relation between denitrification and the percentage of aerated pores in a soil.

Effect of Temperature

As temperature is increased, denitrification increases; the relation is exponential. Rates tend to increase up to 37°C. Effects of temperature are similar to those for biological systems. Q_{10} values of 1.4 to 2.0 have been reported (Focht, 1974).

Effect of Organic Materials

Denitrification is directly related to oxidative respiration. Hence, with higher amounts of readily oxidized organic materials, denitrification will be higher.

Since denitrification is a process causing loss of nitrogen, we must minimize the amount denitrified for efficient use of nitrogen fertilizers. One approach to minimization is using nitrification inhibitors. As long as nitrogen remains as NH_4^+, no denitrification will occur.

Leaching

Nitrogen as NH_4^+ in soils with appreciable cation exchange capacity will only leach in proportion to the quantity in solution. When nitrogen is in the nitrate form, however, it is all in soil solution and can be readily leached. The amount of leaching per unit of water in excess of field capacity will vary with the volumetric moisture content at field capacity.

Volatilization

Nitrogen may also be lost to the atmosphere as NH_3; which may occur whenever NH_4^+ is present at the soil surface, especially at high pH (above 7.0) and high temperature. Ammonium or ammonium-forming fertilizers may suffer loss to the atmosphere by volatilization when surface applied.

NITROGEN-UPTAKE MODELS

Modeling of nitrate uptake has been described by Phillips et al. (1976) and NaNagara et al. (1976); who modeled nitrate flux to the root by both mass flow and diffusion. They made some of the same assumptions used in the model of Claassen and Barber (1976), as described in Chapter 6. However, unlike Claassen and Barber (1976), they used a constant for the relation between nitrate concentration in solution and nitrate influx per unit of root. They estimated this constant from field data for nitrogen uptake; an approach that tends to force agreement between observed and predicted nitrate uptake values.

Bhat et al. (1979) used a model developed by Nye et al. (1975) to compare nitrate uptake by rape (*Brassica napus*), with nitrate uptake predicted by the model. The rape was grown in a controlled-climate facility with relative humidity about 95%. Under these conditions, transpiration and mass flow would be minimized. Bhat et al. used two initial levels of nitrate in solution and calculated that mass flow supplied a maximum of 38 and 25% of total uptake for the respective conditions. They predicted both plant growth and nitrate uptake. Observed nitrogen uptake agreed well with predicted uptake over the entire growth period, which lasted 24 days. Values for nitrogen-influx kinetics were obtained by growing plants in solution culture at the same temperature and day lengths used to predict uptake from the soil.

Barber and Cushman (1981) described a mathematical model for nitrogen uptake, where nitrogen supply to the root was by mass flow and diffusion. The model has not been validated experimentally for nitrogen; however, its validation for potassium and phosphorus was described in Chapter 5.

In modeling nitrogen uptake, the variation of influx with concentration outside the root is different for nitrogen than for phosphate and potassium. The K_m value for nitrogen is very low in comparison to concentrations in solution. Furthermore, with nitrate there is no buffering from the soil solid phase. Hence, nitrogen influx remains at I_{max} until the nitrogen is almost all gone. Therefore, the main parameters of significance in the model are root surface area, I_{max}, and the quantity of nitrogen in the soil. Because of a high D_e value for nitrate and competition between roots, roots can absorb a large fraction of the nitrate present in the soil. The finite amount in solution becomes the limiting factor in nitrogen uptake.

Sensitivity Analysis for Nitrogen Uptake

Results of a sensitivity analysis using the Barber–Cushman model are shown in Figure 8.3; the initial values used for the analysis are shown in Table 8.5. Each parameter was varied between 0.5 and 2.0 of its initial value, while all other parameters were held constant at initial levels. Results indicate that predicted NO_3^- uptake increased rapidly with increases in k, r_0, and I_{max}. The increase in k represents an increase in total NO_3^- available for the

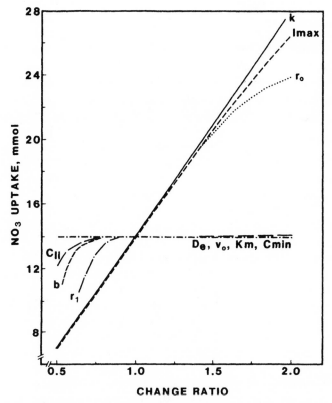

FIGURE 8.3   Sensitivity analysis for predicted nitrate uptake from soil by corn, using the Barber–Cushman (Barber and Cushman, 1981) model.

TABLE 8.5   Soil and Plant Parameters for Running a Sensitivity Analysis Predicting Nitrogen Uptake with the Barber-Cushman Model[a]

Symbol	Initial value
C_{li}	5 mmol/L
b	0.4
D_e	2.5×10^{-6} cm^2/s
v_0	1×10^{-7} cm/s
r_1	0.3 cm
r_0	0.02 cm
I_{max}	100 nmol/m$^2 \cdot$ s
K_m	0.025 mmol/L
C_{min}	0.002 mmol/L
L_0	20 cm
k	0.03 cm/s

Source: Barber, unpublished data.
[a] Values are estimates for NO$_3^-$ uptake by corn from Raub silt loam.

plant, since if r_1 is held constant, increasing k increases the amount of soil by a similar amount. However, increasing r_0 also increases root surface area for absorption, with soil volume, $r_1 - r_0$, decreasing slightly. Increasing I_{max} increases uptake almost linearly. The latter two effects are because C_{li} is high relative to K_m, 5 versus 0.025 mmol/L. Hence, there was enough NO$_3^-$ present to enable the root to absorb at the I_{max} rate even as root surface area was increased by increasing r_0. Consequently, the rate of absorption by the plant, rather than the supply by the soil, was limiting influx.

Values for D_e, v_0, K_m, and C_{min} were without effect or had little effect, because absorption was mainly controlled by I_{max}. Only as C_{li} was reduced below 0.75 of its initial value did it began to have an effect on predicted uptake. Reducing b also reduced the supply of NO$_3^-$ and reduced uptake, while root-to-root competition effects did not occur until r_1 was less than 80% of its initial value.

The type of sensitivity diagram obtained will be influenced by the level of NO$_3^-$ in the soil and the length of the cropping period. In this experiment, plants were grown for 10 days. Values for parameters were those that were observed experimentally. Because of higher D_e values for NO$_3^-$, and the fact that most of the NO$_3^-$ can also move by mass flow, most of it has an opportunity to reach the root and be absorbed. Consequently, plant recoveries of applied NO$_3^-$ are usually much higher than recoveries of potassium or phosphorus.

Nitrogen can also be present in the ammonium form. Ammonium and potassium are adsorbed with similar strengths by the soil, so their b and D_e values are comparable. Absorption parameters of these nutrients by plant roots are also similar. Hence, the sensitivity analysis for NH$_4^+$ should be similar to that shown for potassium in Figure 5.5.

REFERENCES

Alexander, M. 1965. Nitrification. In W. V. Bartholomew and F. E. Clark, Eds. *Soil Nitrogen*. American Society of Agronomy, Madison, WI. Pp. 307–343.

Barber, S. A. 1974. The influence of the plant root on ion movement in soil. In E. W. Carson, Ed. *The Plant Root and Its Environment*. University Press of Virginia, Charlottesville. Pp. 525–564.

Barber, S. A. 1979. Corn residue management and soil organic matter. *Agron. J.* **71**:625–628.

Barber, S. A., and J. H. Cushman. 1981. Nitrogen uptake model for agronomic crops. In I. K. Iskandar, Ed. *Modeling Wastewater Renovation-Land Treatment*. Wiley-Interscience, New York. Pp. 382–409.

Barber, S. A., and M. Silberbush. 1984. Plant root morphology and nutrient uptake. In S. A. Barber and D. R. Bouldin, Eds. *Roots, Nutrient and Water Influx and Plant Growth*. American Society of Agronomy, Madison, WI.

Bartholomew, W. V. 1965. Mineralization and immobilization of nitrogen in the decomposition of plant and animal residues. In W. V. Bartholomew, and F. E. Clark, Eds. *Soil Nitrogen*. Monograph No. 10. American Society of Agronomy, Madison WI. Pp. 285–306.

Bhat, K. K. S., A. J. Brerton, and P. H. Nye. 1979. The possibility of predicting solute uptake and plant growth response from independently measured soil and plant characteristics. VIII. The growth and nitrate uptake of rape in soil at two nitrate concentrations and a comparison of the results with model predictions. *Plant Soil* **53**:169–191.

Claassen, N., and S. A. Barber. 1976. Simulation model for nutrient uptake from soil by a growing plant root system. *Agron. J.* **68**:961–964.

Edwards. J. H., and S. A. Barber. 1976a. Nitrogen flux into corn roots as influenced by shoot requirement. *Agron. J.* **68**:471–473.

Edwards, J. H., and S. A. Barber. 1976b. Nitrogen uptake characteristics of corn roots at low N concentration concentration. *Agron. J.* **68**:973–975.

Focht, D. D. 1974. The effect of temperature, pH, and aeration on production of nitrous oxide and gaseous nitrogen—a zero order kinetic model. *Soil Sci.* **118**:173–179.

Fried, M., F. Zsoldos, P. B. Vose, and I. L. Shatokhim. 1965. Characterizing the NO$_3$ and NH$_4$ uptake process of rice roots by use of ^{15}N labelled NH$_4$NO$_3$. *Physiol. Plant.* **18**:313–320.

Gregory, P. S., D. V. Crawford, and M. McGowan. 1979. Nutrient relations of winter wheat. 2. Movement of nutrients to the root and their uptake. *J. Agric. Sci.* **93**:495–504.

Hassan, M. M., and T. van Hai. 1976. Ammonium and potassium uptake by citrus roots. *Physiol. Plant.* **36**:20–22.

Jungk, A. 1970. Interactions between the nitrogen concentration (NH₄, NH₄NO₃, and NO₃) and the pH of the nutrient solution, and their effects on the growth and ion balance of tomato plants. *Gertenbauwissenschaft* **35**:13–28.

Larson, W. E., C. E. Clapp, W. H. Pierre, and Y. B. Morachan. 1972. Effects of increasing amount of organic residues on continuous corn: II. Organic carbon, nitrogen, phosphorus, and sulfur. *Agron. J.* **64**:204–208.

Lycklama, J. C. 1963. The absorption of ammonium and nitrate by perennial ryegrass. *Acta Bot. Neerl.* **12**:361–423.

McLaren, A. D. 1969. Steady state studies of nitrification in soil: Theoretical considerations. *Soil Sci. Soc. Am. Proc.* **33**:273–276.

McLaren, A. D. 1976. Comments on nitrate reduction in unsaturated soil. *Soil Sci. Soc. Am. J.* **40**:698–699.

Minotti, P. L., D. C. Williams, and W. A. Jackson. 1969a. Nitrate uptake by wheat as influenced by ammonium and other cations. *Crop Sci.* **9**:9–14.

Minotti, P. L., D. C. Williams, and W. A. Jackson. 1969b. The influence of ammonium on nitrate reduction in wheat seedlings. *Planta* **86**:267–271.

Mugwira, L. M., S. M. Elgawhary, and A. E. Allen. 1980. Nitrate uptake effectiveness of different cultivars of triticale, wheat, and rye. *Agron. J.* **72**:585–588.

NaNagara, T., R. E. Phillips, and J. E. Leggett. 1976. Diffusion and mass-flow of nitrate nitrogen into corn roots grown under field conditions. *Agron. J.* **68**:67–72.

Nye, P. H., J. L. Brewster, and K. K. S. Bhat. 1975. The possibility of predicting solute uptake and plant growth response from independently measured soil and plant characteristics. I. The theoretical basis of the experiments. *Plant Soil* **42**:161–170.

Nommik, H., and K. Vahtras. 1982. Retention and fixation of ammonium and ammonia in soils. In F. J. Stevenson, Ed. *Nitrogen in Agricultural Soils.* Monograph No. 22. American Society of Agronomy, Madison, WI. Pp. 123–171.

Olsen, C. 1950. The significance of concentration for the rate of ion absorption by higher plants in water culture. *Physiol. Plant.* **3**:152–164.

Phillips, R. E., T. NaNagara, R. E. Zartman, and J. E. Leggett. 1976. Diffusion and mass-flow of nitrate nitrogen to plant roots. *Agron. J.* **68**:63–66.

Reisenauer, H. M. 1964. Mineral nutrients in soil solution. In P. L. Altman and D. S. Dittmer, Eds., *Environmental Biology.* Fed. Am. Soc. Exp. Biol., Bethesda, MD. Pp. 507–508.

Reisenauer, H. M. 1978. Absorption and utilization of ammonium nitrogen by plants. In D. R. Neilsen and J. G. MacDonald, Eds., *Nitrogen in the Environment.* Academic Press, New York. Pp. 157–170.

Smith, J. L., R. R. Schnabel, B. L. McNeal, and G. S. Campbell. 1980. Potential errors in the first-order model for estimating soil nitrogen mineralization potentials. *Soil Sci. Soc. Am. J.* **44**:996–1000.

Smith, S. J., L. B. Young, and G. E. Miller. 1977. Evaluation of soil nitrogen mineralization potentials under modified field conditions. *Soil Sci. Soc. Am. J.* **41**:74–76.

Stanford, G., and S. J. Smith. 1972. Nitrogen mineralization potentials of soils. *Soil Sci. Soc. Am. J.* **36**:465–472.

Stevenson, F. J. 1959. Distribution of fixed ammonium in soils. *Soil Sci. Soc. Am. Proc.* **23**:121–125.

Stevenson, F. J. 1965. Origin and distribution of nitrogen in soil. In W. V. Bartholomew and F. E. Clark. Eds. *Soil Nitrogen.* Monograph No. 10. American Society of Agronomy, Madison, WI. Pp. 1–42.

Stevenson, F. J. 1982. *Nitrogen in Agricultural Soils.* American Society of Agronomy, Madison, WI.

Stevenson, F. J. 1986. *Cycles of Soil Carbon, Nitrogen, Phosphorus, Sulfur, Micronutrients.* John Wiley, New York.

Teo, Y. H., C. A. Beyrouty, and E. E. Gbur. 1992. Nitrogen, phosphorus and potassium influx kinetic parameters of three rice cultivars. *J. Plant Nutr.* **15**:435–444.

Van den Honert, T. H., and J. J. M. Hooymans. 1955. On the absorption of nitrate by maize in water culture. *Acta Bot. Neerl.* **4**:376–384.

Warncke, D. D., and S. A. Barber. 1973. Ammonium and nitrate uptake by corn (*Zea mays* L.) as influenced by nitrogen concentration and NH_4^+/NO_3^- ratio. *Agron. J.* **65**:950–953.

Warncke, D. D., and S. A. Barber. 1974. Nitrate uptake effectiveness of four plant species. *J. Environ. Qual.* **3**:28–30.

Wijler, J., and C. C. Delwicke. 1954. Investigations on the denitrifying process in soil. *Plant Soil* **5**:155–169.

CHAPTER **9**

Phosphorus

Phosphorus is an important nutrient in crop production, since many soils in their native state do not contain sufficient available phosphorus to maximize crop yield. The total phosphorus content of the earth's crust is about 0.12% (Cathcart, 1980).

This chapter deals with phosphorus in relation to the factors used in the mechanistic nutrient uptake model, levels of C_{li} and C_{si} of phosphorus in the soil in relation to soil properties determining them, and use of the model for predicting phosphorus uptake and sensitivity of the model parameters.

SOIL PHOSPHORUS

Phosphorus content of soils may vary from 0.02 to 0.5%, with an average of approximately 0.05%. The phosphorus present may be divided into four general categories: (1) phosphorus as ions and compounds in the soil solution; (2) phosphorus adsorbed on the surfaces of inorganic soil constituents;

(3) phosphorus minerals, both crystalline and amorphous; and (4) phosphorus as a component of soil organic matter. The total amount of phosphorus in top soil, an average of 1000 kg P/ha, is not large compared to annual crop removals of 10 to 40 kg P/ha. This is especially true since a large fraction of the phosphorus present is in a mineral form not readily available for absorption by the plant.

Soil Solution Phosphorus, C_{li}

Phosphorus concentrations of C_{li} are low compared to nitrogen, potassium, calcium, and magnesium. Represented values are shown in Table 9.1. Reisenauer (1964) surveyed values in western U.S. soils. Barber et al. (1962) measured values from U.S. soils in the central part of the United States. Kovar and Barber (1988) measured values from samples collected from throughout the United States. The increase in concentration in recent years reflects increases due to phosphorus fertilization. Whether soils are alkaline, neutral, or acidic, phosphorus concentrations are usually low. The dominant phosphorus ion form in solution will be either $H_2PO_4^-$ or HPO_4^{2-} depending on soil pH, since soils are normally in the pH range 4.0–8.5. The relationship between the ionic form of phosphorus and pH is shown in Figure 9.1. At pH 7.2 one-half of the phosphate would be present as $H_2PO_4^-$ and one-half as HPO_4^{2-}. In addition to the ionic phosphate in solution, as much as half the phosphate may be present as soluble organic compounds, particularly in soils containing appreciable organic matter.

When phosphate is added to the soil, values of C_{li} increase; the increase varies with the soil. Kovar and Barber (1988) studied the relation between phosphate added to samples of the 0- to 20-cm layer of 33 diverse undried

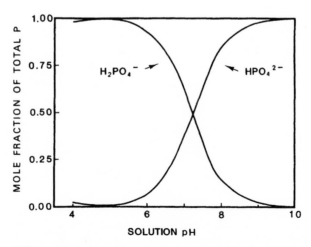

FIGURE   9.1   Effect of solution pH on ionic forms of dissolved phosphate.

TABLE 9.1 Levels of Phosphorus in the Soil Solution

Phosphorus concentration (mg/L)	Reisenauer (1964)[a] (% of samples)	Phosphorus concentration (mg/L)	Barber et al. (1962) (% of samples)	Kovar and Barber (1988) (% of samples)
0–0.03	25.5	0–0.02	4	0
0.03–0.06	18.8	0.02–0.04	23	6
0.06–0.10	16.8	0.04–0.06	31	12
0.10–0.15	12.1	0.06–0.08	28	6
0.15–0.20	2.7	0.08–0.12	19	30
0.20–0.25	2.0	0.12–0.16	17	9
0.25–0.30	4.0	0.16–0.20	2	6
0.30–0.40	6.0	0.20–0.40	11	15
0.40–0.50	4.0	0.41–0.50	5	15
>0.50	8.1	0.81–1.20	2	0
Number of samples	149		135	33

soils and measured soil solution phosphorus, C_{li}, and labile phosphorus, C_{si}. Seven rates of phosphorus ranging from 0 to 655 mg P/kg of soil (oven-dry basis) were equilibrated with the soil. The phosphate source was 11–16–0 clear solution used in liquid fertilizers and containing NH_4 ortho- and polyphosphates. The fertilizer was diluted with water and sprayed on the soil and the soil and fertilizer were mixed thoroughly. Sufficient water was added to increase the soil water content to field capacity (–33 kPa) and the soils were kept at 25°C for 3 weeks. Soil water level was maintained. After 3 weeks the soils were dried to –100 to –300 kPa water tension and sieved through a 2-mm sieve in preparation for determining C_{li} and C_{si}. The procedures used for determining C_{li} and C_{si} for phosphorus are given in Chapter 2. Soils were not allowed to dry below wilting point prior to use. The effect of rate of phosphorus added on C_{li} is shown for four soils in Figure 9.2 giving the range obtained. The increase in C_{li} with phosphorus added was curvilinear and could be described by the relation $C_{li} = ax^c + d$, where x is the amount of phosphorus added and a, c, and d are constants and are given for the 33 soils investigated by Kovar and Barber (1988). The greatest increase in C_l per unit of added P was for Bonifay sand, while the least was for Malabon silt loam. The coefficient c describes the degree of curvilinearity. Its values

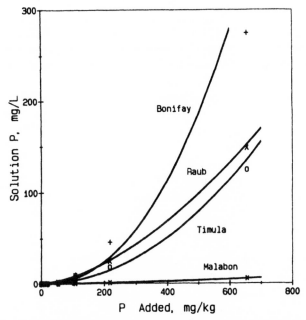

FIGURE   9.2   Relation between phosphorus added, x, and phosphorus in soil solution, C_{li}, for four soils representing the range found for 33 soils after 3 weeks equilibration curves fit with the Freundlich relation $C_{li} = ax^c + d$. Reproduced from Kovar and Barber (1988) by permission of the Soil Science Society of America.

ranged from 1.03 to 3.15. The a coefficient describes the degree of linear increase in C_i with P addition; values ranged from 2.2×10^{-7} to 2.3×10^{-2}, however, two-thirds of the values were between 0.13×10^{-4} and 4.95×10^{-4}. The values are small because only a small fraction of added phosphate ends up in soil solution. The d value, the intercept with the y axis, is the C_i value with no phosphate addition. The high rates of phosphorus application represent what occurs when fertilizer is applied in a concentrated band for row crops.

Adsorbed Phosphorus

Controversy exists over the amount of phosphorus that may be adsorbed on the surfaces of such soil constituents as iron and aluminum oxides versus the amount that is precipitated as discrete mineral forms. Since phosphorus may be adsorbed on mineral oxide surfaces, the chemical composition with respect to phosphate can be highly variable. In uncultivated soils, where no phosphate fertilizer has been applied, phosphorus will be recycled in the system. It is absorbed from the soil by plants, and when the plants die, decomposition of plant residues releases phosphorus into the soil where it can be adsorbed or precipitated once more. This causes surface soils to have higher phosphorus contents than subsoils. Phosphate may be adsorbed or precipitated by soils to which phosphate fertilizer has been applied. When added phosphorus only increases the phosphorus level in the soil solution to less than 5.0 mg/L, added phosphate is commonly considered to be adsorbed by oxide or carbonate surfaces at first rather than precipitated in mineral forms. The adsorbed phosphate has been postulated to result from surface reactions with calcium carbonates and aluminum and iron oxides. When higher phosphate concentrations are maintained in the soil solution, such as near phosphate fertilizer particles, precipitation of phosphate as various calcium, iron, and aluminum phosphates is believed to occur.

Equilibrium of phosphate between solution and solid forms has been described on the basis of rate of reaction (Barrow, 1980). Rate of equilibration between dissolved and adsorbed phosphate can be determined using ^{32}P. There is an initial, relatively rapid rate of equilibration and then a subsequent slow equilibration process. The absorbed phosphate that equilibrates rapidly, usually within 24 or 48 h, is arbitrarily termed labile phosphate; the slowly equilibrating phosphate is termed nonlabile phosphate. We can thus consider two divisions of solid-phase phosphate. One involves labile and nonlabile forms; the other, adsorbed and crystalline forms. These two divisions are not synonymous, but there is overlap between the two (Olsen and Khasawneh, 1980). From the standpoint of phosphate supply to the plant root, the kinetic approach of labile and nonlabile forms is more useful. When phosphate is removed from solution, the equilibrium is disturbed and labile phosphate goes into solution rapidly. This also disturbs the equilibrium between labile and nonlabile forms, with phosphate moving very slowly from nonlabile to labile pools.

Labile phosphorus can be readily measured from ^{32}P-equilibration experiments; since it varies with time of equilibration, standardized conditions are used to obtain relative values among soils. An alternative procedure involves extraction with chloride-saturated anion-exchange resin. Sufficient resin is used to provide a concentration of solution phosphate approaching zero; which results in phosphorus desorption. As with labile phosphate measurements, removal depends on time and experimental conditions. It can be argued that resin values might be more closely related to phosphate uptake by plant roots, where the root acts as a phosphate sink, since the exchange resin serves as a sink for phosphorus just as the root does.

In Kovar and Barber's (1988) study with 0–20 cm samples of 33 soils, as described in this chapter under soil solution, C_{si} was also measured. The effect of rate of phosphorus added on C_{si} is shown for four soils in Figure 9.3, giving the range obtained. For all soils a linear relation (r^2 of 0.91 to 1.00) was obtained that was described by the equation $C_{si} = g + hx$, with the intercept g being the C_{si} level in the untreated soil and h the proportion of added phosphorus that was extracted with Cl-saturated anion-exchange resin in a 24-h extraction period. Values of g varied from 3.5 to 190.7 mg kg^{-1} and

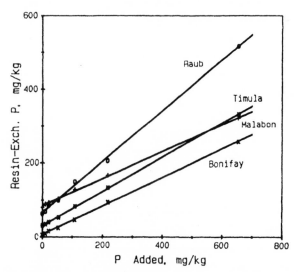

FIGURE   9.3   Relation between phosphorus added and phosphorus removed after 3 weeks equilibration by anion-exchange resin, C_{si}, for four soils representing the range found for 33 soils. Reproduced from Kovar and Barber (1988) by permission of the Soil Science Society of America.

those of h from 0.34 to 0.86 and averaged 0.69. In the range of phosphate addition used the fraction found as C_{si} was not concentration dependent.

In the Kovar and Barber (1988) research phosphate was equilibrated with the soil for 3 weeks. Long-term field experiments (Barber, 1979a) provide a study of the relation between added phosphate, phosphorus removed, and C_{si} that occurs over many years. A four-crop rotation experiment was conducted for 25 years, during which phosphorus was added at average rates of 0, 11, 22, and 44 kg P/ha/year. After 17 to 21 years of application equivalent to 22 and 44 kg P/ha annually, addition of phosphorus was stopped, and the change in levels of resin-exchangeable phosphorus was measured each year. Values were obtained for up to 8 years without phosphorus addition; the amounts of phosphorus removed by the crop were also measured. Data for the effect of phosphorus depletion from the soil on labile phosphorus levels are shown in Figure 9.4. The soil with a higher phosphorus content decreased 0.33 mg/kg for each kg/ha of phosphorus removed by cropping. Since there are about 2000 metric tons of soil/ha, the slope of the upper line suggests that 6 kg/ha of phosphorus removed would reduce the resin-extractable phosphorus level by 1 kg/ha; 5 kg/ha must have come from nonlabile phosphorus. For the lower line, the value is close to 20 kg/ha of phosphorus removal to reduce resin-

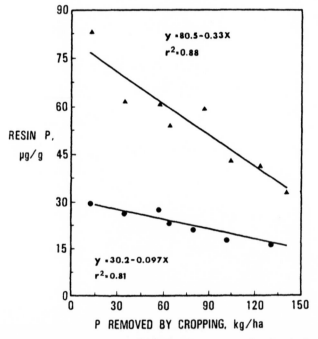

FIGURE   9.4   Effect of crop removal of phosphorus on reduction in level of resin-extracted phosphorus for field soil at two initial phosphorus levels. Reprinted from Barber (1979a) by courtesy of Marcel Dekker, Inc.

extractable phosphorus levels by 1 kg/ha. Hence, there is a considerable quantity of phosphorus, not immediately exchangeable, which comes into solution only slowly over time during cropping. A large portion of total phosphorus in this Raub silt loam (Aquic Argiudoll) was apparently in slow equilibrium with resin-extractable phosphate. Results of this nature will undoubtedly vary with soil. In soils with high amounts of iron and aluminum oxides, less of the total soil phosphorus may move to labile forms as phosphorus is removed. Few experimental results are available from which to evaluate this postulate.

Barrow (1980) found that the rate of phosphate desorption from soil in laboratory experiments decreased with time and was approximately proportional to $t^{0.3}$. Release was described by a curve with a gradually decreasing slope. The rate was such that most of the desorption occurred within 24 h. There is usually no obvious point where rate of release changes from rapid to slow, so that we can only characterize labile and nonlabile types of phosphate empirically.

Results from laboratory studies with resin-extraction suggest that less phosphate may become available than calculated from field experiments where cropping removed phosphorus. This was probably because of much longer time intervals in the field.

Mineral Phosphorus

Many potential phosphorus minerals may exist in the soil; some of the more common minerals are listed in Table 9.2. They are mainly minerals of phosphorus combined with calcium, aluminum, and iron. In soils above pH 7, calcium phosphates should be dominant, while in acid soils iron and aluminum phosphates are the dominant forms. Whether crystalline phosphate minerals or

TABLE 9.2   Some Common Soil Phosphate Minerals

Calcium phosphates	Aluminum phosphates	Iron phosphates
$Ca_5(PO_4)_3F$ Fluoroapatite	$AlPO_4 \cdot 2H_2O$ Variscite	$FePO_4 \cdot 2H_2O$ Strengite
$Ca_5(PO_4)_3OH$ Hydroxyapatite	$H_6K_3Al_5(PO_4)_8 \cdot 18\ H_2O$ Potassium taranakite	
$Ca_3(PO_4)_2$ Tricalcium phosphate	$AlPO_4$ Berlinite	
$Ca_4H(PO_4)_3 \cdot 2.5\ H_2O$ Octacalcium phosphate		
$CaHPO_4$ Dicalcium phosphate		
$CaHPO_4 \cdot 2H_2O$ Dicalcium phosphate dihydrate, brushite		

amorphous phosphate minerals are predominant is difficult to determine. The quantities of each vary with soil.

The solubility-product principle has been used to evaluate the presence of possible crystalline phosphate minerals. The appropriate constants may be expressed as solubility isotherms as a function of pH. To calculate the isotherms, it is also necessary to know the effect of pH on the activity of aluminum and iron in solution. The activity of calcium can be set at a constant arbitrary level consistent with values found for soil solutions, since pH usually does not influence solution calcium activity in noncalcareous soils. Example isotherms are shown in Figure 9.5 for selected iron, aluminum, and calcium compounds. When the pH and pH_2PO_4 of the soil solution are measured, values can be compared with the solubility product line for each mineral. If a point is above a line, the solution will be supersaturated relative to that mineral. If it is below a line, the solution will be undersaturated.

A major limitation in using solubility isotherms is that they are constructed assuming that pure crystalline compounds are at equilibrium with the phosphate in solution. Since attaining equilibrium may be very slow, the system may be either oversaturated or undersaturated. Phosphate compounds in soil are probably not pure crystalline forms but rather impure and have a solubility that is not known. In addition, the fit of points to a particular line is not proof that a particular compound is controlling phosphate solubility. Solubility isotherms are useful for determining the kinds of phosphate compounds that could be controlling phosphate levels in solution.

Soil Organic Phosphorus

One-half or more of the total phosphorus in the A horizon of soils may be present as organic phosphorus; the amount depends on the organic matter content of the soil. The carbon/phosphorus ratio in soil organic matter is approximately 100. Hence, the phosphorus concentration of organic matter is about 0.5%. The organic phosphorus compounds are principally esters of orthophosphoric acid (Anderson, 1980). Phosphate esters are present in both plants and animals. There are many with varying stabilities, which may be added to soil by plants and animals. The most resistant to decomposition is inosital phosphate; which makes up more than half the organic phosphorus in many soils. The most common form is myo-inosital hexaphosphate, also known as phytic acid; chemical formula $C_6H_6(OHPO_3)_6$.

Other common organic phosphate groups in soil organic matter are the nucleic acids. Organic materials recently added to the soil are usually higher in nucleic acids than inosital phosphates, but they are broken down quite rapidly, so there are smaller residual amounts. In addition to inosital phosphates and nucleic acids, phospholipids and other esters are also present.

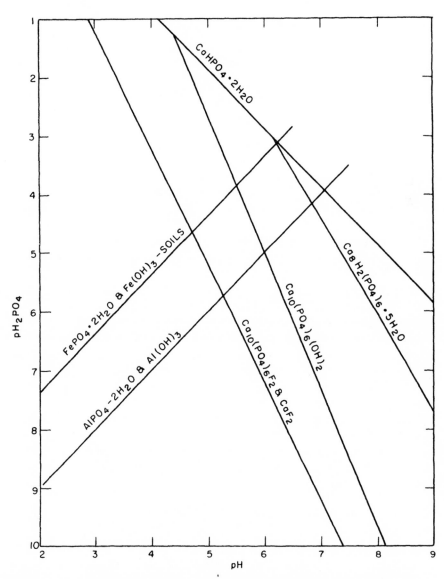

FIGURE 9.5 Solubility isotherms for indicated crystalline phases. Activity of calcium was arbitrarily set at pCa = 2.5. The activities of Al³⁺ and Fe³⁺ were controlled by the solubility of their oxides. Reprinted from Olsen and Khasawneh (1980) by permission of the American Society of Agronomy, Crop Science Society of America, and Soil Science Society of America.

Mineralization of Organic Phosphorus

The release of organic phosphorus into the soil solution where it can move to the root and be absorbed, is controlled by the rate of organic matter mineralization. The rate of mineralization of organic matter in temperate-climate soils depends on soil texture and environmental conditions. Barber (1979b) reported organic matter mineralization of 2.4% per year when an Indiana silt loam soil was fallowed for 6 years. Calculations were expressed as a percentage of total organic matter present. Larson et al. (1972) reported 1.9% per year decomposition when an Iowa soil was fallowed for 10 years. Where crops are grown, new organic matter is produced, so that phosphorus released in organic matter decomposition is recycled into new organic forms. The Indiana soil, Raub silt loam, had an organic phosphorus content of 340 kg/ha (assuming 0.5% phosphorus in the organic matter and 3.4% organic matter). If we assume that phosphorus is released at the same rate as organic matter decomposition, then 8.1 kg/ha of phosphorus would be released per year. Similar calculations for the Iowa soil, Marshall silty clay loam, a Typic Hapludoll, indicated that 7.9 kg/ha of phosphorus would be released per year.

Decomposition of organic phosphorus adds to the general level of labile inorganic phosphorus in the soil just as though this amount of fertilizer phosphorus were mixed with the soil during crop growth. However, since only a small part of the labile phosphorus is positionally available for absorption by plant roots, only a small part of the phosphorus released will be absorbed by current growing plants.

The greatest effects of organic phosphorus have been observed in the tropics (Anderson, 1980), where organic phosphorus can apparently supply a large part of the crop phosphorus requirement. In tropical soils, where temperatures are high, organic matter decomposition is rapid and thus supplies more phosphorus than does organic matter in temperate soils. When organic phosphorus in the soil is above 700 mg/kg, the response to added fertilizer phosphorus is small. Release of phosphorus from organic sources becomes an important factor in producing crops using "bush fallow" systems, where native plants are allowed to grow without harvest for several years in order to build up the level of soil organic phosphorus as well as nitrogen. Then when the bush is cleared and the soil cropped, rapid decomposition of the organic matter supplies the phosphorus needed by the crop.

Tropical soils with an organic phosphorus content of 700 or more kg P/ha frequently do not respond to added phosphorus. Soil organic matter decomposition rates will vary. Rates in tropical soils are undoubtedly much higher, probably by a factor of two or more than in temperate soils. Hence, if 5% decomposition per year occurred on a soil with 700 kg P/ha, this would release 35 kg P/ha. This, along with that already present, would likely be sufficient for crop production.

While 50% of the phosphorus in the A horizon of a temperate soil may be organic phosphorus, the organic phosphorus is not so readily available for absorption by the plant as inorganic phosphorus that only needs to be desorbed. To become available for supply to the root, organic phosphorus must first be converted into inorganic phosphorus. Some organic phosphorus exists as soluble organic compounds in soil solution. Little is known of the relative absorption of these compounds by plant roots. Anderson (1980) provides evidence both for and against their absorption. If the soluble organic phosphorus were absorbed by the root, the kinetics of its replacement would likely depend on microbiological action, so that its total significance may not be great. Much more research is needed in this area to define the mechanisms operating.

Some investigators (Boero and Thien, 1979) have postulated that release of phosphatase by plant roots will increase the release of phosphorus from organic forms and hence render it more available for plant use. Little information is available to test this postulate. Experimental evidence has not shown significant increases in soil phosphorus release when phosphatase amounts were higher.

Fractionation of Soil Phosphates

Soil phosphorus can be separated into organic and inorganic forms. Separation of the various organic phosphorus and inorganic forms is difficult and often imprecise, since not all of the organic phosphorus compounds are known.

Separating inorganic phosphorus into calcium, iron, and aluminum forms plus occluded phosphates, has been attempted by Chang and Jackson (1957). This separation is based on consecutive extractions with different extractants, each designed to remove a particular form of phosphorus. All calcium phosphates are combined, since the fractionation scheme cannot separate within a class. The separation was satisfactory for stable reaction products, such as varasite, strengite, and hydroxyapatite. However, when metastable products, such as amorphous iron and aluminum phosphates were used, the procedure could not distinguish adequately between them. With these reservations, some data are given here to illustrate probable variation in general forms of phosphorus present in soils.

Chang and Jackson (1958) analyzed three soils varying widely in degree of weathering and obtained the data shown in Table 9.3. Less weathered soils were high in calcium phosphates, whereas highly weathered soils were high in occluded phosphates. In moderately weathered soils, which represent large areas of soils used for agricultural production under nonirrigated conditions, there were relatively similar amounts of the various forms delineated by the extraction procedure.

TABLE 9.3   Distribution of Extractable Inorganic Phosphorus among Four Forms in Three Soils Representing Different Degrees of Weathering

Degree of weathering	Soil and horizon	Soil group	Form of phosphate (mg/kg)			
			Ca	Al	Fe	Occluded
Slight	Rosebud B_2	Mollisol	155	18	8	Trace
Moderate	Miami A_2	Alfisol	45	48	100	79
Strong	Wahiawa A	Oxisol	10	20	123	601

Source: Chang and Jackson (1958) by permission of Blackwell Scientific Publications Ltd.

TABLE 9.4   Effect of Position in a Catena on Phosphorus Form Along a Brookston–Crosby Transect Catena in Indiana

	Soil			
	A	B	C	D
	(Brookston)		(Crosby)	
Soil pH	6.4	6.5	6.7	6.5
Organic matter %	5.3	2.7	2.7	1.8
Soluble P, mg/kg	2.9	11.2	10.3	2.0
Al-P, mg/kg	64.2	30.7	26.2	15.8
Fe-P, mg/kg	106.8	52.8	63.2	47.2
Ca-P, mg/kg	84.9	37.8	32.8	23.1
Occluded Al-P, mg/kg	11.5	9.3	8.6	8.5
Occluded Fe-P, mg/kg	221	263	249	279
Organic P, mg/kg	409	241	52	52
Total P, mg/kg	900	646	442	428

Source: Reproduced from Al Abbas and Barber (1964) by permission of the Soil Science Society of America.

Al-Abbas and Barber (1964) used the same procedure to evaluate forms of phosphate in samples taken on a transect across a Crosby-Brookston catena in Indiana. Data are shown in Table 9.4; organic phosphorus showed a wide range of values related to the organic matter content of the soil. Total phosphorus increased as organic matter increased. In the Brookston soil, organic phosphorus constituted 45% of total phosphorus, while at the other end of the spectrum, where organic matter was low, organic phosphorus represented only 12% of the total. As total phosphorus increased, all forms of phosphorus also increased except for occluded forms, which remained relatively constant. While such values indicate the forms of phosphorus that may be present in the soil, they do not evaluate the rate of phosphorus supply to the root for subsequent phosphorus absorption by the plant.

BIOAVAILABLE PHOSPHORUS

Phosphorus flux through the soil to plant roots depends on the values of C_{li}, b, and D_e for soil phosphorus. These are the soil parameters used in the simulation model in Chapter 5. The values usually found for C_{li} were described at the beginning of this chapter, so only b and D_e are discussed here.

Buffer Power

The buffer power of the phosphate associated with the solid phase of the soil for phosphate in solution is important in determining the supply of phosphate moved to the root by diffusion. Because the level of phosphate in the soil solution is low and phosphate is readily adsorbed on soil surfaces, the buffer power for soil phosphorus is usually much larger than that for exchangeable cations, such as potassium. The buffer power also varies indirectly with the level of soil phosphate present. Data in Table 9.5 from Anghinoni and Barber (1980) show the relation between the amount of phosphorus added as $Ca(H_2PO_4)_2$ to two Indiana soils and b and D_e; soils used were Raub silt loam and Wellston silt loam. The sample of Raub soil came from the Ap horizon of a plot that had not received phosphate fertilizer

TABLE 9.5   Effect of Phosphorus Addition on Soil Properties Influencing the Rate of Phosphorus Supply to Plant Roots

Soil	P added (mg/kg)	C_{li} (μmol/L)	b	Mean D_e (cm$^2 \cdot$ s $\times$ 10^9)
Raub silt loam	0	2.9	239	1.79
	60	6.2	234	1.83
	120	14.1	190	3.29
	240	53.9	79	5.41
	480	182.1	38	11.24
Wellston silt loam	0	1.9	1072	0.43
	140	3.1	1072	0.43
	280	9.0	778	0.60
	560	62.8	229	2.02
	1,120	611.8	45	10.28

Source: Reproduced from Anghinoni and Barber (1980) by permission of the Soil Science Society of America.

for 25 years. The Wellston soil came from the A horizon of an uncultivated soil that had no previous history of phosphate application. Phosphate was added to the soil by spraying phosphate solution while mixing. The soils were first incubated at 0.3 bar moisture for 2 weeks at 20°C, then for 4 days at 70°C, and finally for 1 week at 20°C. The incubation at 70°C rapidly equilibrated phosphate, as described by Barrow et al. (1977). Soil solutions were displaced for determination of C_{li}. Soils were then equilibrated with ^{32}P to determine the amount of phosphate from the solid phase equilibrating with phosphorus in solution, in order to determine $\Delta C_s/\Delta C_i$. Values for b were the average of C_s/C_i for the range from 0 to the value of C_{li} shown.

As phosphate was added, C_{li} increased and b decreased. Wellston soil was lower in phosphate initially and adsorbed a greater portion of the added phosphate, so that its values for b were higher. Because of the higher adsorption of phosphate in Wellston soil, rates of application used were higher than for the Raub soil. Hence, as the rate of phosphate application to the soil was increased, the phosphorus concentration in the soil solution increased and b decreased; values for b are therefore concentration dependent. Values for b can also be measured by using desorption isotherms (Brewster et al., 1975).

Phosphorus C_{si} vs. C_{li} Isotherms

The relation between C_{si} and C_{li} in the soil is influenced by the rate of phosphorus added to the soil. The value of b is calculated from $\Delta C_s/\Delta C_i$ so b is concentration-dependent. Values of C_{si} and C_{li} obtained in the Kovar and Barber (1988) research can be used to study the relation of C_{si} to C_{li} for 33 soils. Figure 9.6 shows the relation for the four soils used in Figures 9.2 and 9.3 and gives the range of values obtained. Data were related using the Freundlich equation $C_{si} = m\,C_{li}^n$ where m and n are regression constants. The data were represented more closely by the Freundlich equation than by the Langmuir adsorption equation. Values of n, representing the degree of curvilinearity, varied from 0.34 to 0.62. Values of m varied from 33 to 299 with the majority (25) of the values between 33 and 87. The value of m represents the linearity. Figure 9.6 illustrates the dissociation of adsorbed phosphate into solution. Caldwell et al. (1992) prepared a figure showing the distribution of the curves of C_{si} vs C_{li} for the 33 soils of the experiments of Kovar and Barber (1988) plus their results for a Typic Haplorxeroll. The soils used by Kovar and Barber (1988) were analyzed without drying after collection or during treatment.

Adsorption curves have commonly been determined by using phosphate added minus C_i for the adsorbed phosphate value. This value also includes tightly adsorbed (fixed) phosphate that is not removed with anion-exchange resin. For calculating phosphorus uptake with the mechanistic uptake model C_{si} values are needed to determine b, since C_{si} values do not include the tightly absorbed phosphate.

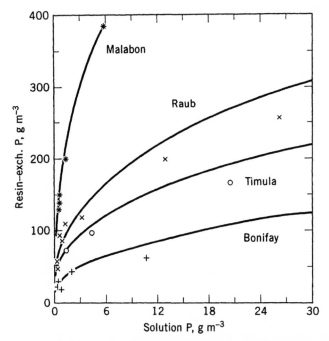

FIGURE   9.6   Relation between C_{li} and C_{si} of phosphorus for the four soils shown in Figures 9.2 and 9.3. Curves calculated with the Freundlich equation. Reproduced from Kovar and Barber (1988) by permission of the Soil Science Society of America.

Effective Diffusion Coefficient, D_e

Because of the high buffer power of phosphorus in soil, D_e for phosphorus is usually much smaller than D_e for nutrients such as nitrate and potassium. As indicated in Chapter 5, D_e can be estimated from the relation $D_e = D_l f_1 \theta/b$ where D_l is the diffusion coefficient in water at the same temperature, θ is the volumetric soil moisture content, and f_1 is the impedance factor largely due to tortuosity of the diffusion path. Values for D_e for phosphorus are usually in the range of 1×10^{-8} to 1×10^{-10} cm²/s. The range of values as affected by phosphate application to two soils is shown in Table 9.5.

PHOSPHORUS-UPTAKE KINETICS

Phosphate absorption by plant roots has been investigated using excised roots, intact plants grown in flowing nutrient culture at several levels of maintained phosphate, and intact plants that were allowed to deplete phosphate from stirred nutrient solution. These methods permit calculating phosphate-absorption kinetics. Asher and Loneragan (1967) studied phosphorus uptake by using flowing solution culture. Volumes as high as 2800 L were used at the low concentrations to reduce variability of the phosphorus level maintained in solution. Phosphorus concentrations ranged from 0.04 to 25 μmol/L. Solutions were analyzed every day, with rates of phosphorus addition used to maintain the phosphorus concentration in solution. The researchers measured the growth and phosphorus-uptake rate of 24 species; even at 0.04 μmol/L P, plants grew slowly and took up phosphorus. Maximum growth rate was reached at relatively low phosphate concentrations; results for a few species are shown in Table 9.6. There was similarity between species in the relation between plant phosphorus concentration and solution phosphorus level, and some species responded to higher levels of solution phosphorus more than others. Lupin, with a large seed, had a lower relative growth rate. The low levels of phosphate in these flowing culture solutions were similar to the levels in soil solution at the root surface. In the soil, the solution concentration at the root is maintained by diffusion and mass flow of phosphate to the plant root from a highly buffered soil.

Phosphate influx can be described by Michaelis–Menten kinetics, and more than one phase may occur, as discussed in Chapter 3. Edwards (1968),

TABLE 9.6   Effect of Phosphorus Concentration in Flowing Solution Culture on Phosphorus Concentration in Tops and Roots and Relative Growth Rate (RGR) of Four Species

| | P concentration in solution (μmol/L) | | | | |
	0.04	0.2	1.0	5.0	25
Flatweed (*Hypochoeris glabra* L.)					
P concentration (%)	0.07	0.24	0.80	0.76	0.81
RGR	7.0	11.5	13.3	15.6	16.4
Bromegrass (*Bromus rigidus* Roth)					
P concentration (%)	0.10	0.24	0.64	0.81	0.74
RGR	6.6	9.8	11.3	12.7	12.8
Clover (*Trifolium subteraneum*, L.)					
P concentration (%)	0.07	0.26	0.66	1.00	1.02
RGR	6.8	0.5	10.9	11.5	10.9
Lupin (*Lupinus digitatus*, Forsk)					
P concentration (%)	0.10	0.20	0.43	0.97	1.79
RGR	5.7	5.7	5.5	6.1	6.0

Source: Asher and Loneragan (1967). Reproduced by permission of Williams & Wilkins Co.

using a wide range of solution phosphorus concentrations, demonstrated dual kinetics. Phase 1 kinetics for subterranean clover was in the range of 0 to 300 µmol/L, while phase 2 kinetics was in the range of 300 to 1000 µmol/L. Soil–solution phosphorus levels are in the range Edwards describes as phase 1; he found a K_m value of 7.4 µmol/L for subclover when calcium was present at 600 µmol/L. This compares with a value of 4 µmol/L obtained by Leggett et al. (1965) for excised barley roots.

Jungk and Barber (1975) investigated the kinetics of phosphate uptake by corn roots using intact plants and solution phosphate depletion for evaluation. The absorption studies were done in solutions of pH 6.0. Influx was measured on corn grown for 12, 14, 28, 35, 43, 52, and 80 days. The K_m values varied from 2.1 to 8.1 µmol/L, with most of the values between 2.1 and 2.9 µmol/L; K_m did not change with plant age, but I_{max} values decreased with plant age. The relation between solution concentration and phosphorus-uptake rate for corn of various ages grown in the growth chamber and greenhouse is shown in Figure 9.7. As the plants increased in age, mean phosphorus influx decreased; C_{min} values decreased from 0.3 to 0.1 µmol/L as plant age increased.

Phosphate Influx Parameters

Influx parameters for a number of plant species grown in solution culture for up to 4 weeks are shown in Table 9.7. The value of I_{max} varied widely among species, but values for K_m were usually 2 to 3 µmol/L and for C_{min}, below 0.2 µmol/L.

Cultivar Variation in Phosphorus-Influx Kinetics

Cultivars differ in their root morphology and phosphorus-influx kinetics. Nielsen and Barber (1978) evaluated a number of corn genotypes. Twelve inbreds were grown in nutrient solution, giving a range of I_{max} of 12 to 36 nmol/m² · s. The authors did not obtain values for K_m and C_{min} on these 12 inbreds. Nielsen and Barber also determined root length per g of root and per g of shoot; there was a range of 141 to 290 m/g (d.w.) roots, indicating a

TABLE 9.7   Influx Kinetics for Phosphorus Absorption by Several Species from Solution Culture

Species	Age (days)	I_{max} (nmol/m² · s)	K_m (µmol/L)	C_{min}	Reference
Corn	16	32.6	5.8	0.09	Schenk and Barber (1979a)
Soybean	18	6.4	2.7	0.04	Silberbush and Barber (1982)
Lettuce	30	10.6	2.0	0.24	Itoh and Barber (1983)
Tomato	26	49.9	6.1	0.12	Itoh and Barber (1983)

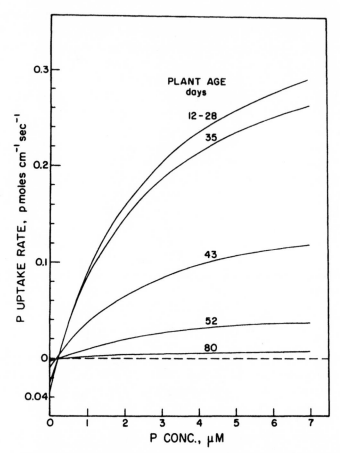

FIGURE   9.7   Effect of corn plant's age on P uptake rate as influenced by P concentration in solution. Reproduced from Jungk and Barber (1975) by permission of Martinus Nijhoff Publishers B.V.

variation in fineness of roots. Root length per g of plant ranged from 18 to 46 m/g (d.w.). Greater root length per g of shoot gives the plant the opportunity to absorb more phosphorus per unit of shoot. There was a negative correlation between I_{max} and root length per g of shoot when plants were growing in soil. Hence, genotypes with longer, finer roots absorbed phosphorus more slowly per cm of root. The results also suggest that phosphorus-influx kinetics were heritable.

Baligar and Barber (1979) compared six corn hybrids developed in Indiana with six hybrids developed in Florida. They found that I_{max} values for the Florida genotypes averaged 40 nmol/m^2 · s compared with an average of 15 nmol/m^2 · s for Indiana genotypes. Respective average K_m values were 6.3 and 2.0 μmol/L, with average C_{min} values of 3.2 and 0.4 μmol/L.

All of these differences were significantly different at the 1% level of significance. Florida genotypes had much finer roots, averaging 260 m/g, while Indiana genotypes averaged 179 m/g. The variability found in both of these studies indicates that it should be possible to develop corn hybrids with properties that would increase the efficiency of phosphorus uptake from soil.

Modeling Phosphorus Uptake

The mechanistic nutrient uptake model of Barber and Cushman (1981), described in Chapter 5, uses a mathematical relation describing the flux of nutrients from soil into the root and calculates the amount of nutrient absorbed by the plant. Table 5.1 gives the correlation and regression coefficients for the relation between observed and predicted phosphorus uptake obtained in pot experiments where the plants were grown in the growth chamber. For these experiments r^2 was between 0.85 and 0.99 and regression coefficients between 0.79 and 0.99 with an average of 0.93. This indicates satisfactory agreement considering that there were seven parameters describing plants and five describing soil supply. Figure 5.3 shows the relation between predicted and observed phosphorus uptake for five soils varying in volumetric water relationships (Cox, 1991). The correlation coefficient of $r^2 = 0.98$ and the regression coefficient of 1.02 indicate close agreement. Closer agreement between predicted and observed phosphorus uptake has been obtained in recent years due to improved measuring procedures for obtaining the parameters used in the model.

In an experiment by Itoh and Barber (1983) phosphorus uptake by root hairs also contributed to uptake. This was discussed in Chapter 7 where the effect was illustrated in Figures 7.9 and 7.10. Significant influence of root hairs depends on the relation between size of D_e and root hair length. Contribution of root hairs to nutrient uptake occurs more frequently for phosphorus because of its relatively low D_e. The mean linear diffusion distance of a phosphorus ion is $(2D_e t)^{1/2}$. Hence if D_e for phosphorus is 2×10^{-9} cm²/s the mean linear diffusion distance in 3 days will be 0.32 mm and in 9 days the distance will be 0.56 mm. Root hair length on wheat (Table 7.2) averaged 0.29 mm and on corn they averaged 0.33 mm. Hence if the root hair length does not appreciably exceed the depletion distance about the root, it will have little effect on uptake. On soils very low in phosphorus, D_e is very low, so root hairs can affect uptake. Some plant species have long root hairs such as 0.6 mm for Russian thistle and these can affect uptake even at higher D_e values. Inclusion of root hair uptake will be needed in such instances.

The relations between predicted and observed phosphorus uptake obtained by Schenk and Barber (1979a) for uptake by five corn genotypes grown at two soil-phosphorus levels are shown in Figure 9.8. At the low phosphorus level (D_e of 5.76×10^{-9} cm²/s) the mean distance phosphorus would diffuse in 4 days is 0.63 mm while at the high phosphorus level (D_e of 11.2×10^{-9}

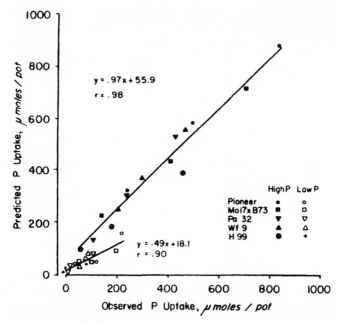

FIGURE   9.8   Relationship between observed phosphorus uptake and phosphorus uptake predicted by the Claassen–Barber model for five corn genotypes grown at two soil-phosphorus levels. Reprinted from Schenk and Barber (1979b) by permission of American Society of Agronomy.

cm²/s) the mean diffusion distance in 4 days would be 0.88 mm. Hence the regression coefficient between observed and predicted in the low phosphorus level was 0.49 and at the high phosphorus level it was 0.97. When the plants become deficient in phosphorus root hairs are sometimes longer, however, in Figure 9.8 the plants grown on the low phosphorus soil still absorbed only about one-fourth the amount of phosphorus of those grown at the higher phosphorus level, hence were deficient in phosphorus.

Mycorrhizae may also increase phosphorus uptake. Where the soil has been disturbed before planting, mycorrhizae usually do not affect phosphorus uptake until the plants are 2 weeks old. Hence they usually do not affect phosphorus uptake in pot experiments with rapidly growing annual plants only grown until root density reaches that observed in the field since they are usually harvested within 2 weeks. Competition between adjacent roots usually is not a large factor in phosphorus uptake because of the low D_e in the soil. Because of this the use of the Claassen–Barber model used in earlier experiments gave the same result as the Barber–Cushman model, however, it was not suitable for nutrients with higher D_e because of competition between adjacent roots.

When observed phosphorus uptake is greater than predicted phosphorus uptake while on the same experiment observed potassium uptake agrees with predicted potassium uptake, the comparison indicates that mycorrhizae or root hairs or both are increasing phosphorus uptake. Root hairs or mycorrhizae usually do not increase potassium uptake primarily because D_e for potassium is usually 10 to 100 times larger than for phosphorus. The effects of mycorrhzae and root hairs are discussed in Chapter 7.

Mycorrhizae may be a greater factor for phosphorus uptake where crops are planted without soil tillage (see Chapter 7). The mycorrhizal effect on uptake will begin much sooner on these soils than where soil has been disturbed.

SENSITIVITY ANALYSIS FOR PHOSPHORUS UPTAKE

The Barber–Cushman model was used with the parameters shown in Table 9.8 to predict phosphorus uptake as each parameter was varied from 0.5 to 2.0 times the value shown, while the other parameters remained at their initial levels. The values in Table 9.8 were obtained for soybeans growing in pots of Raub silt loam in a controlled-climate facility. Results of the sensitivity analysis are shown in Figure 9.9. Predicted phosphorus uptake was most sensitive to changes in root surface area, since changes in k and r_0 gave the greatest changes in uptake. Predicted phosphorus uptake was more sensitive to changes in soil supply, as described by C_{li}, b, and D_e, than to root-uptake kinetics, as described by I_{max}, K_m, and C_{min}. Hence for this soil, phosphorus supply by the soil was the rate-limiting step. Water flux v_0 had little effect, since phosphorus supply was mostly by diffusion; r_1 also had little

TABLE 9.8   Plant and Soil Parameters for Simulation of Phosphorus Uptake by Cushman's Mechanistic Mathematical Model[a]

Parameter	Initial Value
D_e	2.3×10^{-9} cm^2/s
b	163
C_{li}	0.0136 mmol/L
v_0	5.0×10^{-7} cm/s
r_0	0.015 cm
r_1	0.2 cm
I_{max}	6.43 nmol/m$^2 \cdot$ s
C_{min}	0.2 μmol/L
Km	5.45 μmol/L
L_0	250 cm
k	0.02 cm/s

Source: Silberbush and Barber, 1983. Table reproduced by courtesy of Martinus Nijhoff Publishers B.V.
[a]Values are for Williams soybeans growing on Raub silt loam.

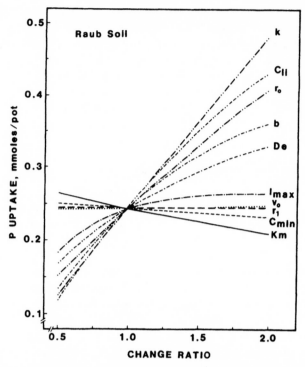

FIGURE   9.9   A sensitivity analysis for the effect on calculated P uptake of separately changing mathematical-model uptake parameters. All other parameters were maintained at the values shown Table 9.8. Figure reprinted from Silberbush and Barber (1983) by permission of Martinus Nijhoff Publishers B.V.

effect, since the D_e value for phosphorus is small and there was little root-to-root competition for phosphorus.

Usually, for soil-grown plants, changing the size of one parameter affects the size of other parameters as well; increasing C_{li} usually reduces b, which in turn increases D_e. Figure 9.10 gives the sensitivity of predicted phosphorus uptake to changes in C_{li} when (A) only C_{li} was changed, as in Figure 9.9; (B) b was changed as C_{li} was changed in order to keep C_{si} constant, D_e was kept constant; and (C) where both b and D_e were changed according to $b = \Delta C_{si}/\Delta C_{li}$ and $D_e = f_i \theta D_l / b$. Curve C represents a more realistic case: Even though b and D_e were varied, there remained a large effect of C_{li} on predicted phosphorus uptake.

The influence of many additional combinations of parameters on predicted phosphate uptake can be measured with the mechanistic uptake model. For example having finer longer roots with the same root volume can show that this will increase predicted phosphorus uptake, hence it may not be necessary for the plant to spend additional energy on producing longer roots.

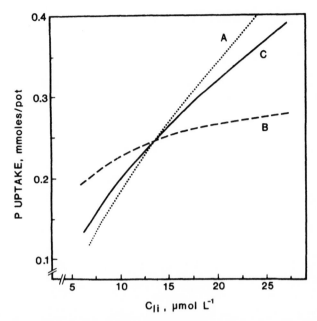

FIGURE   9.10   Calculated P uptake as affected by C_{li}, initial P concentration in soil solution, where (A) only C_{li} was changed, (B) buffer power b was changed to keep C_{si}, diffusible P, constant; D_e, the effective diffusion coefficient, was also kept constant; and (C) both b and D_e were changed according to their dependance on C_{li} as C_{li} was changed. Figure reprinted from Silberbush and Barber (1983) by permission of Martinus Nijhoff Publishers B.V.

Increased uptake occurred because of the greater root surface area as well as improved radial geometry for increased phosphorus supply to the root. Since the significance of this change may vary with phosphorus C_{li} level it needs to be evaluated with the C_{li}, b and D_e values of the soil and the k and r_o values for the plant of interest.

The preceding sensitivity discussion pertains to roots without root hairs; when root hairs are present, there is usually competition between root hairs for phosphorus in the soil. Because of this, C_{si} and hence b become significant factors in determining phosphorus uptake. A sensitivity analysis of the Itoh and Barber simulation model, which includes root hair effects, is given in Chapter 7. Increasing root hair length has a considerable effect on predicted phosphorus uptake because of the radial geometry of the root. In Figures 9.9 and 9.10, C_{li} was a highly significant factor in determining predicted phosphorus uptake by some species. When root hairs were an appreciable factor in uptake, b also became an important factor; presumably, the longer the root hairs, the greater the significance of b relative to C_{li}.

ROOT'S EFFECT ON RHIZOSPHERE SOIL PHOSPHORUS

The plant root can affect the soil supply of phosphorus by changing the pH of the rhizosphere soil. When they adsorb more cations than anions they release H^+ ions into the soil. This can occur for nonlegumes supplied with NH_4^+ fertilizer. The balance between the supply of NH_4^+ and NO_3^- may have a large effect on the rhizosphere soil pH (see Table 6.1). An excess of equivalents of cation uptake due to most of the nitrogen being adsorbed as NH_4^+ causes proton, hence H^+, release in order to balance the cations and anions. When the excess is anions where much of the nitrogen is adsorbed as NO_3^- then OH^- or HCO_3^- are released and the rhizosphere soil pH increases (see Figures 6.2 and 6.3). In legume crops that have a higher calcium, magnesium, and potassium content there is usually an excess of cations adsorbed, hence H^+ is released and soil rhizosphere pH decreases. A decrease in soil pH usually increases C_{li} for phosphorus (Chen and Barber, 1990). The degree of rhizosphere soil pH reduction by several legume crops was calculated by Li and Barber (1991) (see chapter 21).

When soils contain saturated soil solutions of $CaSO_4$, mass flow of calcium increases the calcium content of the rhizosphere soil. The increased calcium level may decrease phosphorus C_{li} where calcium phosphates are responsible for much of the phosphorus C_{si} level in the soil.

Hoffmann and Barber (1971) demonstrated the effects of calcium accumulation on phosphorus uptake by wheat seedlings. Four soils were used with and without addition of $CaSO_4$ each at high and low levels of transpiration induced by changing relative humidity of the air. Where calcium compounds were present, the accumulation of calcium reduced phosphorus uptake.

SOIL MEASUREMENTS OF BIOAVAILABLE PHOSPHORUS

The mechanistic uptake model is a research model useful for investigating the mechanisms involved. The predicted phosphorus uptake is useful for developing methods for measuring bioavailable phosphorus where arbitrary procedures are used to extract a portion of the soil phosphorus. The amount extracted is then correlated with the predicted phosphorus uptake.

Sensitivity analysis of the parameters of the model on predicted phosphorus uptake shown in Figure 9.9 indicates that changes in C_{li} were more sensitive than b or D_e. Measures of C_{li} or C_{li} plus C_{si} may be useful for a simple test of soil phosphorus availability that would be applicable over a wide range of soils. The most commonly used tests extract some portion of the labile soil phosphate pool. They may also extract nonlabile phosphate, but the amount extracted is usually correlated with labile phosphorus if the soil test is useful for predicting soil phosphorus availability. In the model describing phosphate flux from the soil into the plant root, three soil parameters are included, C_{li}, b, and D_e. The product of $C_{li} \times b$ approximates labile soil phosphate. If soil-test procedures are developed on similar soils, it is possible that C_{li}, b, and D_e are correlated; hence, using one of the three values should give a result correlated with results from the prediction equation. However, when different soils are used, the three soil factors may not be correlated, therefore any one of the values would not be correlated with predicted P uptake, and the simpler soil test would not be reliable.

The correlation of phosphorus soil tests for P with the ability of the plant to get phosphorus from the soil may also vary with plant species. This may sometimes be due to the length and density of root hairs present. For species with few or no root hairs, C_{li} may be correlated with observed phosphorus uptake, particularly where there is little competition between roots for phosphorus. For species with many long root hairs, however, C_{si} may be correlated more closely with observed phosphorus uptake because of competition between root hairs for phosphorus. Differences between soils in their effect on the extent and morphology of plant roots could also affect the ability to get close correlations between phosphorus uptake and a single measurement of the soil's phosphorus levels. Hence, development of soil tests is likely to be affected by both the nature of the roots of the test species and characteristics of soils used in the study.

Phosphate-Fertilizer Additions

Phosphate fertilizers are commonly added to the soil as granules, which go into solution before reacting with the soil. Liquid phosphate fertilizers can also be used, and their initial distribution will show less localization. Calcium or ammonium phosphates are the common chemical forms. Reactions with the soil over time and distance from the granule have been reviewed by Barrow (1980). When a monocalcium phosphate granule is added to soil, phosphate goes into solution and moves into the surrounding soil. Three zones may be recognized: A central zone contains the residue of the granule; for monocalcium phosphate, the phosphate remaining would be largely dicalcium phosphate. Around the central zone is a zone into which concentrated solution moves; this solution may dissolve calcium, iron, and aluminum from the soil. Phosphate reacts with these ions and forms precipitates of phosphate minerals. The third zone that grades out from the concentrated solution zone has been subjected to lower phosphorus concentrations, with the phosphorus probably adsorbed on soil particle surfaces. Hence, precipitation reactions based on solubility may be important in zone 2, whereas adsorption reactions may be more important in zone 3. As the soil equilibrates for an even longer period, movement will extend further into the soil. The significance of phosphate reactions in determining the supply of available phosphorus following phosphorus fertilizer additions will vary with fertilizer material, soil, moisture, temperature, and time. Additional information on phosphate fertilizers is given in Chapter 21.

REFERENCES

Al-Abbas, A. H., and S. A. Barber. 1964. A soil test for phosphorus based upon fraction of soil phosphorus. I. Correlation of soil phosphorus fractions with plant available phosphorus. *Soil Sci. Soc. Am. Proc.* **28**:218–221.

Anderson, G. 1980. Assessing organic phosphorus in soils. In F. E. Khasawneh, E. C. Sample, and E. J. Kamprath, Eds. *The Role of Phosphorus in Agriculture.* American Society of Agronomy, Madison, WI. Pp. 411–431.

Anghinoni, I., and S. A. Barber. 1980. Phosphorus application rate and distribution in the soil and phosphorus uptake by corn. *Soil Sci. Soc. Am. J.* **44**:1041–1044.

Asher, C. J., and J. F. Lonergan. 1967. Response of plants to phosphate concentrations in solution culture. I. Growth and phosphorus content. *Soil Sci.* **103**:225–233.

Baligar, V. C., and S. A. Barber. 1979. Genotypic differences of corn in ion uptake. *Agron. J.* **71**:870–873.

Barber, S. A. 1979a. Soil phosphorus after 25 years of cropping with five rates of phosphorus application. *Commun. Soil Sci. Plant Anal.* **10**:1459–1468.

Barber, S. A. 1979b. Corn residue management and soil organic matter. *Agron. J.* **71**:625–628.

Barber, S. A., and J. M. Walker, and E. H. Vasey. 1962. Principles of ion movement through the soil to the plant root. *Proc. Int. Soil Conf.*, New Zealand 121–124.

Barrow, N. J. 1980. Evaluation and utilization of residual phosphorus in soils. In F. E. Khasawneh, E. C. Sample, and E. J. Kamprath, Eds. *The Role of Phosphorus in Agriculture.* American Society of Agronomy, Madison, WI. Pp. 333–360.

Barrow, N. J., N. Malajczuk, and T. C. Shaw. 1977. A direct test of the ability of vesicular-arbuscular mycorrhiza to help plants take up fixed soil phosphate. *New Phytol.* **78**:269–276.

Boero, G., and S. Thien. 1979. Phosphatase activity and phosphorus availability in the rhizosphere of corn roots. In J. L. Harley and R. S. Russell, Eds. *The Soil-Root Interface.* Academic Press, New York. Pp. 231–242.

Brewster, J. L., A. N. Gancheva, and P. H. Nye. 1975. The determination of desorption isotherms for soil phosphate using low volumes of solution and an anion exchange resin. *J. Soil Sci.* **26**:364–377.

Caldwell, M. M., L. M. Dudley, and B. Lilieholm. 1992. Soil solution phosphate, root uptake kinetics and nutrient acquisition: Implications for a patchy environment. *Oecologia* **89**:305–309.

Cathcart, J. B. 1980. World phosphate reserves and resources. In F. E. Khasawneh, E. C. Sample, and E. J. Kamprath, Eds. *The Role of Phosphorus in Agriculture.* American Society of Agronomy, Madison, WI. Pp. 1–18.

Chang, S. C., and M. L. Jackson. 1958. Soil phosphorus fractions in some representative soils. *J. Soil Sci.* **9**:109–119.

Chen, J.-H., and S. A. Barber. 1990. Soil pH and phosphorus and potassium uptake by maize evaluated with an uptake model. *Soil Sci. Soc. Am. J.* **54**:1032–1036.

Claassen, N., and S. A. Barber. 1977. Potassium influx characteristics of corn roots and interaction with N, P, Ca, and Mg influx. *Agron. J.* **69**:860–864.

Cox, M. S., 1991. Predicting soil phosphorus levels needed for equal uptake on soils with different water levels. MSc Thesis, Purdue Univ.

Edwards, D. G., 1968. The mechanism of phosphate absorption by plant roots. *Trans. Ninth Int. Cong. Soil Sci.* Adelaide, Australia **2**:183–190.

Hoffman, W. E., and Barber, S. A. 1971. Phosphorus uptake by wheat (*Triticum aestivum*) as influenced by ion accumulation in the rhizocylinder. *Soil Sci.* **112**:256–262.

Itoh, S., and S. A. Barber. 1983. Phosphorus uptake by six plant species as related to root hairs. *Agron J.* **75**:57–461.

Jungk, A., and S. A. Barber. 1974. Phosphate uptake rate of corn roots as related to the proportion of the roots exposed to phosphate. *Agron. J.* **66**:554–557.

Jungk, A., and S. A. Barber. 1975. Plant age and the phosphorus uptake characteristics of trimmed and untrimmed corn root systems. *Plant Soil* **42**:227–239.

Jungk, A., C. J. Asher, D. G. Edwards, and D. Meyer. 1990. Influence of phosphate status on phosphate uptake kinetics of maize (*Zea mays* L.) and soybean (*Glycine max*). *Plant Soil* **124**:175–182.

Kovar, J. L., and S. A. Barber. 1982. Phosphorus supply characteristics of 33 soils as influenced by seven rates of phosphorus addition. *Soil Sci. Soc. Am. J.* **52**:160–165.

Larson, W. E., C. E. Clapp, W. H. Pierre, and Y. B. Morachan. 1972. Effects of increasing amount of organic residues on continuous corn: II. Organic carbon, nitrogen, phosphorus, and sulfur. *Agron. J.* **64**:204–208.

Leggett, J. E., R. A. Galloway, and H. G. Gauch. 1965. Calcium activation of orthophosphate absorption by barley roots. *Plant Physiol.* **40**:897–902.

Li, Y., and S. A. Barber. 1991. Calculating changes of legume rhizosphere soil pH and soil solution phosphorus from phosphorus uptake. *Commun. Soil Sci. Plant Anal.* **22**:955–973.

Nielsen, N. E., and S. A. Barber. 1978. Differences among genotypes of corn in the kinetics of phosphorus uptake. *Agron. J.* **70**:695–698.

Olsen, S. R., and F. E. Khasawneh. 1980. Use and limitations of physical-chemical criteria for assessing the status of phosphorus in soils. In F. E. Khasawneh, E. C. Sample, and E. J. Kamprath, Eds. *The Role of Phosphorus in Agriculture.* American Society of Agronomy, Madison, WI. Pp. 361–410.

Reisenauer, H. 1964. Mineral nutrients in soil solution. In P. L. Altman and D. S. Dittmer, Eds. *Environmental Biology.* Fed. Am. Soc. Exp. Biol., Bethesda, MD. Pp. 507–508.

Riley, D., and S. A. Barber. 1971. Effect of ammonium and nitrate fertilization on phosphorus uptake as related to root-induced pH changes at the root-soil interface. *Soil Sci. Soc. Am. Proc.* **35**:301–306.

Schenk, M. K., and S. A. Barber. 1979a. Phosphorus uptake by corn as affected by soil characteristics and root morphology. *Soil Sci. Soc. Am. J.* **43**:880–883.

Schenk, M. K., and S. A. Barber. 1979b. Root characteristics of corn genotypes as related to phosphorus uptake. *Agron. J.* **71**:921–924.

Silberbush, M., and S. A. Barber. 1983a. Sensitivity of simulated phosphorus uptake to parameters used by mechanistic-mathematical model. *Plant Soil* **74**:93–100.

Silberbush, M., and S. A. Barber. 1983b. Prediction of phosphorus and potassium uptake by soybeans with a mechanistic mathematical model. *Soil Sci. Soc. Am. J.* **47**:262–265.

CHAPTER **10**

Potassium

Potassium is a major component of the earth's crust. The lithosphere contains an average of 1.9% potassium (Rich, 1968). The average potassium concentration in the soil, 1.2%, is lower than the lithosphere's, due to potassium loss through weathering. Organic soils are low in potassium because of their low mineral contents, and average values may be less than 0.03% potassium. Young soils, having little weathering, have higher than average potassium contents.

SOIL POTASSIUM

Soil potassium can be placed in four categories: (1) soil solution potassium, (2) exchangeable potassium, (3) difficultly exchangeable potassium, and (4) mineral potassium. Potassium ions move from one category to another whenever equilibrium is disturbed by removing or adding potassium. However, the rates of equilibration can vary. Equilibration between solution and exchangeable potassium is rapid and usually complete within a few minutes. Equilibration between difficultly exchangeable potassium and exchangeable or solution potassium is much slower, requiring days or even months to reach equilibrium. Conversion of potassium from the mineral form is extremely slow and varies with mineral. It is frequently so slow that mineral forms have little significance in supplying plants with potassium during a single season's growth.

Soil Solution Potassium

Potassium in the soil solution, C_{li}, is usually considered the primary source of potassium absorbed by the plant root. Potassium concentrations in soil solutions vary with soil weathering, past cropping, and potassium fertilization practices. Results of two surveys of potassium concentrations in displaced soil solution are given in Table 10.1; Reisenauer (1964) found 155 values in the literature. Displaced soil solutions reported in the literature have most commonly been obtained from arid soils, however, where salinity is a problem; fewer have been obtained from leached soils. Hence, many of the soil solution potassium concentrations found by Reisenauer appear relatively high, ranging up to 100 mg/L. Barber et al. (1962) reported displaced soil solutions from 142 midwestern U.S. soils and Kovar and Barber (1990) 33 U.S. soils. Many of these soils had been collected from soils low in exchangeable potassium, since they came from sites selected for potassium experiments. In this study, 41% of the displaced soil solutions were in the concentration range of 2 to 5 mg/L. Values from Barber et al. (1962) and Kovar and Barker (1990) in Table 10.1 are about an order of magnitude less than those from Reisenauer's survey. The three surveys indicate the range of values that may be expected in soils.

TABLE 10.1   Results from Three Summaries of Potassium Concentrations in Displaced Soil Solutions[a]

| Reisenauer (1964[a]) | | | | |
Potassium concentration (mg/L)	% of samples	Potassium concentration (mg/L)	Barber et al. (1962) (% of samples)	Kovar and Barber (1990) (% of samples)
0–11	7.7	0–2	2.1	—
11–20	11.0	2–3	9.9	3.0
21–30	12.9	3–4	16.2	—
31–40	12.9	4–5	14.8	—
41–60	18.0	5–6	8.5	12.1
61–80	11.6	6–8	9.2	6.1
81–100	10.3	8–10	9.2	15.2
101–200	10.3	11–40	23.3	54.5
>200	5.2	>40	7.0	9.1

[a]Reprinted from Reisenauer (1964) by permission of Federation of American Societies for Experimental Biology.

Potassium in the soil solution is assumed to exist as K^+, since potassium forms few ion pairs or soluble organic coordination complexes. There could be potassium in some soil solutions attached to soluble organic compounds.

Exchange Potassium

Exchangeable potassium is held by the negative charge on soil clay and organic matter exchange sites, as described in Chapter 2. Potassium is one of the major exchangeable cations, along with calcium, magnesium, sodium, aluminum and hydrogen. The amount of exchangeable potassium in soils usually ranges from 40 to 500 mg/kg. A value of 150 mg/kg is frequently considered high enough to ensure that plants growing on the soil receive sufficient potassium so that plant growth is not reduced. The strength of the bond of exchangeable potassium to the soil varies with type of exchange site and the nature of other cations present. As described in Chapter 2, there are both permanent-charge and pH-dependent charge cation-exchange sites. Potassium is usually the dominant monovolent exchangeable cation (except for hydrogen on acid soils); however, there are normally much larger amounts of exchangeable calcium and magnesium ions present. When many of the exchange sites are filled with calcium and magnesium, potassium moves to the more weakly bonded sites. Exchangeable potassium equilibrates within one hour with solution potassium in most soils. In some soils, equilibrium is almost instantaneous, while as long as 24 h may be required for some vermiculitic clays to reach equilibrium (Sparks et al., 1980). Amounts of exchangeable potassium are usually measured by displacement with 1 mol/L NH_4OAc, pH 7.

Nonexchangeable Potassium

Nonexchangeable potassium is held between clay plates in positions that are not readily accessible for exchange with solution cations. When exchangeable and solution potassium are removed from the soil by plant uptake, potassium that was initially nonexchangeable moves into the exchangeable and solution forms. Conversely, when potassium fertilizer is added to a soil, potassium moves from solution and exchangeable to nonexchangeable forms. This potassium may not all move back into exchangeable and solution positions as potassium is removed by cropping; some potassium may remain nonavailable for crop uptake. The potassium balance between different forms, and its rate of equilibration, vary with soil.

The amount of nonexchangeable potassium can be measured by exhaustive cropping of the soil, with many experiments conducted in greenhouse pots. Barber and Humbert (1963) reviewed experiments reported in the literature and found amounts of nonexchangeable potassium removed by crops that varied from 0 to 750 mg/kg; most of the soils released less than 200 mg/kg. Alluvial soils tended to release the largest amounts of potassium from nonexchangeable positions. Binnie and Barber (1964) compared the release of potassium from alluvial soils with potassium release from soils collected from the watershed that presumably was the origin of the alluvial soils. Comparisons were made between the soils at the same exchangeable potassium levels by selecting soils with a range of exchangeable potassium values. The soils were exhaustively cropped using Neubauer techniques. The relation between the decrease in exchangeable potassium and potassium removed by cropping is shown for both sets of soils in Figure 10.1. There was a much greater release of nonexchangeable potassium from the alluvial soils. Binnie and Barber (1964) speculated that this was due to extra potassium added to these soils from flood waters over hundreds of years. This potassium was assumed to be held less tightly than nonexchangeable potassium in nonalluvial soils.

Chemical estimates of difficultly available potassium (nonexchangeable potassium that may be released with exhaustive cropping) have also been made by boiling the soil with 1 mol/L nitric acid. This treatment decomposes most of the clay-sized particles releasing their potassium. The amount released has usually been of the same magnitude as that released by repetitive cropping until little additional release occurs.

Talibudeen et al. (1978) studied the kinetics of potassium release from soils sampled from the Rothamstead classical experiments by shaking the soil with cation-exchange resin that reduced solution potassium concentrations to approximately 1 μmol/L. They found three simultaneous rate

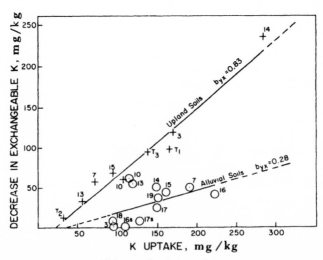

FIGURE 10.1 Relation between potassium uptake by rye (*Secale cereale* L.) seedlings and reduction in exchangeable potassium in alluvial and upland soils. Reproduced from Binnie and Barber (1964) by permission of Soil Science Society of America.

processes: The initial process was complete within 24 h, the second type of release ceased after 35 days; while the third, the release from the soil matrix itself, continued at a rate that would indicate a diffusion-controlled process involving a diffusion coefficient of 3×10^{-19} cm^2/s. This coefficient is of the same magnitude reported for potassium diffusion out of illitic clay minerals. The latter two processes both involve releasing nonexchangeable potassium. Plant roots can also reduce the soil solution potassium levels to approximately 1 to 2 μmol/L.

Jungk et al. (1982) found that potassium released from the soil near the root was about twice the initial exchangeable level when plant roots reduced C_{li} for potassium to 2 to 3 μmol/L. Release of nonexchangeable potassium was restricted to the zone within the radius of root hair length. If a soil has a root density of 4 cm/cm^3 and root hairs 1 mm long, the volume in the root hair zone will be about 2.5% of total soil volume. Beyond the root hair zone, the level of solution potassium will probably be too large for significant release.

Binnie and Barber (1964) found that in alluvial soils there was uptake of potassium in addition to that measured as C_{si} (Figure 10.1). There measurement was made with high root density in the soil, hence the release occurred in the rhizosphere zone where C_{l0} would be low. A review by Barber (1963) found that of 30 investigations with successive cropping of 386 soils only 14 released more than 300 mg potassium/kg of soil and of these eight were alluvial or lacustrine.

Mineral Potassium

Soils commonly contain potassium-bearing minerals. The main minerals found in soils are given in Table 10.2. Those present in a soil depend primarily on the source of the parent material. Minerals release potassium slowly, and the rate varies with the mineral. Studies of the relative rate of

TABLE 10.2   Main Potassium-Bearing Minerals in Soils

Mineral	Formula	K content (%)
Micas		
Biotite	$K_2Al_2Si_6(Fe^{2+},Mg)_6O_{20}(OH)_4$	8.7
Muscovite	$K_2Al_2Si_6Al_4O_{20}(OH_4)_4$	9.8
Feldspars		
Orthoclose	$KAlSi_3O_8$	13.7
Microcline	$KAlSi_3O_8$	13.8
Clays		
Illite		~7

release from similar-sized particles show that the order of release is biotite > muscovite > orthoclase > microcline (Rich, 1968). Biotite is seldom present in soil, because it has a high rate of release and hence is removed rapidly by weathering.

Potassium in micas occurs between the silicate layers as described in Chapter 2. Removing potassium without clay decomposition requires separating the clay sheet so the potassium can be exchanged by other cations; this release occurs along the edges of the mica. Potassium can also be released when Fe^{2+} is oxidized to Fe^{3+}, so that the additional positive charge of iron in the mica structure balances the charge previously balanced by potassium. With prolonged weathering and removing the potassium ions binding the mica sheets together, mica will become vermiculite or montmorillonite. It must be emphasized that this is an extremely slow process under most conditions. Feldspars, orthoclase, and microcline are three-dimensional minerals, so that there are no cleavage planes for weathering. Hence, release of potassium from these forms is usually slower and arises primarily from weathering of surfaces and decomposition of the mineral.

POTASSIUM ADSORPTION BY SOILS

Potassium is the largest in size of the mineral cations required by plants; it has a dehydrated radius of 0.133 nm. The number of oxygen that can coordinate around it is high, 8 to 14. As a result, the potassium-oxygen bond is relatively weak. Potassium has a lower hydration energy than sodium, calcium, or magnesium, so it is not highly hydrated in mineral structures. Table 10.3 compares the relative dissociation of exchangeable potassium, calcium, and magnesium from three widely different soils. The degree of dissociation of the three cations varied greatly between soils. Potassium was held less

TABLE 10.3   Dissociation of Calcium, Magnesium, and Potassium from Three Soils

Soil	Exchangeable			Dissociation		
	Ca^{2+}	Mg^{2+}	K^+	Ca^{2+}	Mg^{2+}	K^+
	[cmol(p^+)/kg]			($C_h/C_u \times 100$)		
Toronto silt loam (Udollic Ochraqualf)	4.9	3.9	0.36	1.2	4.0	7.6
Raredon silt loam (Aquic Hapludult)	6.3	5.3	0.42	0.33	0.71	1.35
Edwards muck (Limnic Medesaprists)	41.0	25.0	3.3	4.5	10.2	14.2

Source: Baligar and Barber (unpublished data)

tightly than divalent calcium or magnesium, and magnesium was held less tightly than calcium. The degree of dissociation was calculated by assuming a linear relation between exchangeable and solution ions, as concentration changed from zero to the level measured in the soil.

Adsorption curves have been used to study the adsorption of potassium by soils and clays. Soil samples are commonly shaken with 0.01 mol/L $CaCl_2$ solutions containing graded amounts of potassium at 10:1 solution to soil ratio. After equilibrating these systems for 24 hs, potassium is measured in the solution phase, and the adsorbed potassium is calculated. The solution potassium level is then plotted versus the adsorbed potassium level; an example of such a plot is shown in Figure 2.6. The slope of the linear section depends on potassium-adsorption characteristics of the soil. In general, soils having a higher cation-exchange capacity will have a steeper slope, since there are more exchange sites for adsorption. The relation cannot be solely related to cation exchange-capacity, however, because for differing cation-exchange sites, the slope is also affected by the strength of bonding of potassium to the exchange material. This in turn can be affected by the strength of bonding of the companion cations.

Potassium Adsorption into Nonexchangeable Positions

When potassium is added to soil, some of it goes into exchangeable positions and some into nonexchangeable positions. The potassium may or may not move readily back into solution from nonexchangeable positions as soil potassium is removed by plant uptake. Movement of potassium into nonexchangeable positions has often been termed fixation. Fixation of potassium frequently occurs in soils containing weathered micas and vermiculites; the extent of fixation is greater the higher the charge density of the clay involved (Rich, 1968). Measuring fixation is accomplished by measuring the added potassium that is not exchangeable by ammonium.

The mechanism of potassium fixation is usually assumed to be fitting nonhydrated potassium into the ditrigonal holes on the surface of the silicate sheets of clay minerals. When the charge density of the clay is high [above 150 cmol(p$^+$)/kg], potassium ions can be fixed in ditrigonal holes of adjacent clay plates. The high charge leads to enough potassium fixation to stabilize the structure; which, in turn, reduces the subsequent rate of release of potassium from the structure. Rich (1968) indicated that fixation is most common for dioctahedral vermiculite. Potassium fixation is greater in soils that have not formerly been acid and do not contain iron and aluminum oxides precipitated on their surfaces. Oxide precipitates can prop the clay plates apart, so that they cannot collapse about potassium and trap it in the ditrigonal holes (Figure 10.2).

Potassium fixation varies with soil type because of the nature of the clays present. Soils whose clays are primarily kaolinitic do not fix potassium. Horton (1959) compared moist fixation of potassium by several Indiana

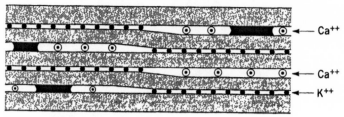

K not exchangeable to large cations

Hydroxy – Al (or Fe) "islands"

FIGURE 10.2 Diagrammatic illustration of locations in illitic-clay minerals where potassium is not exchangeable (layers collapsed) and islands of hydroxy Al (or Fe) prop layers apart so that potassium is exchangeable. Reproduced from Rich (1968) by permission of American Society of Agronomy, Crop Science Society of America, and Soil Science Society of America.

soils. Potassium was added to samples taken at five different locations for each of six soil types. The soil was incubated moist at room temperature, was sampled at intervals over a 6-month period and exchangeable potassium measured. The amount of potassium that became nonexchangeable increased with time. There was a difference between soil types that was related to clay mineralogy. The amount of fixation was correlated with the amount of clay that collapsed to 1 nm when treated with KCl. Fixation was greatest on Brookston silt loam, a Typic Argiaquoll. This soil type had very few iron and aluminum oxides between the clay layers; hence, the clay plates could collapse around the added potassium.

The rate of potassium fixation of potassium by four of these soils studied by Horton (1959) is illustrated in Figure 10.3; data are means for five different samples of each soil. The amount of fixation varied with soil. When fixation is large, efficiency of added potassium fertilizer is reduced. In this experiment, fixation was a major factor in reducing exchangeable potassium levels after adding potassium to the Brookston silt loam.

Potassium Release on Drying

Some soils release nonexchangeable potassium to exchangeable positions when air-dried or oven-dried at 105°C. This release seldom occurs in nature; it is an artifact occurring when sampled soils are dried prior to analysis. Subsoil samples frequently release large amounts of potassium on drying, so resultant values for exchangeable potassium measured on an air-dried sample may be three to four times that measured on a moist, undried sample. Surface soils subjected to potassium fertilization for many years usually show less change on drying. Rich (1968) reported that for soils with high exchangeable potassium levels, potassium fixation may occur on drying, while release may occur for soils having low exchangeable potassium soil levels. Several theories for the release have been suggested. One is that the

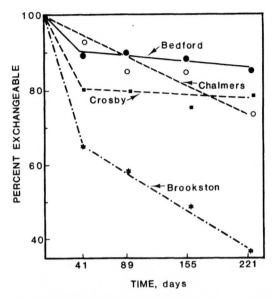

FIGURE   10.3   The change in exchangeable-potassium level with time after applying 400 mg/kg of potassium as KCl to four soils. Values are the average for five different samples of each soil type (Horton, 1959).

clay plates warp on drying, causing more ready access of ions and water at clay plate edges and easier exchange between solution cations and potassium held between the clay layers.

In illite, potassium is fixed between the clay layers; which expand at the edges and allow H_3O^+, Ca^{2+}, or Mg^{2+} to replace potassium and expand the lattice. When all of the potassium has been replaced, we may have a vermiculite like clay. However, adding potassium may change the clay into a more illite like structure. Aluminum and iron oxide islands between the clay layers can prevent the clay from collapsing and fixing the added potassium (see Figure 10.2).

POTASSIUM-UPTAKE KINETICS

Potassium was one of the first ions used in studying ion-absorption mechanisms by plant roots, so many experiments have been conducted on potassium uptake. Uptake, as influenced by concentration of potassium in solution, has usually been described by Michaelis–Menten kinetics: this approach was outlined in Chapter 3. Such kinetics describe the curves of solution potassium concentration versus potassium influx. Epstein et al. (1963) observed separate influx curves for low and high ranges of solution potassium concentration. They labeled them as due to uptake by mechanisms I and

II; their results are illustrated in Figure 10.4. Mechanism I operated over the concentration range up to 0.5 mmol/L; the value of K_m was 18 µmol/L. This range includes potassium concentrations found in most soil solutions. Mechanism II operated in the range up to 50 mmol/L; the value of K_m was 16 mmol/L. The maximum influx for mechanism II was approximately double that for mechanism I during 30-min uptake by excised barley roots. The concentration range where mechanism II operates would occur only near bands of recently applied potassium fertilizer.

Uptake by intact plant roots is affected by both absorption and translocation of potassium within the plant. Claassen and Barber (1974) studied potassium influx by intact 18-day-old corn plants. They measured potassium uptake by the solution–depletion procedure and obtained a K_m value of 27.9 µmol/L and an I_{max} (in terms of root length) of 40 nmol/m · s, measured between 0 and 0.20 mmol/L. This was in the concentration range where only mechanism I was operating. The authors also found that there was still potassium in solution when net influx was zero. Hence, in order to accurately describe the curve of potassium influx versus potassium concentration, it was necessary to include an efflux, E, of 0.55 µmol/g_{FW} · h. In later studies, Claassen and Barber (1976) found a K_m of 16.7 µmol/L for 17-day-old corn and an E value of 0.07 I_{max}. Instead of using efflux as one of the three para-

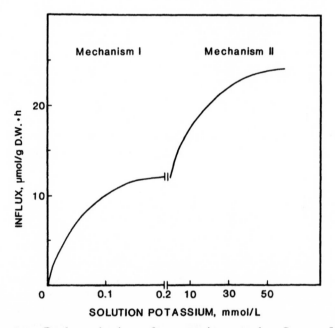

FIGURE 10.4 Dual mechanisms for potassium uptake. System I. Uptake between 0 and 500 µmol/L. System II. Uptake for the range 0–50,000 µmol/L. Reprinted from Epstein et al. (1963) by permission of the authors.

meters describing influx versus concentration, the solution potassium concentration below which potassium influx ceases, C_{min}, can also be used. Values of 1 to 2 μmol/L are common C_{min} values for potassium uptake by young corn plants. An illustration of the relation between influx and concentration, as described by I_{max}, K_m, and C_{min} in the Michaelis–Menten equation was shown in Figure 3.6.

Potassium influx versus solution potassium has also been studied in flowing culture experiments where a range of potassium levels is maintained. Asher and Ozanne (1967) grew 14 crop and pasture species for 3 to 6 weeks in flowing solutions ranging from 1.0 μmol/L to 1.0 mmol/L potassium. All species were healthy at 24 μmol/L, with eight reaching maximum yield; the remaining six reached maximum yield at 95 μmol/L. Increasing concentrations above 95 μmol/L to 1.0 mmol/L had little effect on growth and gave only a small increase in plant potassium concentrations. These results are in agreement with potassium influx measurements made during short-term experiments.

Hassan and van Hai (1976) measured potassium influx by intact citrus seedlings using a continuous-flow technique. They observed a K_m value of 3 ± 1.7 μmol/L for 120-day-old seedlings, with a transition to phase II uptake at 26 μmol/L. These K_m values are much lower than those observed by others.

Wild et al. (1979) studied potassium uptake by ryegrass (*Lolium perenne*) and radish (*Raphanus sativus*). Their lowest constant potassium concentration in solution was 1.2 μmol/L. Influx at this concentration was always at least 70% of that at higher potassium concentrations, indicating that these species have very low K_m and C_{min} values for potassium uptake. This may explain why ryegrass is very efficient in absorbing potassium from the soil. Growing plants in solutions of low potassium concentrations for long periods may also allow plants to adapt to low potassium concentrations.

Factors Affecting Potassium Influx

Plant Age and Potassium Influx

Mengel and Barber (1974) measured the mean potassium influx by roots of corn plants growing in the field. Potassium uptake decreased from 52.9 μmol/m/day at 20 days of age to 0.16 at 100 days. Although the uptake rate per unit of root decreased, root growth increased more rapidly, so uptake rate per plant increased to 60 days. Barber (1978) measured potassium uptakes by soybean roots and found that they also decreased with plant age from 10.2 μmol/m/day ($\sim$ 120 nmol/m^2 $\cdot$ s) at 13 to 24 days to 0.65 at 85 to 106 days. The rate was lower than for corn at 20 days but more rapid at 100 days. Soybeans have fewer roots than corn, so that a higher mean rate of potassium uptake is required during growth in order to supply the potassium requirements of the plant. The decrease in rate with increased plant age was much less for soybeans than for corn.

Plant Potassium Status and Potassium Influx

Potassium influx by plant roots is influenced by the potassium status of the plant. Claassen and Barber (1977) used several procedures to vary potassium levels in corn shoots; they included growing plants at several levels of potassium in the soil solution, using split-root systems where only a part of the root system received potassium, and growing plants for portions of the growth period in solutions devoid of potassium. After these varying pretreatments, the authors found the relation between potassium concentration in the corn shoot and subsequent potassium influx shown in Figure 3.8. Potassium influx doubled when shoot potassium was reduced to a low level. Wild et al. (1979) obtained similar results with ryegrass and radish when they compared influx with plants grown at 1.2 μmol/L potassium to plants grown at 50 μmol/L. With the species they used, influx was increased sixfold. Apparently there is a difference among species in the degree to which starving plants for potassium increases the potassium influx when potassium is subsequently supplied.

Effect of Temperature on Potassium Influx

Ching and Barber (1979) investigated the effect of temperature on potassium influx by corn plants; the values they obtained for two temperatures and two plant ages are given in Table 10.4. Shoot temperature was held constant at 29 and 15°C. I_{max} at the higher root temperature was double that at the low root

TABLE 10.4   Effect of Potassium Addition and Soil Temperature on Parameters Used in a Simulation Model to Predict Potassium Uptake by Corn Growing in Raub Silt Loam

| Parameters | K applied (mg/kg of soil) | | | |
| | 0 | | 500 | |
	15°C	29°C	15°C	29°C
C_{li}, mmol/L	0.046	0.090	6.26	8.10
b	39	23	1.2	1.2
$D_e \times 10^7$, cm^2/s	0.15	0.39	5.0	7.5
I_{max}, nmol/m $\cdot$ s	56	112	56	112
K_m, μmol/L	14	28	14	28
C_{min}, μmol/L	1.6	0.89	1.6	0.89
L_0, cm	3070	3070	2260	2260
k, 10^6s	1.29	2.38	1.46	2.49
r_0, cm	0.021	0.018	0.026	0.023
v_0, $\times 10^6$ cm^3/cm$^2 \cdot$ s	0.70	1.2	0.76	1.3

Source: Reproduced from Ching and Barber (1979) by permission of American Society of Agronomy.

temperature; K_m also doubled, but C_{min}, the potassium concentration in solution where influx ceased, was less at the higher temperature. The effect of temperature will depend on plant species and the temperature range investigated, since species vary in their temperature adaptation.

Effect of Supplying Potassium to Part of the Root System on Potassium Uptake per Plant

The significance of shoot demand in determining potassium influx can be evaluated by varying the proportion of the root system supplied with potassium. Claassen and Barber (1977) compared potassium uptake by corn plants having different portions of the root system supplied with potassium. Their results are shown in Table 10.5. The plants were grown in split-root systems in solution culture for 7 to 17 days of age. Using a split-root system increased potassium influx in proportion to the reduction of potassium concentration in the shoot, as was shown in Figure 3.8. Potassium-absorption rate for the root system was determined to a greater extent by the root uptake mechanisms involved rather than the demand of the plant for potassium. Shoot potassium levels decreased as a smaller fraction of the roots was supplied with potassium.

Potassium Influx as Affected by Root Potassium Concentration

We might assume there would be a relation between the potassium status of the root and potassium influx. Claassen and Barber (1977) found that this was not the case. Their data, in Table 10.6, show that I_{max} was only marginally greater in roots that had not been supplied with potassium, even though

TABLE 10.5   Average Influx Characteristics of the Proportion of 17-Day-Old Corn Plant Roots Supplied with Potassium as Influenced by the Proportion Supplied with Potassium

Proportion of roots supplied with K (%)	Shoot K concentration	K_m (μmol/L)	$I_{max_{F\,W}}$ (μmol/g $_{F.W.}$ · s × 10^4)	E
100[a]	7.64	17	16	2
75	6.24	17	26	2
50	5.89	18	27	2
25	3.84	17	39	3
2–15	3.26	10	38	4

Source: Reproduced from Claassen and Barber (1977) by permission of American Society of Agronomy.
[a]Values for 100 and 50 are the average of four experiments; for 75 and 25, three experiments; and 2 or 15, two experiments.

TABLE 10.6   Relation of Potassium Level of Corn Roots versus Level in the Shoot on Potassium Influx

Proportion of root (%)	Presence of K	K (%) Shoots	K (%) Roots	$I_{max_{FW}}$[b]	K_m (μmol/L)
100	+	8.00	4.69(5.85)[a]	15.8	9.6
50	+	6.45	3.90(5.55)	28.0	13.4
50	−		2.23(1.88)	30.8	5.7
25	+	4.35	3.00(4.99)	33.8	14.2
75	−		1.64(1.24)	39.8	7.7
15	+	4.13	3.90(5.51)	36.8	7.0
85	−		1.55(1.20)	38.0	8.2

Source: Reproduced from Claassen and Barber (1977) by permission of American Society of Agronomy.
[a]I_{max} in μmol/g$_{FW}$ · s × 10^4.
[b]Values after running depletion curve. Values in parentheses are for roots treated similarly but not having a K-depletion curve run on them.

untreated roots had potassium concentrations of 1.20 to 1.88% as compared to 4.99 to 5.85% for the potassium supplied roots. Differences in potassium level in the shoot had a much larger effect on I_{max} for potassium uptake. Hence, shoot potassium levels for corn appeared to have more effect on I_{max} for potassium uptake than did root potassium levels. Potassium influx appeared to be regulated by a feedback mechanism between the root and shoot.

Potassium Influx as Related to Light

Light is necessary for plant leaves to convert carbon dioxide and water into photosynthate; this is required, in turn, for root respiration, in order to provide energy for active potassium uptake. After lights are turned off, carbohydrates in the plant are still available for several hours. Claassen and Barber (1974) measured the effect of day versus night on the potassium uptake rate of four 20-day-old corn plants grown in solution culture. Potassium was added continuously to a pot containing 3 L of solution at 54 μmol/h. When plant uptake equaled 54 μmol/h, the concentration in solution remained constant. When uptake rate decreased, solution potassium concentration increased; results are shown in Figure 10.5. When the lights were turned off, the uptake rate was decreased to about 35 μmol K/h. When the lights came on, uptake rapidly increased until the solution concentration returned to its previous level. Results suggest that the effect of transpiration in translocating potassium was the primary reason for the difference in rates of potassium uptake in dark and light periods, since transpiration resumed when the lights were turned on.

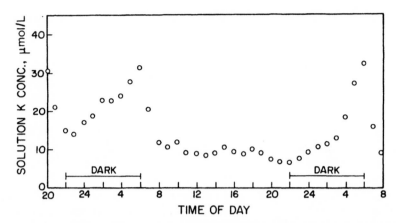

FIGURE   10.5   Relationship of solution potassium to time of day when potassium was added at a constant rate of 54 µmol/h to a 3-L pot containing 10 20-day-old corn plants growing in a growth chamber during day and night cycles. Reprinted from Claassen and Barber (1974) by permission of American Society of Plant Physiologists, copyright 1969.

Differences between Genotypes in Potassium Influx

Baligar and Barber (1979) compared potassium influx in six Florida corn genotypes with six genotypes used in Indiana. They found no significant difference in I_{max}, though K_m and C_{min} for the Florida genotypes were both significantly lower than for the Indiana genotypes. Florida genotypes had more root length per plant, and because of this, average potassium influx during growth was less than for the Indiana genotypes, indicating that differences in the physiology of potassium uptake exist even between genotypes of the same species. Because of this variation, it should be possible to develop genotypes having the desired influx characteristics needed to maximize yield and increase the fertilizer potassium use efficiency.

Root Morphology and Potassium Influx

Peterson and Barber (1981) studied the effect of root morphology on potassium influx. They changed the morphology of soybean roots by growing plants in sand instead of growing them in nutrient solution. Roots grown in sand were larger in radius—0.22 versus 0.17 mm—than those grown in solution (Figure 6.7). Increase in radius was due to an increase in the radial size of the cortical cells; the size of the stele was not affected, nor were the number of cortical cells in the root cross section. The larger radius was due to a shortening and broadening of the cortical cells. I_{max} for potassium influx was 55% greater for sand-grown roots than for their solution-grown counterparts; K_m was not affected. The greater influx per unit of root surface was probably due to the greater area of plasmalemma surface per unit of root surface that occurred with the sand-grown roots of larger radius.

Significance of Growing Plants in Soil vs. Solution Culture on Potassium Influx

Many of the preceding factors were evaluated for plants that were grown in solution culture. When plants are grown in soil rather than solution, the rate of potassium supply to the root by mass flow and diffusion is a factor in rate of nutrient uptake. The sensitivity analysis of predicted potassium uptake obtained by individual variation of each parameter shown in Figure 5.5 shows that predicted uptake is much more sensitive to increasing C_{li}, b, and D_e than to increasing I_{max}. A sensitivity diagram of predicted phosphorus uptake for the interaction of C_{li} and I_{max} is shown in Figure 5.10; a similar diagram can be obtained for potassium. The significance of increasing influx kinetics is important with high levels of soil potassium but has little effect at low soil potassium levels. Hence if factors are used to increase I_{max} in order to correct potassium deficiency for plants growing in soils low in C_{li}, they are not likely to be successful.

MODELING POTASSIUM UPTAKE

A mechanistic mathematical uptake model accurately describing potassium uptake was given in Chapter 5. The Barber–Cushman model used 11 parameters. Three, C_{li}, b, and D_e, describe potassium supply to the root surface. Four, L_0, k, r_0, and r_1, describe root geometry and growth rate. Four, v_0, I_{max}, K_m, and C_{min}, describe water and potassium uptake kinetics. In an experiment by Shaw et al. (1983) four 6-day-old seedlings were transplanted into 3-L pots of four soils and grown for 14 days. The four soils differed in cation-exchange capacity from 119 to 247 mmol(p^+)/kg. Four replicates were used. The root density at harvest varied from 2.67 to 3.34 cm/cm³. The predicted vs. observed potassium uptake values for the four replicates of four soils are shown in Figure 10.6. The relation $y = 1.11x + 0.37$, $r^2 = 0.93$ indicates relatively close agreement between observed and predicted values; while the C_{si} values were similar for all four soils, the CEC value was not related to potassium uptake.

In the potassium uptake experiment of Li and Barber (1991) reported in Chapter 5, four plant species were grown on one soil with soil pH adjusted to levels of 5.6, 7.3, 7.6, and 8.3. The relation of predicted potassium uptake to observed potassium uptake shown in Figure 5.1 had an r^2 of 0.99 and a regression coefficient of 0.99 indicating close agreement using the Barber–Cushman model.

Silberbush and Barber (1984) investigated potassium uptake by soybeans growing in the field. Five cultivars of soybeans were grown with four repli-

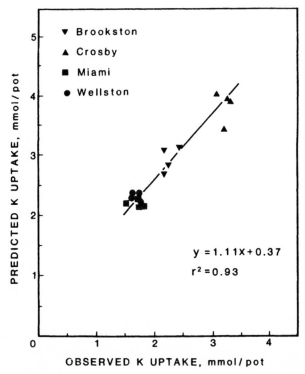

FIGURE   10.6   Relation between potassium uptake calculated by the Barber-Cushman model and observed-potassium uptake for four soils varying in cation-exchange capacity. Reproduced from Shaw et al. (1983) by permission of Marcel Dekkar Inc.

cations on Raub silt loam. The soil varied in potassium with depth, so that the model's soil parameters were evaluated for depths of 0 to 15, 15 to 30, and 30 to 76 cm. An impervious layer at 76 cm restricted deeper root growth. Soybeans were sampled at the R_4 to R_6 maturity stage, depending on the maturity group of the cultivar, since all cultivars were sampled on the same day. The relation between observed and predicted potassium uptake is shown in Figure 10.7. There was close agreement, indicating that the model satisfactorily described potassium uptake for soybeans growing in the field. In a similar experiment with Chalmers silt loam (Figure 5.4) predicted potassium uptake agreed more closely with observed uptake than obtained for Raub silt loam (Figure 10.7) due to less experimental variation.

SENSITIVITY ANALYSIS FOR POTASSIUM UPTAKE

A sensitivity analysis of the effect of parameters used in the Barber–Cushman simulation model on predicted potassium uptake was con-

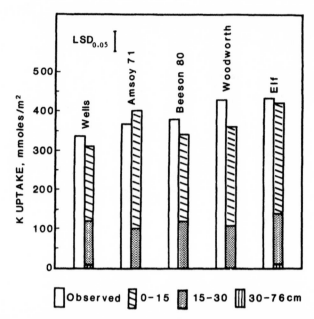

FIGURE 10.7 Comparison of predicted with observed potassium uptake by five soybean cultivars grown in the field on Raub silt loam. Predicted-potassium uptake is shown for each soil depth. Silberbush and Barber, unpublished data.

ducted by Silberbush and Barber (1983). This analysis was discussed in Chapter 5 where the model was introduced, so only a summary of the results are given here. Each parameter of the model was varied from 0.5 to 2.0 times the initial value, with predicted potassium uptake calculated while holding remaining parameters at the initial values given in Table 5.2. Results were shown in Figure 5.5. Predicted potassium uptake was most sensitive to k and r_0, which affect root surface area. Since r_1 remained constant, the amount of soil increased with an increase in k. The next three most sensitive parameters were the soil supply parameters C_{li}, b, and D_e. Potassium uptake approached a maximum as each of these parameters was increased, because they increase C_{l0}. As C_{l0} increases In increases to I_{max}.

Decreasing r_1 greatly decreased uptake, because it reduced the amount of soil supplying potassium and increased root-to-root competition. Increasing r_1 increased predicted potassium uptake asymptotically, because the extent of the zone supplying potassium to the root increased as r_1 increased and root-to-root competition decreased concurrently. Changing v_0 had little effect, because most of the potassium was reaching the root by diffusion in this situation. The influx parameters I_{max}, K_m, and C_{min} had smaller effects, because influx was limited more by soil supply than by the plant's ability to absorb potassium.

When the level of one parameter is changed, it can often affect the level of other parameters. Sensitivity analysis was also conducted where interdependent parameters were changed according to their interdependence. Three interdependent parameters of root morphology are k, r_0, and r_1; sensitivity analysis of these three variables was shown in Figure 5.6. It showed that where root volume and soil volume were held constant, predicted potassium uptake still increased as k increased, because the roots were longer.

Soil parameters C_{li}, b, and D_e are interdependent, though the degree of their interdependence depends on the soil. If we take one soil, hold C_{si}, the level of diffusible potassium, constant, and then increase C_{li}, b (which is $\Delta C_s/\Delta C_l$) will decrease, and D_e (from the relation $D_e = D_l\theta f_l/b$) will increase. A sensitivity analysis where these three soil variables were varied was shown in Figure 5.7. In one case, b and D_e were held constant as C_{li} varied and in the other, b and D_e were varied so that C_{si} remained constant. There was not a large difference between the two curves obtained, although there might have been more if there had been more root competition for soil potassium.

SOIL MEASUREMENTS OF BIOAVAILABLE POTASSIUM

Soil potassium bioavailability is often measured by growing plants in soils, measuring the amount of potassium they absorb, and using this value to characterize potassium availability. Such measures of potassium uptake are usually used as the dependent variable in correlations with empirical laboratory measurements of amounts of potassium extracted from the soils. In this presentation, it is assumed that the primary parameters determining potassium uptake by a soil-grown plant are those used in the mathematical model described in Chapter 5. The mathematical framework is a mechanistic approach to determining soil potassium bioavailability. When plant parameters are held constant, the soil parameters C_{li}, b, and D_e interact to determine the rate of potassium supply to the root. In this section, I will discuss proposed empirical methods for determining the rate of soil potassium supply to plant roots. These methods are evaluated in terms of the rate of supply described by the mathematical model.

Exchangeable Potassium

Exchangeable potassium measures the amount of soil potassium that is in a form immediately available for movement into the soil solution as the plant root absorbs potassium. Because of the usual measurement procedure, exchangeable potassium also includes that in soil solution; however, the amount in solution is usually small compared to truly exchangeable potassium. In systems where C_{li}, b, and D_e are closely correlated with each other, as would happen if different levels of C_{si} existed in a single soil, values of C_{si}

would correlate closely with observed potassium uptake by the plant. However, when soils of widely different types and amounts of cation-exchange sites are used to evaluate the correlation between potassium uptake and C_{si}, agreement may be poor. In this case, C_{li}, $1/b$, and D_e may not be closely correlated with one another as the level of C_{si} changes.

A large factor in determining whether C_{si} is correlated with potassium uptake is root density in the soil, reflecting subsequent competition between roots for potassium. The degree of competition depends on both root density and the size of the diffusion coefficient, which reflects the distance potassium can diffuse through the soil to the root. Root density in the field depends on the crop grown. For corn, a root density L_v, of 3 to 4 cm/cm³ may occur in the topsoil (Mengel and Barber, 1974). For soybeans, L_v may reach only 1.0 cm/cm³ in the topsoil, while grasses may have L_v above 20 cm/cm³. Hence, roots of grass crops will compete for potassium sooner than soybean roots. The mean half-distance between root axes is $1/(\pi L_v)^{1/2}$.

The D_e for potassium in the soil is usually between 1×10^{-7} and 1×10^{-8} cm²/s. Using the model parameter values in Table 5.2 and keeping all constant except r_1, which was varied, the uptake model was used to calculate C_l/C_{li} in the soil perpendicular to the root surface; the results are shown in Figure 10.8. Values of r_1 were 0.14, 0.28, and 0.56 cm, which correspond to L_v values of 16, 4, and 1 cm/cm³. As root density in the soil increases, the value of C_l/C_{li} at the half-distance decreases, which indicates competition between adjacent roots. Hence, for plants with high root density and for which diffusion is important in supplying potassium to the root, root competition for potassium will occur. The greater the root competition, the closer the relation between C_{si} for potassium in the soil and potassium uptake by the plant, because the quantity of potassium present becomes more important than the rate of movement to the root. The plant takes up a large portion of the exchangeable potassium present under these conditions.

Frequently potassium availability in the soil is evaluated by growing plants in pots of soil in the greenhouse or plant-growth chamber. Plants may be grown long enough that root density in the soil is high. As a result, a high correlation usually occurs between exchangeable potassium and potassium uptake by the plant. In addition nonexchangeable potassium may be absorbed by plants grown on some soils as shown in Figure 10.1.

When evaluations are made in the field, however, the rooting densities may be much lower. In this case, results can differ because of less root competition for potassium. The crop used for evaluation will also make a difference. A crop with low root density may give a different result from a crop with high root density; typical results are illustrated in Table 5.3. In this table, the mathematical model is used to predict potassium uptake as affected by root density. In this case, root length remained constant, so that there was more soil present with the low root density of 1.6 cm/cm³ than the high root density of 45 cm/cm³. At high root density, changes in C_{li}, b, and D_e had

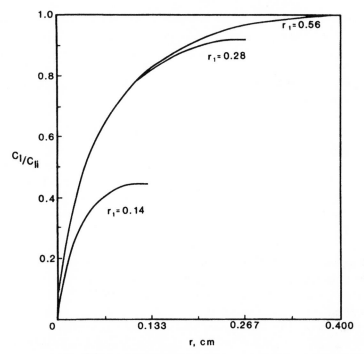

FIGURE   10.8   Calculated C_1:C_{li} distribution perpendicular to the root using L_v values of 1, 4, and 16 cm/cm³ (r_1 = 0.56, 0.28, and 0.14 cm). All parameters other than r_1 were those given in Table 5.2; time was 5 days.

little effect on predicted potassium uptake if C_{si} remained constant. Values as high as 45 cm/cm³ are likely to occur in greenhouse pots.

Potassium Potential

The potassium potential ΔG has been described as the change in free energy from the exchange of potassium for calcium and magnesium in the soil.

$$\Delta G = - nRT \ln \frac{K}{(Ca + Mg)^{1/2}} \qquad (10.1)$$

where n is valence, R is the gas constant, and T is absolute temperature. This relation suggests that there is an effect of the activity of calcium and magnesium on potassium uptake by plant roots. No evidence exists for interaction between the level of calcium and magnesium in solution and the amount of potassium uptake. However, measurements of soil potassium potential are usually made at a uniform level of solution calcium and magnesium, so potassium is measured under relatively uniform conditions of ionic strength. The measurement is usually made using 0.01 mol/L of $CaCl_2$ in equilibrium with soil at a ratio of 10 mL of solution to 1 g of soil. Hence, the measurement represents partial displacement of potassium by $CaCl_2$. This partial displacement is essentially a measure of soil solution potassium C_{li} at a uniform salt concentration. Relative bonding strengths of different soils for potassium compared to calcium and magnesium can be described in terms of the energy of exchange.

One reason for not using the ratio of calcium to potassium in soil to evaluate potassium uptake is that calcium primarily moves to the root by mass flow and potassium primarily by diffusion. The ratio of these two cations in equilibrium soil solution is thus entirely different from the ratio that will occur at the surface of an actively absorbing root after mass flow and diffusion have been operating for a few hours. In some cases, calcium will accumulate at the root as more calcium moves to the root by mass flow than is absorbed, so the concentration may be more than double that in the initial solution. For potassium moving to the root by diffusion, the potassium concentration near the root may be reduced to 10% or less of that in the equilibrium solution. Hence, the calcium to potassium ratio in solution at the root surface could differ by a factor of 20 from that in the initial equilibrium solution.

Measurement of the potassium potential itself may describe the availability of potassium for uptake during the first few minutes of absorption into the root, since it describes potassium level in solution. However, without other values to determine how the potassium concentration at the root changes with time, the potassium potential is of doubtful value as a measure of potassium availability.

Capacity/Intensity Measurements

Investigators have used capacity/intensity relations or Q/I values to describe the status of potassium in the soil (Beckett, 1964; Barrow, 1966). This relation acknowledges the inability of the potassium potential (solution potassium concentration) or intensity to adequately describe potassium availability; and it includes a value for the soil's capacity to replenish its potassium in solution. The capacity value of many soils is essentially the level of exchangeable potassium.

As exchangeable potassium is commonly measured, solution potassium is included, hence the value represents C_{si}. If we take the ratio of the change in C_{si} to the change in solution potassium, we have the soil buffer power; one of the parameters used in the mathematical description of uptake (Chapter 5). Barrow (1966) measured potential and capacity values for soil potassium and used them in a multiple regression expression with potassium uptake by the plant as a measure of potassium availability. He found that "as uptake of potassium became progressively larger, the potential became increasingly unsuitable as a single index of availability, but a multiple regression containing terms for potential and for buffering capacity continued to account for a large proportion (89%) of the variation." This result, where a large part of potassium is absorbed, agrees with what would be expected from calculations made with the mathematical model.

The Q/I relations are obtained from a potassium adsorption-desorption curve with ionic strength held constant; an example of such a curve was shown in Figure 2.6. The curve characterizes a given soil with respect to potassium adsorption. From the adsorption curve, it is also possible to obtain a value for C_{li} and calculate an accurate value for b, both of which are used in the Claassen–Barber and Barber–Cushman models to describe potassium uptake by plant roots.

Percent Saturation

Since cation-exchange capacity varies among soils, there is a possibility that it may influence the availability of potassium. McLean (1976) suggested modifying soil test procedures that use exchangeable potassium as an index by adjusting the exchangeable potassium level needed for an adequate supply by a factor related to the cation-exchange capacity of the soil. With higher cation-exchange capacity, the level of exchangeable calcium for a similar pH will be greater. Hence, to have similar calcium to potassium ratios in the soil solution, a higher level of exchangeable potassium would be needed.

Barber (1981) investigated the probable effect of cation-exchange capacity on potassium supply to the root by calculating uptake predicted by the Claassen–Barber mathematical model as the measure of available potassium. Using a curve for adsorption of potassium by the soil, Barber determined values for C_{li}, b, and D_e for the case where C_{si} for potassium was 0.4 cmol/kg. He did this for Chalmers silt loam and Wellston silt loam, which had cation-exchange capacities of 31.5 and 12.9 cmol(p$^+$)/kg, respectively;

TABLE 10.7   Values of CEC, C_{li}, b, D_e, and Calculated Uptake for Chalmers Silt Loam and Wellston Silt Loam, where C_{si} Was Constant[a]

Soil property	Chalmers	Wellston
CEC, cmol (p$^+$)/kg	31.5	12.9
C_{li}, mmol/L	0.14	0.52
C_{si}, cmol/kg	0.4	0.4
b	18	7.9
D_e, cm^2/s $\times$ 10^8	6.6	7.5
Uptake/week, μmol/plant	70	119

Source: Reproduced from Barber (1981) by permission of American Society of Agronomy and Soil Science Society of America.

[a]L_v, 3 cm/cm^3; I_{max}, 415 nmol/m^2 · s; K_m, 20 μmol/L; C_{min}, 2.7 μmol/L; L_0, 200 cm; k, 1 $\times$ 10^{-6}/s; r_0, 0.02 cm. Calculated uptake allowed for root competition for potassium.

data are shown in Table 10.7. Calculated uptake, assuming no root competition for potassium, was much higher for the Wellston soil. Predicted uptake, assuming root competition with a root density of 3 cm/cm³ showed much less difference between soils because predicted uptake was more closely related to C_{si} when root competition increased. In a greenhouse study of potassium uptake affected by cation-exchange capacity, where potassium level was approximately constant, Shaw (1979) found a greater difference in potassium uptake among soils of the same cation-exchange capacity than soils differing in cation-exchange capacity. There was close agreement ($r^2 = 0.93$) between predicted and observed uptake using the Barber–Cushman model (Figure 10.6). Root density at harvest varied between 2.65 and 3.25 cm/cm³ and thus was similar to root densities observed in the field. Table 10.8 gives the relation among soil cation-exchange capacity, exchangeable potassium, and potassium flux into the root; there was no apparent effect of cation-exchange capacity on potassium uptake. For these soils, differences in strengths of bonding, independent of cation-exchange capacity, were apparently the more important effect.

Cation-exchange capacity *may* affect potassium uptake if the dissociation of bonded potassium into solution becomes less with an increase in cation-exchange capacity. However, if rooting density in the soil is high enough so that roots compete for potassium, effects on potassium uptake other than the level of exchangeable potassium are relatively smaller. For many crops, potassium uptake is equal to 30 to 50% or more of the amount of exchangeable potassium present in the soil. Because uptake is such a large proportion of that present, root competition for potassium occurs. Hence, exchangeable potassium level can be an important factor in determining potassium uptake. When comparing soils low in cation-exchange capacity, such as 2 to 4 cmol(p$^+$)/kg, with those high in cation-exchange capacity, 30 to 50 cmol/kg, it is likely that some effect of cation-exchange capacity on potassium uptake may occur. However, when smaller differences in cation-exchange capacity occur, factors such as the type of

TABLE 10.8   Cation-Exchange Capacities for Four Soils as Related to Potassium Uptake by Corn Growing in These Soils

Soil	Cation-exchange capacity (cmol(p$^+$)/kg)	Exchangeable K (cmol(p$^+$)/kg)	Potassium uptake (mmol/pot)
Brookston siltly clay loam	24.8	0.17	2.11
Crosby silt loam	13.2	0.21	2.93
Miami silt loam	11.9	0.16	1.62
Wellston silt loam	13.2	0.18	1.59

Source: Reproduced from Shaw et al. (1983) by permission of Marcel Dekker Inc.

exchange site may be more important. Hence, reliance on a single factor to increase the ability to predict potassium uptake is not likely to have broad significance.

ENVIRONMENTAL FACTORS AFFECTING POTASSIUM UPTAKE

Such soil properties as soil moisture, aeration, and pH may influence potassium uptake by the root. These properties can affect the kinetics of potassium influx by the root or the rate of potassium supply through the soil to the root. In addition, other soil properties may affect the morphology and size of the root system, which would reduce potassium uptake in proportion to the reduction in root size. The discussion presented here is confined primarily to the effect on root physiology and soil supply of potassium rather than on root morphology and growth rate.

Soil Water

Volumetric water content influences the diffusive supply of potassium to the root, so that increasing volumetric water content increases diffusion due to an increase in cross-sectional area for diffusion and a decrease in tortuosity of the diffusion path. Place and Barber (1964) investigated the effect of soil water content on diffusion, using ^{86}Rb to indicate the relation for soil potassium. They measured D_e on a Raub silt loam that had been incubated at three levels of rubidium and five levels of soil water; results were shown in Figure 4.10. As water level increased, D_e increased, the degree of increase being greater the higher the level of exchangeable rubidium.

Place and Barber (1964) also measured rubidium uptake by corn as affected by water level and rubidium concentration. Their results were shown in Figure 4.11. Uptake increased linearly with increases in soil moisture in a manner similar to the increase in D_e that was shown in Figure 4.10. Danielson and Russell (1957) also obtained a linear relation between soil water level and uptake of ^{86}Rb of corn seedlings. Hence, soil water level is important in supplying potassium to plant roots.

Aeration

Soil aeration has a greater effect on potassium uptake than on uptake of other ions. When oxygen levels drop below 10% potassium uptake decreases; Danielson and Russell (1957) presented data illustrating this effect. Oxygen level in the soil pore space seldom drops below 10% unless the soil becomes saturated with water. The relation between oxygen content of the soil atmosphere and rubidium uptake is illustrated in Figure 10.9. Aeration may affect potassium uptake from the soil.

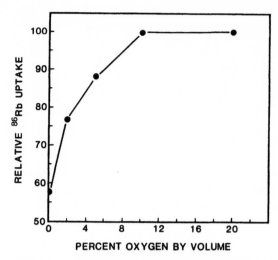

FIGURE 10.9 Influence of oxygen concentration on ^{86}Rb uptake by corn seedlings. Each point is the mean for six levels of osmotic pressure varying from 0.005 to 1.2 MPa. Reproduced from Danielson and Russell (1957) by permission of Soil Science Society of America.

Temperature

Soil temperature influences both potassium uptake by the root and potassium diffusion through the soil. The effects of temperature on potassium influx, mentioned earlier in this chapter, were discussed in Chapter 3 and shown in Figure 3.9.

Temperature affects the parameters D_e, b, and C_{li} for soil potassium. These, in turn, influence supply of soil potassium to the root. The degree of the effect is influenced by the level of soil potassium. Results of the investigations of Ching and Barber (1979) on Raub silt loam are shown in Table 10.9; values for C_{li} and D_e were lower at 15 than at 29°C. Since both plant and soil factors are influenced by temperature, each will have an effect on potassium uptake by the plant. In order to estimate the relative effects of temperature on soil parameters compared to temperature effect on plant parameters, potassium uptake was calculated from the mathematical model using all combinations of temperature–soil and temperature–plant parameters; results are given in Table 10.10. At the low soil potassium level, a C_{li} value of 0.046 mmol/L at 15°C, increasing soil temperature increased uptake due to soil factors more than plant factors. At the high soil potassium level, a C_{li} value of 6.25 mmol/L at 15°C, there was no effect of temperature on uptake due to soil parameters, while a large effect occurred due to changes in plant parameters. Hence, raising soil temperature in the range below 29°C will increase potassium uptake a variable amount, depending on the effect of temperature on soil and plant parameters involved in potassium uptake.

TABLE 10.9   Effect of Soil Temperature and Potassium Level on C_{li}, b, and D_e in Raub Silt Loam

K added (mg/kg)	Soil parameters					
	C_{li} (μmol/ml)		b		$D_e \times 10^7$ (cm^2/s)	
	15°C	29°C	15°C	29°C	15°C	29°C
0	0.046	0.089	39	23	0.15	0.39
50	0.174	0.256	12	9.5	0.50	0.94
100	0.355	0.516	8.7	3.3	0.69	2.7
300	1.97	2.66	2.2	1.8	2.7	5.0
500	6.26	8.10	1.2	1.2	5.0	7.5
700	11.90	13.90	1.2	1.2	5.0	7.5

Source: Reproduced from Ching and Barber (1979) by permission of American Society of Agronomy.

Soil pH

The effect of soil pH on potassium uptake includes the effects of aluminum, hydrogen, calcium and magnesium on potassium uptake by the root and also on the equilibrium between C_{si} and C_{li}. Only extremely acid soils, pH below 5.2, contain appreciable amounts of exchangeable aluminum. Little information is available to indicate whether changing the levels of any of these cations affects potassium uptake by roots, assuming that systems always have sufficient calcium to maintain the integrity of the plant root membrane. The effect on soil parameters suggests that the presence of aluminum causes more displacement of potassium into solution. Replacing exchangeable hydrogen with calcium also increases C_{li} for potassium.

TABLE 10.10   Predicted Potassium Uptake (mmol/pot) by Corn from Raub Silt Loam for All Combinations of Two Soil Potassium Levels, Two Soil Temperatures, and Two Air Temperatures

Air temperature	Soil potassium level			
	Low		High	
	Soil temperature		Soil temperature	
	15°C	29°C	15°C	29°C
15°C	0.44	1.0	3.1	3.1
29°C	0.77	1.8	8.9	8.9

Source: Reproduced from Ching and Barber (1979) by permission of American Society of Agronomy.

Salt Concentration in the Soil

As the anion concentration of the soil is increased, more potassium becomes present in the soil solution; however, there will be an equilibrium between calcium, magnesium, and potassium, the principal cations in the system. The equilibrium will be such that

$$\frac{K_1}{(Ca_1 + Mg_1)^{1/2}} = \frac{K_2}{(Ca_2 + Mg_2)^{1/2}} \qquad (10.2)$$

where subscripts 1 and 2 refer to low and high salt concentrations, respectively. Calcium and magnesium must increase as the square of their concentrations to keep these ratios constant. Hence, the increase in potassium will be only the square root of the increase in calcium and magnesium concentration. Baligar and Barber (unpublished data) measured the effect of salts added to Raub silt loam on the soil parameters for calculating potassium availability; their results are shown in Table 10.11. Predicted potassium uptake was increased but much less than the increase in salt concentration. Influx increased with increasing concentration of potassium at the root surface, so the initial level of potassium in the soil should also influence results. For these calculations, it was assumed that increasing calcium concentration had no effect on potassium absorption kinetics. Clark (1978) found that changing calcium levels in seven steps from 0 to 25.4 mmol/L had very little effect on the potassium concentration in the shoots of 21-day-old corn, even though calcium concentration in the plant increased from 0.4 to 1.8%.

TABLE 10.11   Effect of KCl, KH₂PO₄, and CaCl₂ Additions on Potassium-Uptake Parameters and Calculated Potassium Uptake by Corn[a]

Treatment salt added (mmol/kg)	Potassium Parameters			
	C_{li} (μmol/L)	b	D_e (cm^2/s $\times 10^7$)	Uptake (mmol/pot)
10 KH₂PO₄	2.1	7.5	1.5	7.01
10 KCl	3.3	4.6	2.5	9.04
10 KH₂PO₄, 15 CaCl₂	5.4	3.0	3.9	9.49
10 KH₂PO₄, 45 CaCl₂	8.2	1.9	6.2	9.52

Source: Baligar and Barber (unpublished).

[a]Uptake calculated using the Claassen–Barber (1976) model with values for C_{li}, b, and D_e from this table together with common values for I_{max}, 400 nmol/m² · s; K_m, 1.5 × 10⁻² mmol/L; C_{min}, 2 × 10⁻³ mmol/L; v_0, 7 × 10⁻⁷ cm/sec; r_0, 4 × 10⁻² cm; L_0, 500 cm; k, 1.6 × 10⁻⁶; t, 8.64 × 10⁵s.

SUMMARY

The amount (w/w) of potassium absorbed by crops is often greater than any other mineral nutrient. Nitrogen uptake may exceed potassium uptake for some crops. The fraction of available soil potassium absorbed by a crop in a given year is greater than any other major nutrient with the exception of nitrogen. Because plants may absorb such a large fraction of the available soil potassium, potassium fertilization of crops is an important factor in crop production.

Potassium concentration of the grain is usually low, with most of the absorbed potassium remaining in the leaves and stems. When only the grain is removed from the field, there is usually a large amount of potassium cycled from the plant back into the soil. Cycling subsoil potassium to the surface through plant uptake may also occur. In many soils, potassium cycling over the centuries has apparently moved much of the available potassium to the topsoil. Because of this, Schenk and Barber (1980) found that the subsoil of a Raub silt loam contributed less than 10% of the potassium absorbed by the crop, even though more than half the root system was growing in the subsoil.

Potassium is an important nutrient in crop production, therefore it is important to understand the kinetics of potassium flux within the soil–plant root system. Diffusion usually is the greatest contributor to soil supply of potassium. The Barber–Cushman mechanistic nutrient uptake model can be used to provide an understanding of why there are differences between plant species and soils in potassium uptake. Many of the parameters of the model can play a significant role in determining potassium uptake. The sensitivity diagram shown in Figure 5.5 will vary with soil–plant conditions; at high root densities, low r_1, and low soil supply, C_{li}, increasing r_1 will have a large effect on predicted uptake. At high C_{li} values increasing I_{max} will have a large effect. At high C_{li} values increasing r_0 will have a greater effect. Hence, as noted in this chapter, the important factors in determining potassium availability vary with soil potassium level, root density, uptake kinetics of the plant, and amount absorbed during the growth season, so that drawing broad conclusions on the basis of one experiment should not be attempted.

REFERENCES

Asher, C. J., and P. G. Ozanne. 1967. Growth and potassium content of plants in solution cultures maintained at constant potassium concentrations. *Soil Sci.* **103**:155–161.

Baligar, V. C., and S. A. Barber. 1980. Ion equilibria, selectivity and diffusion of cations in soils. Unpublished data. Agronomy Dept., Purdue University, West Lafayette, Indiana.

Baligar, V. C., and S. A. Barber. 1979. Genotypic differences of corn in ion uptake. *Agron. J.* **71**:870–873.

Barber, S. A. 1978. Growth and nutrient uptake of soybeans under field conditions. *Agron. J.* **70**:457–461.

Barber, S. A. 1981. Soil chemistry and the availability of plant nutrients. In *Chemistry in the Soil Environment.* Spec. Pub. No. 40. American Society of Agronomy, Madison, WI. Pp. 1–12.

Barber, S. A., and J. H. Cushman. 1981. Nitrogen uptake model for agronomic crops. In I. K. Iskandar, Ed. *Modeling Wastewater Renovation-Land Treatment.* Wiley-Interscience, New York. Pp. 382–409.

Barber, S. A., and R. P. Humbert. 1963. Advances in knowledge of potassium relationships in the soil and plant. In M. H. McVicker, G. L. Bridger, and L. B. Nelson. Eds. *Fertilizer Technology and Usage.* Soil Science Society of America, Madison, WI. Pp. 231–268.

Barber, S. A., J. M. Walker, and E. H. Vasey. 1962. Principles of ion movement through the soil to the plant root. *Proc. Int. Soil Conf.*, New Zealand 121–124.

Barrow, N. J. 1966. Nutrient potential and capacity. II. Relationship between potassium potential and buffering capacity and the supply of potassium to plants. *Aust. J. Agric. Res.* **17**:849–861.

Beckett, P. H. T. 1966. Studies on soil potassium. I. Confirmation of the ratio law: Measurement of potassium potential. *J. Soil Sci.* **15**:1–8.

Binnie, R. R., and S. A. Barber. 1964. Constrasting release characteristics of potassium in alluvial and associated upland soils of Indiana. *Soil Sci. Soc. Am. Proc.* **28**:387–309.

Ching, P. C., and S. A. Barber. 1979. Evaluation of temperature effects on potassium uptake by corn. *Agron. J.* **71**:1040–1044.

Claassen, N., and S. A. Barber. 1974. A method for characterizing the relation between nutrient concentration and the flux into roots of intact plants. *Plant Physiol.* **54**:564–568.

Claassen, N., and S. A. Barber. 1976. Simulation model for nutrient uptake from soil by a growing plant root system. *Agron. J.* **68**:961–964.

Claassen, N., and S. A. Barber. 1977. Potassium influx characteristics of corn roots and interaction with N, P, Ca, and Mg influx. *Agron. J.* **69**:860–864.

Clark, R. B. 1978. Differential response of corn inbreds to calcium. *Commun. Soil Sci. Plant Anal.* **9**:729–744.

Danielson, R. E., and M. B. Russell. 1957. Ion absorption by corn roots as influenced by moisture and aeration. *Soil Sci. Soc. Am. Proc.* **21**:3–6.

Epstein, E., D. W. Rains, and E. O. Elzam. 1963. Resolution of dual mechanisms of potassium absorption by barley roots. *Natl. Acad. Sci. Proc.* **49**:684–692.

Hassan, M. M., and T. van Hai. 1976. Ammonium and potassium uptake by citrus roots. *Physiol. Plant.* **36**:20–22.

Horton, M. M. 1959. Influence of Soil Type on Potassium Fixation. M.S. thesis, Purdue Univ.

Jungk, A., and N. Claassen, and R. Kuchenbuch. 1982. Potassium depletion of the soil-root interface in relation to soil parameters and root properties. In A. Scaife, Ed. *Plant Nutrition 82; Proc. 9th Int. Plant Nutr. Colloq.* **1**:250–255.

Kovar, J. L., and S. A. Barber. 1990. Potassium supply characteristics of thirty-three soils as influenced by seven rates of potassium. *Soil Sci. Soc. Am. J.* **54**:1356–1361.

Li, Y., and S. A. Barber. 1991. Calculating changes of legume rhizosphere soil pH and soil solution phosphorus from phosphorus uptake. *Commun. Soil Sci. Plant Anal.* **22**:955–973.

McLean, E. O. 1976. Exchangeable K levels for maximum crop yields on soils of different cation exchange capacities. *Commun. Soil Sci. Plant Anal.* **7**:823–828.

Mengel, D. B., and S. A. Barber. 1974. Rate of nutrient uptake per unit of corn root under field conditions. *Agron. J.* **66**:399–402.

Peterson, W. R., and S. A. Barber. 1981. Soybean root morphology and K uptake. *Agron. J.* **73**:316–319.

Place, G. A., and S. A. Barber. 1964. The effect of soil moisture and rubidium concentration on diffusion and uptake of rubidium-86. *Soil Sci. Soc. Am. Proc.* **28**:239–243.

Reisenauer, H. 1964. Mineral nutrients in soil solution. In P. L. Altman and D. S. Dittmer, Eds. *Environmental Biology*. Fed. Am. Soc. Exp. Biol., Bethesda, MD. Pp. 507–508.

Rich, C. I. 1968. Mineralogy of soil potassium. In V. J. Kilmer, S. E. Younts, and N. C. Brady, Eds. *The Role of Potassium in Agriculture*. American Society of Agronomy, Madison, WI. Pp. 79–108.

Schenk, M. K., and S. A. Barber. 1980. Potassium and phosphorus uptake by corn genotypes grown in the field as influenced by root characteristics. *Plant Soil* **54**:65–76.

Shaw, J. K., R. K. Stivers, and S. A. Barber. 1983.Evaluation of differences in Potassium availability in soils of the same exchangeable potassium level. *Commun. Soil Sci. Plant Anal.* **14**:1035–1049.

Silberbush, M., and S. A. Barber. 1983. Sensitivity analysis of parameters used in simulating potassium uptake with a mechanistic mathematical model. *Agron. J.* **75**:851–854.

Silberbush, M., and S. A. Barber. 1984. Phosphorus and potassium uptake of field-grown soybean cultivars predicted by a simulation model. *Soil Sci. Soc. Am. J.* **48**:592–596.

Sparks, D. L., L. W. Zelazny, and D. C. Martens. 1980. Kinetics of potassium exchange in a Paleudult from the coastal plain of Virginia. *Soil Sci. Soc. Am. J.* **44**:37–40.

Talibudeen, O., J. D. Beasley, P. Lane, and N. Rajendran. 1978. Assessment of soil potassium reserves available to plant roots. *J. Soil Sci.* **29**:207–218.

Wild, A., P. J. Woodhouse, and M. J. Hopper. 1979. A comparison between uptake of potassium by plants from solutions of constant potassium concentration and during depletion. *J. Exp. Bot.* **30**:697–704.

CHAPTER 11

Calcium

Calcium is a divalent alkaline earth cation. It is the fifth most plentiful element in the earth's crust, which has an average calcium concentration of 3.6%. Noncalcareous, highly weathered soils contain less than 1% calcium (McLean, 1975); calcareous soils may have 50% or more calcium carbonate, so their calcium contents can be above 10%. The calcium content of soil depends on its parent material, degree of weathering, and whether or not calcium has bee added by liming. Approximate calcium concentration for several soils are aridisols, 5%; alfisols, 1%; and oxisols, 0.6%.

FORMS OF CALCIUM IN THE SOIL

Calcium in the soil solution is balanced by soluble anions. Larger amounts are present as exchangeable calcium associated with the soil's cation-exchange capacity. Calcium may also be present in soil minerals having varying degrees of solubility. Solution and exchangeable calcium are the main forms that can move to the plant root and be absorbed.

Mineral Calcium

Two soil calcium minerals having the greatest solubility are calcium sulfate and calcium carbonate. Calcium sulfate (gypsum) usually occurs only in arid soils, where the sulfate concentration in solution exceeds 0.01 mol/L (Lindsay, 1979). Calcium carbonate occurs only in soils above pH 7.0, and frequently controls calcium levels in the soil solution for soils above pH 7.8 that contain free calcium carbonate or calcite. The level of calcium in solution will be affected by the soil solution's CO_2 level, which in turn controls bicarbonate and carbonate concentrations. Calcitic limestone, calcium carbonate, or dolomitic limestone, a combination of calcium carbonate and magnesium carbonate, are added to acid soils to increase soil pH, increase the levels of exchangeable calcium and magnesium in the soil, and reduce exchangeable aluminum and sometimes manganese levels. At soil pH values between 7.5 and 8.0, calcium sulfate and calcium carbonate can coexist, providing a calcium level in solution of approximately 3 mmol/L or 120 mg/kg.

Other soil minerals having lower calcium contents are plagioclase feldspars, augite, hornblende, and epidote. Each occurs in igneous and metamorphic rocks, weathers slowly, and has less significance than either carbonates or sulfates in supplying calcium to the soil.

Exchangeable Calcium

Calcium is the dominant exchangeable cation in many soils. In general, only alkaline soils containing sodium, acid soils containing large amounts of hydrogen and aluminum, and serpentine-derived soils high in magnesium have cations other than calcium as the dominant exchangeable cation. Exchangeable calcium is in equilibrium with soil solution calcium. The relative strength of bonding controls the equilibrium between soluble and exchangeable calcium. Bonding depends on the nature of the cation-exchange site, degree of calcium saturation of soil-exchange sites, complementary cations present, and the anion content of the soil solution. When comparing the bonding of calcium and strontium by soils, Juo and Barber (1969) found that calcium was bonded more tightly than strontium on organic exchange sites, though strontium was bonded more tightly to the permanent charge sites on clays. The difference in bonding is apparently determined by the relative field strength of the exchange site as compared to the energy of hydration of the cation. For organic sites, the field strength was high enough that the order of cation adsorption was related to unhydrated cation size. On the permanent-charge sites of soil clay minerals, the field strength was less, and the order of cation adsorption was related to the hydrated cation size, with the larger ion held less tightly.

Exchangeable calcium is almost always held more tightly on soils than either potassium or magnesium, the next two most plentiful exchangeable cations. Table 10.3 showed the dissociation of calcium relative to potassium and magnesium for three widely different soils. For all three, exchangeable calcium was less dissociated than either exchangeable magnesium or potassium.

Exchangeable calcium levels are increased by soil liming. The material of choice is usually calcium or calcium-magnesium carbonate, added as finely ground limestone. The reaction that occurs is shown in Equation 11.1

$$CaCO_3 + 2H^+exch \rightleftharpoons Ca^{2+}exch + CO_2(g) + H_2O \qquad (11.1)$$

Because the end products of this reaction are exchangeable calcium, carbon dioxide gas, and water, the reaction can proceed to completion. The limestone has to be finely ground in order to provide a large surface area for rapid reaction with the soil. When exchangeable aluminum is present on highly acid soils, it will precipitate as $Al(OH)_3$ as the pH increases. Hydrogen from exchange sites or from H_2O will enter into the reaction with calcium carbonate.

Soil Solution Calcium

The level of calcium in the soil solution is usually regulated by equilibrium with exchangeable calcium. The degree of calcium saturation of cation-exchange sites, complementary cations, the nature of the bonding with exchange sites, and level of anions in solution all interact to determine the soluble calcium level in solution. Table 11.1 gives values for calcium concentrations in the soil solution. Barber et al. (1962) reported calcium concentrations in displaced soil solutions from 135 north central U.S. soils of less than 5 to over 100 mg/L. The most common values were 20 to 40 mg/L (0.5 to 1.0 mmol/L). Reisenauer (1964) summarized data from 979 soils and found data from less than 50 to over 1000 mg/L, with 54.6% of the values in

TABLE 11.1   Range of Calcium Conentrations in Displaced Soil Solutions

Reisenauer survey (1964)[a]		Data from Barber et al. (1962)	
Concentration (mg/L)	Fraction of samples (%)	Concentration (mg/L)	Fraction of samples (%)
0–50	23.1	0–10	7.7
51–100	54.6	11–20	19.0
101–200	8.1	21–30	22.5
201–500	8.1	31–40	23.9
500–1000	5.5	41–90	14.8
		100+	12.0
979 samples		142 samples	

[a]Reprinted from Reisenauer (1964) by permission of Federation of American Societies for Experimental Biology.

the 50 to 100 mg/L range. Many of these soils were from the western United States where more arid conditions occur, consequently soil solution calcium values were high. Adams (1974) reports soil solution values for calcium ranging from 1.7 to 19.4 mmol/L; Al Abbas and Barber (1964) obtained calcium values of 0.6 to 2.3 mg/L for acid soils from South Carolina. Hence, a wide range of calcium concentrations exists for soil solutions, depending on soil type and degree of weathering.

While there will be a relation between levels of exchangeable calcium and the soil solution calcium levels for a particular soil, the relation will not be the same when diverse soils are compared. Soils varying widely in cation-exchange capacity will have a wide range of exchangeable calcium values associated with the same level of solution calcium.

CALCIUM-ADSORPTION ISOTHERMS

Data shown in Figure 11.1 illustrate calcium-adsorption isotherms obtained by Hunsaker and Pratt (1971). They investigated calcium–magnesium exchange equilibria in soils where calcium and magnesium were the only exchangeable cations. Since exchange sites preferred calcium over magnesium, a smaller percentage of the calcium was in solution than the fraction

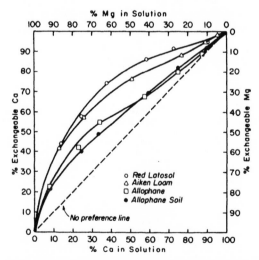

FIGURE   11.1   Relation between exchangeable calcium and magnesium as a percent of CEC and concentrations in soil solution as a percent of the moles of calcium and magnesium for allopane and three soils. The soils are a Brazilian oxisol, Aiken loam (Xeric Haplohumulls) from California and a Mexican volcanic soil. Reproduced from Hunsaker and Pratt (1971) by permission of Soil Science Society of America.

represented by the exchangeable calcium. The relative strength of bonding varied with soil.

As soils are limed, the exchangeable calcium level is increased, which also influences the relative C_{li} level for calcium. The relationship between soil pH and solution calcium for samples taken from two field experiments in Indiana is shown in Figure 11.2. The Chalmers silt loam had a cation-exchange capacity of 23 cmol(p^+)/kg, while the Wellston silt loam had a cation-exchange capacity of 13 cmol(p^+)/kg. In addition, the organic matter level of the Chalmers soil was 5% and the Wellston soil 2%. Hence, as the amount of calcium added to the soil was increased, soil pH increased more rapidly for the Wellston soil.

In Chapter 12, I suggest that magnesium is adsorbed into nonexchangeable forms as soil pH increases above 6.0. There is no corresponding evidence for calcium fixation in soils, except as $CaCO_3$.

Soil Calcium Supply Parameters

The three soil parameters, C_{li}, b, and D_e that are related to adsorption isotherms are important when calculating calcium supply to the plant root by a mechanistic mathematical model.

Soil Solution Calcium

Levels of calcium in soil solution vary with the pH and nature of the soil. Values are commonly between 20 and 40 mg/L, 0.5 to 1.0 mmol/L on leached soils. Corresponding values for arid soils are 50 to 100 mg/L, 2.5 to 5.0 mmol/L (Table 11.1).

Buffer Power

Because of the widely varying nature of cation-exchange capacity, there is a wide range of values for the calcium buffer power. Values on the same soil

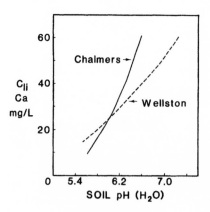

FIGURE 11.2 Relation between soil pH and soil solution calcium for two Indian soils limed in the field at several rates (Barber, unpublished data).

will be higher for calcium than magnesium, because calcium is held more tightly by the exchange sites. Al Abbas and Barber (1964) found b values varying from 19 to 107. Values were higher the lower the fraction of the cation-exchange capacity saturated with exchangeable calcium.

Diffusion Coefficient

The diffusion coefficient for calcium in water is 0.78×10^{-5} cm²/s at 20°C. If we assume a silt loam soil with 20% volumetric moisture, we can assume values of 0.20 for θ and 0.18 for f_i (Warncke and Barber, 1972). Using a buffer power $b = 40$ and the relation $D_e = Df_i\theta/b$, we would calculate an effective diffusion coefficient of 7.0×10^{-9} cm²/s. This value would vary widely with changes in volumetric moisture content and b.

CALCIUM-UPTAKE KINETICS

Moore et al. (1961) reported that calcium uptake by 6-day-old barley (*Hordeum vulgare*) roots was nonmetabolic, since it was insensitive to temperature and the presence of dinitrophenol. Handley and Overstreet (1961), however, found that uptake by the vacuolated portion of the root was metabolic. Dunlop (1973) investigated calcium uptake by intact plants of barley, subterranean clover (*Trifolium subterraneum*), and mung bean (*Phaseolus aureus*) using ^{45}Ca. He found the first increment of ^{45}Ca uptake to be nonmetabolic and displaceable, whereas the rate of additional uptake was constant with time and metabolic. Maas (1969) found that calcium uptake was metabolically controlled at solution calcium concentrations up to 0.05 mmol/L. However, at higher solution calcium concentrations, part of the uptake was nonmetabolic. Drew and Biddulph (1971) studied calcium uptake by bean plants from regular and 1/5-strength Hoagland solutions. The authors found that such uptake was not affected by metabolic inhibitors. Ferguson and Clarkson (1975) found that calcium uptake by maize roots followed the apoplastic pathway and was stopped when the endodermis became suberized. Hence, whether calcium uptake is deduced to be metabolic or not apparently depends on the system measured. Since we are interested in the long-term uptake of calcium, evidence indicates that this uptake is metabolic when calcium concentrations are low. Dunlop measured the kinetics of uptake from solutions ranging in calcium concentration from 0.0005 to 5.0 mmol/L. He obtained a $K_m = 0.039$ mmol/L and a $V_{max} = 0.075$ mol/g$_{F.W.}$ · h.

Calcium Uptake Rate

Loneragan et al. (1968) used a flowing culture experimental system to measure growth and calcium uptake by 30 grasses, cereals, legumes, and herbs; all were grown with six levels of calcium in flowing culture solution. The calcium concentrations used were 0.3, 0.8, 2.5, 10, 100, and 1000 μmol/L; all other nutrient levels were the same for each treatment. There was wide variation among species in response to calcium. For *Lolium perenne*, maximum yield for plants harvested 17 to 19 days after transplanting was obtained at a solution calcium concentration of 2.5 μmol/L, while *Hordeum vulgare* and *Medicago sativa* required 1000 μmol/L for maximum yield. The solution concentration required for maximum growth rate was not related to calcium concentration of the plant. Species vary in their shoot calcium concentrations, with legumes usually having higher values than nonlegumes. Foy (1974) found that solution concentrations over 400 μmol/L were required to maximize yields of soybeans grown in 1/5 Steinberg solution. Calcium-deficiency symptoms were observed with 200 μmol/L calcium.

Calcium level in plant tops varied with solution calcium level as well as among species. The effect of calcium solution level on calcium concentrations in the tops of five species representing the range of values found for the 30 species studied by Loneragan and Snowball (1969a) is shown in Table 11.2. In all instances, calcium concentration continued to increase as the calcium level in solution increased. Loneragan and Snowball (1969b) also measured calcium-uptake rates; uptake rates versus solution calcium concentration for the five species are given in Table 11.3. As with calcium concentration, uptake rate continued to increase with an increase in solution calcium concentration. Data do not give a linear relation when $1/In$ was plotted against $1/C_1$, so there was not a simple Michaelis–Menten relation over the entire range; instead, there were probably two or more phases. Visual inspection of the data would indicate that a K_m value between 100 and 300 μmol/L, depending on the species, would describe the K_m approximately for calcium uptake.

Clarkson (1965) found that uptake increased with increasing solution calcium concentrations to much higher solution calcium concentrations for

TABLE 11.2   Relation between Calcium Concentration in Flowing Solution and Percent of Calcium in the Tops of Five Species Grown in Solution Culture

Solution Calcium (μmol/L)	Avena sativa	Zea mays	Lolium perenne	Lupin alba	Lycopersicon esculentum
0.8	0.05	0.02	0.06	0.14	0.21
10	0.10	0.12	0.15	0.28	0.30
100	0.32	0.43	0.37	0.86	1.29
1000	0.57	0.92	1.08	1.19	2.49

Source: Reproduced from Loneragan and Snowball (1969a) by permission of CSIRO.

TABLE 11.3   Relation between Calcium Concentrations in Solution (μmol/L) and Calcium-Uptake Rates by Roots of Five Species (μmol/g_{FW} · day)

Solution calcium	Avena sativa	Zea mays	Lolium perenne	Lupin alba	Lycopersicon esculentum
0.8	0.1	0.2	0.3	0.2	–
10	0.5	0.9	1.5	1.3	3.8
100	1.7	3.9	5.7	6.3	20.3
1000	3.8	8.2	12.3	11.3	36.3

Source: Reproduced from Loneragan and Snowball (1969b) by permission of CSIRO.

Agrostis stolonifera, a calcicole, than for *Agrostis setacea*, a calcifuge. Halstead et al. (1968) found that tomato took up the greatest amount of calcium; soybean and lettuce were intermediate, and wheat absorbed the least of the four plant species investigated. For all species, calcium uptake increased linearly with increased supply of calcium to the root.

Lazaroff and Pitman (1966) investigated the effect of transpiration rate on calcium uptake by barley seedlings. When the solution was 15 mmol/L in calcium, uptake increased linearly with an increase in transpiration rate. When the solution was only 0.5 mmol/L in calcium, however, transpiration rate had no effect on calcium uptake. It appeared that uptake was metabolically controlled at calcium concentrations of 0.5 mmol/L, with increasing transpiration rate having little effect on calcium-uptake rate. However, when calcium concentration was 15 mmol/L, nonmetabolic uptake occurred and uptake was proportional to transpiration rate; this effect also occurred with magnesium.

Factors Affecting Calcium Influx

Effect of Magnesium Level on Calcium Uptake

The effect of magnesium on calcium uptake by soil-grown plants is influenced by relative levels of calcium and magnesium in solution. When the level of calcium plus magnesium was 20 mmol/L, Lazaroff and Pitman (1966) found that the ratio of calcium to magnesium uptake by barley seedlings was similar to the ratio of the ions in solution. With higher rates of transpiration, magnesium uptake was increased more than calcium uptake. Little information is available on the effect of magnesium on calcium uptake at solution concentrations below 1 mmol/L, where uptake may be metabolically controlled.

Effect of Ammonium and Potassium on Calcium Uptake

The effect of three rates of potassium application and nitrogen form on calcium concentration of young corn plants (Claassen and Wilcox, 1974) is shown in Table 11.4. Increasing potassium decreased corn plant calcium concentrations on both soils, but the effect was less than that for magnesium. Increasing ammonium concentration reduced calcium uptake; this effect was also less than that for magnesium. The effect of ammonium on reducing calcium uptake was greater than for a similar concentration or added potassium. The reason for the depression of calcium uptake when levels of potassium and ammonium are increased is not clear.

Effect of pH on Calcium Uptake

Since calcium is added as lime to reduce the levels of aluminum and hydrogen in the soil system, there is usually an interaction between calcium and pH for plants growing in soil systems. Maas (1969) found that reducing solution pH values below 4.5 reduced calcium uptake by excised maize roots; reducing solution pH at levels above 4.5 had little effect. Leggett and Gilbert (1969) found no effect of solution pH on calcium uptake by excised soybean roots over the pH range 3.8 to 6.5. In soil systems and at pH levels above 5.0, calcium concentrations in solution are high compared to hydrogen concentrations. Hence, pH probably has little direct effect on calcium uptake rate, because hydrogen would not offer much competition for common absorption sites.

TABLE 11.4   Effect of Nitrogen Form and Potassium Rate on Calcium and Magnesium Composition of Corn Plants Grown on Two Soil Types

		\% Composition for Plants Growing on			
		Princeton sand		Fincastle silt loam	
Nitrogen[a]	K rate	(Typic Hapludalfs)		(Aeric Ochraqualfs)	
form	(mg/kg)	Ca	Mg	Ca	Mg
NO_3^-	0	0.86	0.32	0.78	0.71
	50	0.79	0.27	0.66	0.57
	100	0.71	0.24	0.62	0.44
NH_4^+	0	0.67	0.20	0.50	0.33
	50	0.63	0.19	0.45	0.33
	100	0.69	0.18	0.47	0.30
L.S.D.					
(0.05)		0.115	0.033	0.123	0.070

Source: Reproduced from Claassen and Wilcox (1974) by permission of the American Society of Agronomy.
[a]Nitrogen applied at a rate of 100 mg/kg.

SOIL MEASUREMENTS OF BIOAVAILABLE CALCIUM

Calcium is absorbed by plant roots from solution; the uptake rate related to solution calcium may he influenced by formation of soluble ion pairs, such as $CaSO_4$, since these may not be absorbed. Ion-pair formation increases as ionic strength of the solution increases, with such pairs becoming appreciable where concentrations of calcium and sulfate exceed 1 mmol/L. Absorption is assumed to be from only Ca^{2+} in solution; however, exchangeable calcium and ion pairs buffer the Ca^{2+} level of the soil solution.

Bioavailable calcium is usually evaluated by measuring exchangeable calcium in the soil and reporting it as amount present, a percentage saturation of the cation-exchange capacity, or a fraction of the total exchangeable cations other than hydrogen and aluminum that are present. Alternatively, calcium levels in the soil solution can be measured and calculated as a lime potential.

Calcium is added to reduce the acidity of acid soils, with hydrogen activity varying as the reciprocal of calcium as calcium is added. Schofield and Taylor (1955) found that the pH of the solution, along with the calcium activity when a nonsaline soil was mixed with 0.01 mol/L $CaCl_2$, combined as shown in Equation 11.2, gives a useful relation for expressing calcium in the soil. They called the expression pH–1/2 pCa the lime potential.

$$\text{pH} - \frac{1}{2}p\text{Ca} = \log a_{\text{Ca(OH)}_2} + 14.2 \tag{11.2}$$

Since both calcium and magnesium are usually present in the soil, the expression has been expanded to pH $- 1/2p$(Ca + Mg). The authors suggest that results for pH be routinely expressed as pH $- 1/2p$Ca or pH $- 1/2p$(Ca + Mg). The lime potential expresses the ratio between hydrogen and the square root of calcium concentration. Since calcium uptake is not necessarily affected by pH, the utility of the lime potential as a useful measure of calcium availability has yet to he demonstrated. However, the lime potential is useful in calculating the effect of pH on the solubility of soil calcium compounds, such as calcium phosphates.

Calcium Supply by Mass Flow and Diffusion

Mass flow is frequently the primary mechanism for supplying calcium to the root surface because of the balance between soil solution calcium levels and plant requirements for calcium. Assume that a plant transpires 300 L of water per kg of dry matter produced and the plant has a calcium concentration of 0.3% or 3000 mg/kg. If the soil solution has a mean calcium concentration of 10 mg/L, mass flow would just meet the needs of the plant. Most soils have higher calcium concentrations in solution, however, so mass flow would more than supply calcium requirements of this type of plant. With a legume containing 1.2% calcium, the soil solution calcium level would have to be 40 mg/L in order for mass now to meet the plant calcium requirements.

Diffusion of calcium toward the root occurs when mass flow plus root interception does not supply the total calcium requirement. In this case, a concentration gradient decreasing toward the root is established. When mass flow plus root interception supplies *more* than uptake, however, accumulation at the root causes a concentration gradient decreasing away from the root, and calcium diffuses outward along this gradient.

Barber and Ozanne (1970) grew four plant species, ryegrass (*Lolium rigidum*), subclover (*Trifolium subterraneum*), capeweed (*Arctotheca calendula*), and lupin (*Lupinus digitalus*) in sandy soil in the greenhouse. Autoradiographs were made of the soil–root system (Figures 4.3 and 11.3). Autoradiographs indicated that ryegrass, subclover, and capeweed all caused calcium to accumulate around the root, since more calcium moved to the root by mass flow than was absorbed by the root; the greatest accumulation occurred with ryegrass and the least with capeweed. Lupin, on the other hand, absorbed calcium faster than it was being supplied to the root; hence, the calcium level was depleted around the root. These figures illustrate how the rate of calcium uptake by the root influences whether mass flow or diffusion is more important in supplying calcium to the plant root, since all plants were grown at the same time in different portions of the same soil. In this experiment, the soil solution contained approximately 100 mg Ca/L.

The preceding data compare differences among four young plants grown in one soil. Rate of nutrient uptake by roots also varies with plant age. Mengel and Barber (1974) measured calcium-uptake rates for corn plants (*Zea mays*) grown in the field and found a 230-fold reduction in calcium uptake rate per unit of root as the plant grew from 20 to 80 days in age. Hence, depending on the change of water influx with age, concentration around the root will undoubtedly change with plant age.

Barber (1974) found that calcium carbonate can precipitate around roots of perennial plants (Figure 4.5). This accumulation of calcium around the root, combined with bicarbonate formed from carbon dioxide given off during root respiration, or released as bicarbonate by the root in exchange for anion absorption, caused calcium carbonate to precipitate as pH increased.

Malzer and Barber (1975) also found that calcium sulfate can precipitate around corn roots. Both calcium and sulfate accumulated at the root because of a greater rate of supply by mass flow than uptake. Since the solubility of calcium sulfate is not high, precipitation occurred. It is probable that calcium sulfate precipitates around the plant root in many unleached soils. The K_{sp} for calcium sulfate is 1.95×10^{-4} at 10°C. The calcium level in solution in equilibrium with solid calcium sulfate will thus be about 0.0140 mol/L or 560 mg Ca/L, depending on the sulfate level. The calcium sulfate precipitate will maintain a relatively constant level of calcium around the root. Many

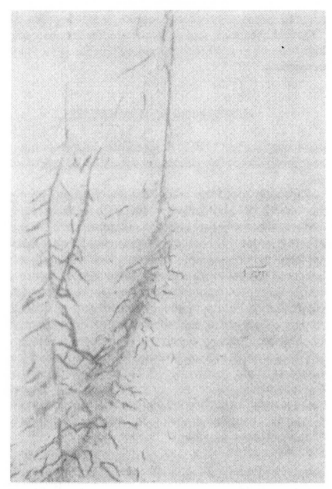

FIGURE 11.3 Depletion of calcium around lupin roots (*left side*) and accumulation of calcium around capeweed roots (*right side*). [45]Ca autoradiograph taken after 24 days of growth. Reproduced from Barber and Ozanne (1970) by permission of Soil Science Society of America.

soils in arid climates probably have soluble calcium concentrations maintained by calcium sulfate solubility or combinations of calcium sulfate and calcium carbonate solubility.

When mass flow supplies less calcium than the plant can absorb, root interception and diffusion supply calcium to the root. Oliver and Barber (1966) used soil and sand mixtures to vary the amount of calcium reaching soybean roots by root interception. They used a system where mass flow was minimized and obtained a correlation between calculated supply by root interception and calcium uptake minus that supplied by mass flow. Uptake was greater than calculated root interception, indicating that diffusion also occurred. The fact that there was a linear relation between calcium uptake and calculated supply is evidence supporting the concept of calcium supply by root interception.

MODELING CALCIUM UPTAKE

The Claassen and Barber (1976) mechanistic model describing nutrient uptake based on soil and plant parameters can also be used to describe calcium uptake.

Three soil parameters of the model describe calcium supply to the root surface; they are C_{li}, the soil solution calcium concentration; b, the buffer power of labile soil calcium for calcium in solution; and D_e, the effective diffusion coefficient for calcium in the soil. Values for C_{li} were reported in Table 11.1. Values for b may vary widely because of the wide range of exchangeable calcium levels found in the soil. Exchangeable calcium is the primary buffer for soil solution calcium in all soils except those containing precipitated $CaSO_4$ or $CaCO_3$. For purposes of illustration, we will assume a soil having a cation-exchange capacity of 20 cmol(p$^+$)/kg and an exchangeable plus solution calcium level of 6 cmol/kg. The soil solution in equilibrium with this soil is assumed to have a calcium level of 0.5 mmol/L, or 20 mg/L. If we assume a soil bulk density or 1.3, the concentration of exchangeable plus solution calcium is 0.06×1.3, or 0.078 mmol/cm^3. The solution concentration was assumed to be 0.0005 mmol/cm^3, hence b would be 156. The value of D_e can be estimated as indicated earlier in this chapter. Using a b value of 156, we obtain a D_e value of 2×10^{-9} cm^2/s.

The levels of C_{li} and D_e can vary in a soil because of the anion concentration. Being a divalent cation, C_{li} will increase directly with an increase in the anion content of the solution. Fertilization with nitrogen fertilizers, or release of nitrogen from the mineralization of organic matter, will greatly increase the C_{li} value for calcium. Increases in C_{li} will reduce b and consequently increase D_e; hence, values for the soil-supply parameters for calcium can vary greatly during the growth of a plant.

Increasing the calcium saturation of the soil's cation-exchange sites will provide calcium that is more highly dissociated from the soil than that present at lower degrees of saturation. Hence, liming the soil increases calcium supply to the root at a rate greater than proportional to the amount of lime applied, particularly when the cation-exchange capacity nears saturation.

Few values are available for calcium-absorption kinetics; the K_m value appears to be between 40 and 300 µmol/L. Values of I_{max} will likely vary with the calcium requirement of the plant; values for C_{min} are of little significance because of the relatively high values of calcium usually present in the soil solution. The rate of water influx v_0 is important, because mass flow is usually the most significant mechanism for calcium supply to the root. Scaife and Clarkson (1976) observed that calcium disorders in plants were often weather related. This may be due to changes in calcium supply by mass flow. Little modeling of calcium uptake has been done, since lack of calcium seldom reduces crop yield. Where it does, there is often a specific situation where demand is high, such as developing tomato or peanut fruits. Calcium may also be needed to balance a high level of an ion that is depressing calcium uptake.

SUMMARY

Wallace et al. (1966) considered that calcium requirements for plants were so small that calcium could be considered a micronutrient. Calcium appears to have its greatest significance in providing the appropriate balance for levels of other nutrients within the plant. Calcium is needed in small amounts to maintain the integrity of the plasma membrane so that ion uptake is facilitated. Since calcium only moves toward the shoot, roots only grow into soil where calcium is present. Because calcium is not retranslocated within the plant once it has been used in plant growth, the calcium needed during reproductive growth must be supplied by uptake. Calcium appears to be absorbed through the apoplastic pathway; such absorption ceases when the endodermis becomes suberized. Hence, only apical roots can absorb calcium, and if any factor interferes with their development, it may affect calcium uptake.

On soils with appreciable cation-exchange capacities [5 cmol(p$^+$)/kg and higher] and for nonlegume crops, calcium deficiency is rarely a problem on soils with pH levels above 5.3, since the amount in most soils is large relative to plant requirements. There are soils that are highly weathered, have a low pH, and a low cation-exchange capacity so they have inadequate supplies of calcium for plant growth. Legumes and other crops with high calcium requirements may need high pH levels in some soils for the soil to supply sufficient calcium.

In modeling calcium uptake, mass flow is a dominant supply factor. Concentration in the soil solution will have an effect on supply, by both diffusion and mass flow.

REFERENCES

Adams, F. 1974. Soil Solution. In E. W. Carson, Ed. *The Plant Root and Its Environment*. University Press of Virginia, Charlottesville. Pp. 441–481.

Al Abbas, H., and S. A. Barber. 1964. The effect of root growth and mass flow on the availability of soil calcium and magnesium to soybeans in a greenhouse experiment. *Soil Sci.* **97**:103–107.

Barber, S. A. 1974. The influence of the plant root on ion movement in soil. In E. W. Carson, Ed. *The Plant Root and Its Environment*. University Press of Virginia, Charlottesville. Pp. 525–564.

Barber, S. A., and P. G. Ozanne. 1970. Autoradiographic evidence for the differential effect of four plant species in altering the Ca content of the rhizosphere soil. *Soil Sci. Soc. Am. Proc.* **34**:635–637.

Barber, S. A., J. M. Walker, and E. H. Vasey. 1962. Principles of ion movement through the soil to the plant root. *Proc. Int. Soil Conf.*, New Zealand 121–124.

Claassen, M. E., and G. E. Wilcox. 1974. Comparative reduction of calcium and magnesium composition of corn tissue by NH_4-N and K fertilization. *Agron. J.* **66**:521–522.

Claassen, N., and S. A. Barber. 1976. Simulation model for nutrient uptake from soil by a growing plant root system. *Agron. J.* **68**:961–964.

Clarkson, D. T. 1965. Calcium uptake by calcicole and calcifuge species in the genus *Agrostis* L. *J. Ecol.* **53**:427–435.

Drew, M. C., and O. Biddulph. 1971. Effect of metabolic inhibitors and temperature on uptake and translocation of ^{45}Ca and ^{42}K by intact bean plants. *Plant Physiol.* **48**:426–432.

Dunlop, J. 1973. The kinetics of calcium uptake by roots. *Planta* **112**:159–167.

Ferguson, I. B., and D. T. Clarkson. 1975. Ion transport and endodermal suberization in the roots of *Zea mays*. *New Phytol.* **75**:69–79.

Foy, C. D. 1974. Effects of soil calcium availability on plant growth. In E. W. Carson, Ed. *The Plant Root and Its Environment*. University Press of Virginia, Charlottesville. Pp. 565–600.

Handley, R. and R. Overstreet. 1961. Uptake of calcium and chlorine in roots of *Zea mays*. *Plant Physiol.* **36**:766–769.

Halstead, E. H., S. A. Barber, D. D. Warncke, and J. B. Bole. 1968. Supply of Ca, Sr, Mn, and Zn to plant roots. *Soil Sci. Soc. Am. Proc.* **32**:69–72.

Hunsaker, V. E., and P. F. Pratt. 1971. Calcium magnesium exchange equilibria in soils. *Soil Sci. Soc. Am. Proc.* **35**:151–152.

Juo, A. S. R., and S. A. Barber. 1969. An explanation for the variability in Sr-Ca exchange selectivity of soils, clays, and humic acid. *Soil Sci. Soc. Am. Proc.* **33**:360–363.

Lazaroff, N., and M. G. Pitman. 1966. Calcium and magnesium uptake by barley seedlings. *Aust. J. Biol. Sci.* **19**:991–1005.

Leggett, J. E., and W. A. Gilbert. 1969. Magnesium uptake by soybeans. *Plant Physiol.* **44**:1182–1186.

Lindsay, W. L. 1979. *Chemical Equilibria in Soils*. John Wiley, New York.

Loneragan, J. F., and K. Snowball. 1969a. Calcium requirements of plants. *Aust. J. Agric. Res.* **20**:465–478.

Loneragan, J. F., and K. Snowball. 1969b. Rate of calcium absorption by plant roots and its relation to growth. *Aust. J. Agric. Res.* **20**:479–490.

Loneragan, J. F., K. Snowball, and W. J. Simmons. 1968. Response of plants to calcium concentration in solution culture. *Aust. J. Agric. Res.* **19**:845–857.

Maas, E. V. 1969. Calcium uptake by excised maize roots and interactions with alkali cations. *Plant Physiol.* **44**:985–989.

Malzer, G. L., and S. A. Barber. 1975. Precipitation of calcium and strontium sulfites around plant roots and its evaluation. *Soil Sci. Soc. Am. Proc.* **39**:492–495.

McLean, E. O. 1975. Calcium levels and availabilities in soils. *Commun. Soil Sci. Plant Anal.* **6**:219–232.

Mengel, D. B., and S. A. Barber. 1974. Rite of nutrient uptake per unit of corn root under field conditions. *Agron. J.* **66**:399–402.

Moore, D. P., L. Jacobson, and R. Overstreet. 1961. Uptake of calcium by excised barley roots. *Plant Physiol.* **36**:53–57.

Oliver, S., and S. A. Barber. 1966. An evaluation of the mechanisms governing the supply of Ca, Mg, K, and Na to soybean roots (*Glycine max*). *Soil Sci. Soc. Am. Proc.* **30**:82–86.

Reisenauer, H. M. 1964. Mineral nutrients in soil solution. In P. L. Altman and D. S. Dittmer, Eds. *Environmental Biology.* Fed. Am. Soc. Exp. Biol., Bethesda, MD. Pp. 507–508.

Scaife, M. A., and D. T. Clarkson. 1976. Calcium related disorders in plants—a possible explanation for the effect of weather. *Plant Soil* **50**:723–725.

Schofield, R. K., and A. W. Taylor. 1955. The measurement of soil pH. *Soil Sci. Soc. Am. Proc.* **19**:164–167.

Wallace, A., E. Frolich, and O. R. Lunt. 1966. Calcium requirements of higher plants. *Nature (London)* **209**:634.

Warncke, D. D., and S. A. Barber. 1972. Diffusion of zinc in soil. II. The influence of soil bulk density and its interaction with soil moisture. *Soil Soc. Am. Proc.* **36**:42–46.

CHAPTER **12**

Magnesium

Magnesium is the eighth most common element in the lithosphere, with an average concentration of 2.1%. However, due to weathering of relatively soluble magnesium minerals, the average magnesium concentration of soils is only 0.5%. It appears that three-fourths of the magnesium has been lost by weathering during soil formation. Because of differences in weathering and parent materials, magnesium contents of soils vary widely.

FORMS OF MAGNESIUM IN THE SOIL

Magnesium in the soil is a constituent of many soil minerals, it is present as exchangeable magnesium on the cation-exchange complex and in soil solution as the soluble magnesium ion. Small amounts of magnesium may also be combined in the soil's organic fraction.

Mineral Magnesium

The principal magnesium-containing soil minerals, their approximate chemical formulae, and appropriate magnesium concentrations are shown in Table 12.1. Weathering of the more soluble magnesium minerals, such as magnesite, dolomite, sulfate, and brucite, has reduced their concentration in the soil compared to that in the lithosphere. Magnesium sulfate is so soluble that it rarely occurs in soils. Magnesium is a constituent of soil clays, such as montmorillonite, vermiculite, chlorite, and illite. The location of magnesium in octahedral coordination in the structure of these clays is described in Chapter 2. In weathered soils, the amount of magnesium released from the remaining magnesium-containing minerals during a crop season is usually small compared with the magnesium requirement of the crop (Salmon and Arnold, 1963). However, Rice and Kamprath (1968) found that corn removed nonexchangeable magnesium from sandy coastal-plain soils. Christenson and Doll (1973) found that fractions separated from soil and used as a magnesium source for plant growth released magnesium from nonexchangeable positions. Release of nonexchangeable magnesium usually occurs where the level of C_{li} for magnesium is very low.

A few total soil magnesium values in the literature indicate a range of 0.015 to 1.02%. Total magnesium usually increases with increasing percentages of soil clay. The sequence of relative ease of weathering of the minerals shown in Table 12.1 is approximately sulfate > dolomite = magnesite > brucite > olivene > serpentine > vermiculite = illite = chlorite = montmorillonite.

Exchangeable Magnesium

Plant roots absorb soluble and ultimately exchangeable magnesium, which is assumed to go into solution before absorption by the root.

TABLE 12.1   Principal Magnesium-Containing Soil Minerals

Mineral	Formula	Magnesium concentration (%)
Dolomite	$CaCO_3 \cdot MgCO_3$	13
Olivine	$Mg_{1.6}Fe_{0.4}SiO_4$	25
Vermiculite	$Mg_3Si_4O_{10}(OH)_2 \cdot 2H_2O$	12–17
Illite	$K_{0.6}Mg_{0.25}Al_{2.3}Si_{3.5}O_{10}(OH)_2$	2
Montmorillonite	$Al_5MgSi_{12}O_{30}(OH)_6$	Up to 6
Chlorite	$Al_2Mg_5Si_3O_{10}(OH)_8$	Up to 23
Brucite	$Mg(OH)_2$	41
Sulfate	$MgSO_4$	20
Magnesite	$MgCO_3$	29
Serpentine	$H_4Mg_3Si_2O_9$	49

Source: Lindsay (1979) and Salmon (1963).

Measurement of exchangeable cations by obtaining those displaced from soil will include both exchangeable and solution cations unless solution ions are subtracted. These measurements give C_{si}. Exchangeable calcium plus magnesium usually accounts for more than 60% of the exchangeable cations on soils with pH 5.5 or higher. Exchangeable hydrogen, aluminum, potassium, and sodium occupy the majority of remaining exchange sites. Exchangeable magnesium is almost always present in smaller quantities than calcium, though on serpentine-derived soils, magnesium may be the dominant exchangeable cation. Since pure dolomitic limestone is $CaCO_3 \cdot MgCO_3$ and liming materials range from pure dolomite to pure calcite, limed soils will always contain as much or more exchangeable calcium as magnesium. Alston (1972) found that exchangeable magnesium levels in Irish soils ranged from 2.6 to 27.9% of the exchangeable bases present.

Baker (1972) investigated 12 Pennsylvania soils and found that exchangeable magnesium varied from 0.05 to 1.85 cmol/kg. This represented a percent magnesium saturation of 1.5 to 35.6%. Barber (unpublished data) found that exchangeable magnesium on 23 cultivated Indiana soils ranged from 0.81 to 8.1 cmol/kg. Mokwunye and Melsted (1972) analyzed samples of 24 tropical and temperate soils; exchangeable-magnesium levels ranged from 0.21 to 14.4 cmol/kg. Total magnesium levels for these soils ranged from 1.44 to 23.4 cmol/kg, and very little magnesium was complexed by the organic matter. Exchangeable magnesium has usually been measured by displacing it with 1 mol/L of ammonium acetate at pH 7.0, C_{si}. Values of C_{si} for magnesium are usually greater in soils having higher cation-exchange capacities. The amount of exchangeable magnesium in soils usually varies from 0.5 to 14 cmol/kg, with magnesium saturation of the exchange capacity usually 10 to 20%; exchangeable magnesium is commonly 20 to 60% of total magnesium.

Soil-Solution Magnesium, C_{li}

Soil-solution magnesium, C_{li}, is in equilibrium with exchangeable magnesium. Magnesium concentrations measured in solutions displaced from soils are reported in Table 12.2. Barber et al. (1962) reported magnesium concentrations in displaced solutions from 137 north central U.S. soils ranging from less than 5 to over 100 mg/L; only one value exceeded 100 mg/L, and five were less than 5 mg/L; 39% were between 16 and 35 mg/L. The range 26 to 30 mg/L, or 1.09 to 1.25 mmol/L, was the mode for the values. Alston (1972) studied 18 Irish soils, where the magnesium concentration in the soil solution ranged from 0.48 to 5.75 mmol/L; the mode of both sets of values had comparable ranges. Reisenauer (1964) reviewed the values reported in the literature prior to 1964 and found magnesium concentrations ranging from less than 25 to over 1000 mg/L; however, 38.7% of his 337 samples were in the 51 to 100 mg/L range. These values are higher than those reported by

TABLE 12.2   Soil Solution Magnesium Levels

Magnesium[a] concentration (mmol/L)	Fraction of samples (%)	Magnesium[b] concentration (mmol/L)	Fraction of samples (%)	Magnesium[c] concentration (mmol/L)	Fraction of samples (%)
0 −0.4	8.0	0 − 1.04	9.2	0 −1.0	24
0.4 −0.8	13.9	1.05− 2.08	21.4	1.1−2.0	28
0.8 −1.25	21.2	2.09− 4.16	38.6	2.1−3.0	12
1.26−1.67	13.1	4.16− 8.33	25.2	3.1−4.0	20
1.68−2.08	5.8	8.34−12.5	0.9	4.1−5.0	8
2.09−2.50	12.4	12.6−25.0	0.6	5.1+	8
2.51−2.92	10.2	25.1−29.2	1.8		
2.92+	15.2	29.2+	2.4		
$n = 137$		337		25	

[a]Barber et al. (1962).
[b]Reisenauer (1964); by permission of Federation of American Societies for Experimental Biology.
[c]Alston (1972).

Alston and Barber et al., because many were from aridic soils where displaced solution measurements have been more common. Aridic soils have soil solutions with much higher solute concentrations, since soluble salts may have accumulated or at least not have been lost by leaching. Data indicate a range in soil solution magnesium values of less than 0.4 to over 30 mmol/L. Hence, a common range for leached soils is 0.5 to 2 mmol/L, while in unleached soils it is 2 to 8 mmol/L.

MAGNESIUM-ADSORPTION ISOTHERMS

The relation between exchangeable magnesium and soluble magnesium in a soil can be described by an adsorption isotherm. Few magnesium-adsorption isotherms have been reported where concentrations in the soil solution and on cation-exchange sites have been measured. They are needed to measure the relative buffering of magnesium in the soil solution by magnesium held on cation-exchange sites.

Magnesium is almost always held less tightly by most exchange sites than calcium, so that the Ca:Mg ratio of the soil solution is less than on the exchange sites. Most of the evaluation of magnesium adsorption has been in comparison with adsorption of calcium, by calculating selectivity coefficients or by comparing Ca:Mg ratios in solution with those on cation-exchange sites. Salmon (1964a) made an extensive investigation of magnesium adsorption by soils; he expressed his results in terms of activity ratios of magnesium to calcium plus magnesium. He investigated 40 British soils and obtained the relation between Mg/(Ca + Mg) in solution and Mg/(Ca + Mg) on the exchange sites shown in Figure 12.1. He also investigated values for a peat soil and two clays, an illite and a bentonite. The values for the soils showed the ratio in solution to be higher than the ratio on the exchange sites by a factor between one and two. Peat adsorbed calcium much more tightly than it did magnesium, while there was little difference between the two ions for the clays used. Barber (unpublished data) compared the Ca:Mg ratios of exchangeable cations to ratios in the displaced solution for 12 Indiana soils; he found the value on the exchange sites to be 2.18 ± 0.5 of that in the soil solution; which agreed with Salmon, in that calcium was held more tightly than magnesium. There was no apparent relation between clay content or organic-matter content in these soils and the selectivity of calcium over magnesium by the exchange sites.

The buffer curve of exchangeable magnesium to solution magnesium will vary among soils because of differences in cation-exchange capacity, the nature of the exchange sites, and variation in other cations competing for the sites. The effect of adding or removing exchangeable magnesium on the activity ratio Mg:(Ca + Mg) in solution for individual soils gave a linear relationship between exchangeable magnesium and the activity ratio in studies reported by Arnold (1967). However, the slope of the linear relation varied widely among soils, indicating differences among soils in buffer power and relative adsorption strengths for calcium and magnesium. Baligar and Barber

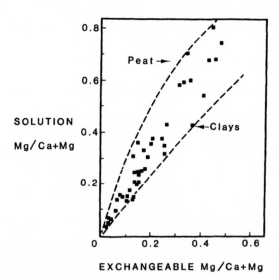

FIGURE 12.1   Relation between magnesium/(calcium + magnesium) in solution and magnesium/(calcium + magnesium) on the exchange sites of 40 British soils. Redrawn from Salmon (1964a) by courtesy of Blackwell Scientific Publications Ltd

(unpublished data) investigated adsorption curves following the addition of relatively large amounts of magnesium chloride to four soils. The curves were steeper, more magnesium adsorbed per unit increase in solution magnesium, the higher the cation-exchange capacity of the soil.

Udo (1978) determined selectivity coefficients for magnesium-calcium adsorption on a kaolinite soil clay; he used a wide range of Mg:Ca ratios. The selectivity coefficient Mg:Ca ranged from 0.520 to 0.558 at 10°C and 0.635 to 0.677 at 30°C, indicating that calcium was adsorbed relatively more tightly compared to magnesium at 10 than at 30°C. Varying the Mg:Ca ratio on the clay had little effect on the selectivity coefficient.

Data from values for exchangeable magnesium and soluble magnesium were obtained by Barber from 18 Indiana soils. Values for the buffer power can be obtained by dividing exchangeable magnesium whose value usually contains solution magnesium, C_{si}, expressed as mmol/dm³ of soil (most values are per kg, but they can be converted to per dm³ by dividing by the soil bulk density) by solution magnesium expressed as mmol/L. Values for b ranged from 1.2 to 61.7 with a mean of 15.3. Values varied with solute concentration of the soil solution as well as with cation-exchange capacity, nature of exchange sites, and complementary cations, as discussed in Chapter 2.

Little information is available on how solution magnesium is affected by the amount of calcium present, the degree of base saturation of the cation-exchange capacity, or the nature of the cation-exchange complex, particularly with respect to differences between permanent charge and pH-dependent charge sites. Information is needed on both the effect on magnesium concentration in solution and the Ca:Mg ratio of the soil solution.

REACTIONS OF MAGNESIUM WITH SOIL

Magnesium may be released from nonexchangeable sites of 2:1 clay minerals. It may also be adsorbed into these positions so tightly that it is referred to as "fixed." Magnesium may also be fixed by the amorphous minerals in Ultisols. Finally, magnesium may come into solution due to the solubility of magnesium-containing soil minerals.

Magnesium Fixation

Chan et al. (1979) showed that for soils with pH-dependent or variable charge, magnesium that was exchangeable at soil pH values below 6.0 became nonexchangeable as soil pH increased above 6.5. At pH 8.4, 25% of the magnesium exchangeable at pH 2.5 was no longer exchangeable. Sumner et al. (1978) found even greater fixation of magnesium in Ultisols from both the United States and South Africa. They observed magnesium fixation at pH values as low as 5.5. After adding calcium hydroxide to replace exchangeable hydrogen and aluminum and raise the soil pH, they observed that more than half of the formerly exchangeable magnesium became nonexchangeable when the pH was raised above 7.5. Even where Sumner et al. added magnesium by applying dolomitic lime, exchangeable-magnesium levels dropped as pH increased above 6.5. The effect of liming with dolomitic limestone on pH and exchangeable-magnesium and calcium levels in these soils is shown in Figure 12.2. Calcium did not appear to be fixed. Chan et al. (1979) suggested that magnesium became nonexchangeable through specific adsorption in the Stern layer. Sumner et al. (1978) theorized that magnesium became nonexchangeable through formation of magnesium silicate precipitates. Information is needed on the effect of pH on solution magnesium levels in soils where magnesium fixation occurs—primarily soils with a pH-dependent charge.

Release of Magnesium from Soil Materials

Intensive cropping of soil in the greenhouse has been used to study the release of nonexchangeable magnesium to crops. Salmon and Arnold (1963) found that exhaustive cropping released negligible amounts of magnesium from English and Welsh soils when cropped with ryegrass (*Lolium perenne*). Rice and Kamprath (1968) cropped Atlantic coastal plain soils with corn. On some soils that were low in exchangeable magnesium, more magnesium was obtained from nonexchangeable than exchangeable sources. The release of magnesium occurred where the exchangeable-magnesium level was below that needed to supply sufficient magnesium to

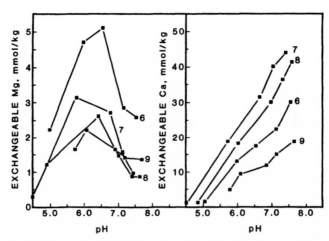

FIGURE 12.2   Effect of soil pH (water) on levels of exchangeable calcium and magnesium in soils. Reprinted from Sumner et al. (1978) by courtesy of Marcel Dekker, Inc.

the plant. Release was believed to be from magnesium-containing soil clay minerals. Kidson et al. (1975) measured magnesium release during exhaustive cropping of 11 New Zealand soils. In four soils, a small release was found, but the remainder showed no or little measurable release. Christenson and Doll (1973) obtained magnesium release from octahedral layers of clay minerals by growing oats (*Avena sativa*) with mineral magnesium as the only magnesium source. While release of mineral magnesium may occur, for most soils the amount released appears to be small relative to the quantity present.

Solubility of Magnesium Compounds in the Soil

Lindsay (1979) has reviewed pH versus solubility characteristics of various soil magnesium minerals. At pH values below 7.0, all of the minerals were sufficiently soluble to maintain a soluble magnesium concentration in excess of 1 mmol/L. Because of their solubility, minerals such as magnesium sulfate, brucite, and magnesite are leached out of weathered soils and not likely to form in these soils even when magnesium is added. Solubility of magnesium minerals does not appear to control the soluble-magnesium level, except where magnesium is precipitated at pH values above 5.5 in Ultisols, as discussed under magnesium fixation.

Soil Supply Parameters

The three soil parameters used in the mathematical model in Chapter 5 to describe the process of nutrient uptake from soil by plant roots are C_{li}, b, D_e. These three factors interact to determine the rate of magnesium supply to the root. Soil solution magnesium concentration is usually 1 or more mmol/L. Sandy soils low in exchangeable magnesium usually have the lowest levels.

Solution Magnesium

Table 12.2 reports data from three summaries of soil solution magnesium concentrations; less than 10% of the values were below 10 mg/L (0.42 mmol/L). The majority of the values on leached soils were from 21 to 30 mg/L (0.8 to 1.25 mmol/L) and from 51 to 100 mg/L (2.1 to 4.2 mmol/L) on aridic soils.

Buffer Power

Few values of magnesium buffer power have been calculated.

The relationship between C_{si} for magnesium and solution magnesium, C_{li}, for 14 Indiana soils sampled by Barber (unpublished data) indicated that there was no correlation between C_{li} and C_{si}. There was variability due to different soil-exchange sites, cation-exchange capacities, and levels of the complementary cations—calcium, hydrogen, aluminum, and potassium.

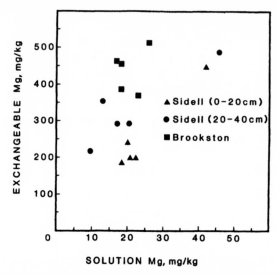

FIGURE   12.3   Relation between exchangeable magnesium and soil solution magnesium for two Indiana soils each at several levels of exchangeable magnesium. Drawn from data of Al Abbas and Barber (1964) by permission of the Williams & Wilkins Co.

Using different calcium and magnesium additions to two surface and one subsoil gave the results shown in Figure 12.3 (Al Abbas and Barber, 1964). Brookston soil had a higher organic matter content and also a higher magnesium buffer power. Soil magnesium buffer-power values for samples evaluated by the author varied from 1.2 to 61.7.

Diffusion Coefficient

The diffusion coefficient can be estimated by the relation $D_e = D_l \theta f_1/b$. The value of D_l, the diffusion coefficient in water, for magnesium is 0.70×10^{-5} cm²/s. θ is the volumetric water content, for which a value of 0.2 is reasonable. At this water content for a silt loam soil, f_1, the impedence factor, is approximately 0.18 (Warncke and Barber, 1972). Using these values and a value for $b = 15$, we get a D_e value of 1.7×10^{-8} cm²/s for magnesium on a midwestern U.S. silt loam soil.

MAGNESIUM-UPTAKE KINETICS

Magnesium uptake by plant roots growing in stirred solution has been studied with both excised and intact roots. Fewer experiments have been conducted with magnesium than other nutrients, because the [28]Mg isotope has a short half-life. Moore et al. (1961) studied magnesium uptake by excised barley roots and found that the absorption rate from solution culture was about one-half that of potassium. Uptake rate was reduced greatly by adding calcium as compared to a system without calcium. Increasing calcium from equal calcium and magnesium levels to four times as much calcium as magnesium reduced magnesium uptake by the excised barley roots to one-third or less. Moore et al. (1960) did not measure the effect of solution magnesium concentration on magnesium-uptake rate.

Ferguson and Clarkson (1976) found that suberization of the endodermis of corn roots restricted magnesium uptake from solution culture just as it did calcium uptake. On the basis of this evidence, they inferred that magnesium was being absorbed passively by the apoplastic pathway. They used solutions with a magnesium concentration of 0.2 mmol/L and a calcium concentration of 0.5 mmol/L.

Factors Affecting Magnesium Influx

The relation between magnesium concentration in solution and magnesium-uptake rate was investigated by Maas and Ogata (1971) using excised corn roots; similar studies using excised soybean roots were conducted by Leggett and Gilbert (1969). Maas and Ogata used magnesium concentrations ranging from 0 to 5 mmol/L and obtained a K_m value of 0.15 mmol/L. Leggett and Gilbert used concentrations from 0 to 5 mmol/L and obtained an uptake by excised soybean roots that indicated a K_m value of 0.4 mmol/L; uptake was from $MgCl_2$ solution. Joseph and Van Hai (1976) used intact soybean plants and calculated a V_{max} of 18.8 μmol/$g_{D.W.} \cdot$ h and a K_m of 13 μmol/L for phase-1 uptake. Above 31 μmol/L, influx increased to phase-2 uptake (Nissen, 1977).

The K_m value of 13 μmol/L reported by Joseph and van Hai (1976) appears extremely low compared to values of 150 μmol/L reported by Maas and Ogata (1977) and 400 μmol/L reported by Leggett and Gilbert (1969). Hence, it would appear that either a different uptake phase is being investigated in each case or plant species and conditions are giving different values. With soil solution values frequently in the range of 1 mmol/L magnesium, it appears that the soil-solution level is either much above K_m or near K_m, depending on K_m values selected.

The maximum magnesium-influx values will depend on the levels of calcium, potassium and ammonium in the soil solution at the root surface. Maximum adsorption rates reported in the literature are Maas and Ogata (1971), excised corn roots, 1.5 μmol/$g_{F.W.} \cdot$ h; Leggett and Gilbert (1969), excised soybean roots, 0.35 μmol/$g_{F.W.} \cdot$ h; Joseph and van Hai (1976), intact soybeans, 18.8 μmol/$g_{D.W.} \cdot$ h. Since root dry weight is about 10–15% of fresh weight, these values would be about 1.2 to 1.9 μmol/$g_{F.W.} \cdot$ h.

Kelly et al. (1992) measured the uptake kinetics of magnesium by loblolly pine (*Pinus taeda* L.) seedlings and obtained an I_{max} of 1.29×10^{-7} μmol/$cm^2 \cdot$ s, a K_m of 9.8 mmol/L, and a C_{min} of 1 mmol/L. Rengel and Robinson (1990a) measured magnesium uptake kinetics of ryegrass

(*Lolium multiflorum* Lam.) and obtained the I_{max} of 1.3 μmol/cm² · s, a K_m of 39 μmol/L, and a C_{min} of 2 μmol/L for 6 to 23 days growth of Gulf ryegrass.

Effect of Transpiration Rate

Lazaroff and Pitman (1966) used intact barley seedlings, 5 to 12 days old, to show how increasing transpiration increased magnesium uptake when the solution concentration was 15 mmol/L but not when 0.5 mmol/L (Figure 12.4). Similar results were observed for calcium. It may be that passive uptake occurs at higher solution concentrations, while magnesium is absorbed by energy-dependent mechanisms at lower solution concentrations. Magnesium concentrations in the soil solution are often about 1 mmol/L, so transpiration rate may have little effect unless magnesium accumulation occurs around the root or the soil solution has a particularly high magnesium concentration.

Effect of Calcium Level on Magnesium Uptake

The level of soluble calcium can affect the rate of magnesium uptake. A few studies have investigated uptake over both a few hours and several days. Lazaroff and Pitman (1966) grew barley plants for 9 days at various calcium/magnesium ratios; they used a constant solution concentration of calcium plus magnesium of 20 mmol/L. This concentration was probably near what may occur under arid-soil conditions but about 10 times that found in leached soils. They found a very marked effect of calcium level on magnesium uptake into the shoot. In contrast, Maas and Ogata (1971), using 5-day-old corn

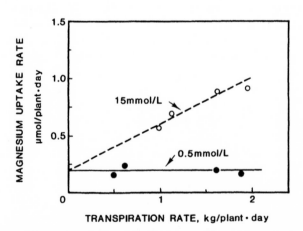

FIGURE   12.4   Magnesium influx into barley as affected by solution magnesium concentration and transpiration rate. Dashed line is for solution concentration of 15 mmol/L, solid line for 0.5 mmol/L. Reproduced from Lazaroff and Pitman (1966) by permission of CSIRO.

roots, varied calcium from 1 to 2.5 mmol/L and only reduced magnesium uptake slightly when magnesium concentration was 0.25 mmol/L. Leggett and Gilbert (1969), using excised soybean roots, found little effect of magnesium/calcium ratio on magnesium uptake in 24 h when the magnesium/calcium ratio was varied through the range of 4 to 0.25. The calcium plus magnesium concentration was 2.5 mmol/L; results are shown in Figure 12.5.

We could speculate that magnesium is absorbed actively when calcium plus magnesium concentrations are below 5 mmol/L, so that rate of transpiration has little effect on uptake. Also under these conditions, calcium concentration may have little effect on magnesium-uptake rate. However, when soluble calcium plus magnesium concentration is 20 mmol/L or higher and intact transpiring plants are used, uptake is largely non-energy-dependent. Hence, the calcium/magnesium ratio of uptake is similar to the calcium/magnesium ratio of the solution, because at this concentration, both calcium and magnesium are largely absorbed passively. Applying this speculation to crop production would suggest that the calcium/magnesium ratio has only a small effect on magnesium uptake from leached soils as long as the magnesium concentration in solution is high enough to maintain magnesium influx at near its maximum rate but not so high that passive uptake becomes important. It is probably also important that calcium concentration be low enough so that transpiration rate does not affect uptake. Under these conditions, it is probable that magnesium absorption is by active uptake and dependent on respiration. On unleached soils from drier areas, the soil solution salt concentration is high, and both calcium and magnesium uptake may be largely passive because of high ion concentrations. The calcium/magne-

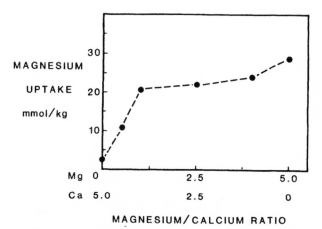

FIGURE 12.5 Effect of calcium/magnesium ratio on magnesium uptake by excised soybean roots. Time 24 hs; calcium plus magnesium concentration remains at a total of 2.5 mmol/L. Reproduced from Leggett and Gilbert (1969) by permission of American Society of Plant Physiologists.

sium ratio of uptake is then more nearly that of the ratio in the soil solution. Hence, a high calcium/magnesium soil solution ratio for an unleached soil may cause low plant magnesium contents when the same ratio on a leached soil may not cause low plant magnesium contents because of the change in uptake mechanism with solution calcium and magnesium concentrations. These speculations await experimental verification.

Effect of Potassium Level on Magnesium Uptake

Increasing potassium level in the soil or solution causes a reduction in both calcium and magnesium uptake; frequently, magnesium is affected more than calcium (Metson, 1974). Claassen and Wilcox (1974) grew corn for 33 days on two soils fertilized with three rates of potassium (0, 50, and 100 mg/kg). Adding potassium decreased shoot magnesium composition from 0.32 to 0.24% on one soil and from 0.71 to 0.44% on the second soil (Table 11.4). Claassen and Barber (1977) grew corn from 7 to 17 days of age in split-root experiments where potassium was supplied to the roots on one side of a barrier and not on the other; data are shown in Table 3.6. These and other data show that adding potassium depresses magnesium uptake to as little as one-half the rate where no potassium is present. Additional data are needed to determine the interaction of potassium with magnesium uptake as the magnesium level is varied. Does potassium affect magnesium uptake when magnesium is absorbed passively?

Mayland and Grunes (1979) reviewed data on the effect of soil potassium on magnesium uptake. They reported that the reduction in magnesium uptake caused by potassium additions is greater for grasses than associated clovers. Potassium additions reduce plant magnesium levels, but magnesium additions do not alter plant potassium levels. However, Roberts and Weaver (1974) found an exception to this in a study where applying magnesium lowered potassium concentrations of sudan grass grown on a slightly calcareous Shano silt loam, Xerollic Camborthids, soil.

Salmon (1964b) found that an increased soil-potassium level decreased magnesium uptake by plants. This factor had to be considered in correlating the activity ratio of calcium and magnesium with magnesium uptake. McLean and Carbonell (1972) found that doubling the soil potassium level from 2.5 to 5.0% saturation of the cation exchange capacity reduced magnesium concentration of millet and alfalfa grown on two soils. This reduction was restored by increasing magnesium saturation from 5 to 15%. Hannaway et al. (1980) studied the effect of potassium concentration in solution on ^{28}Mg uptake by fescue. They found that potassium had a greater depressive effect on magnesium translocation to the shoot than on magnesium uptake by the root.

Effect of Ammonium on Magnesium Uptake

Addition of ammonium fertilizer, particularly where nitrification is inhibited, so that nitrogen is absorbed as ammonium, will reduce magnesium uptake. Claassen and Wilcox (1974) showed that ammonium additions reduced magnesium uptake to a greater degree than potassium additions (Table 11.4). Information is also needed to determine if magnesium level affects this reduction. Can sufficient magnesium be added to overcome the depression? Cox and Reisenauer (1973) showed that adding NH_4 to a solution culture reduced magnesium uptake by wheat.

Effect of Solution pH on Magnesium Uptake

Moore et al. (1960) showed a rapid decrease in magnesium uptake from solution by excised barley roots when pH decreased below 5.0. Maas and Ogata (1971) obtained a reduction in magnesium uptake by excised corn roots when solution pH was below 5.0. Leggett and Gilbert (1969), working with excised soybean roots, showed a reduction in magnesium uptake when pH was below 5.5. All three studies gave similar results; information is needed with intact roots to determine if the same effect occurs when whole plants and longer absorption times are used.

Effect of Temperature on Magnesium Uptake

Grunes et al. (1968) grew ryegrass at two temperatures in sand watered with nutrient solution; temperature regimes were 20°C day, 14°C night and 26°C day, 23°C night. Four levels of magnesium were used in the nutrient solution. The effect of temperature on plant magnesium concentration is shown in Figure 12.6. Magnesium concentration was higher with higher temperature at all levels of magnesium; calcium and potassium concentrations were also increased. Increasing temperature increased rate of magnesium uptake more than it increased plant growth. Since solution magnesium level was in

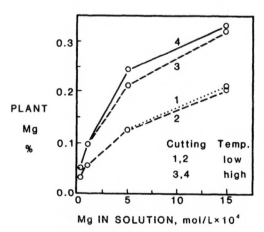

FIGURE   12.6   Effect of magnesium level and temperature on magnesium concentration in perennial ryegrass grown in sand (Grunes et al., 1968).

the range of 0.03 to 1.5 mmol/L, effect of temperature on transpiration probably did not affect uptake.

SOIL MEASUREMENTS OF BIOAVAILABLE MAGNESIUM

Several empirical soil measurements have been suggested for evaluating relative rates of magnesium supply from soil to plants; these measurements include

1. Exchangeable magnesium, usually measured by displacement with 1 mol/L ammonium acetate, pH 7.0, and expressed per unit weight of soil.
2. Expressing exchangeable magnesium as percent saturation of the cation-exchange capacity (McLean and Carbonell, 1972; Alston, 1972).
3. Calcium–magnesium activity ratio in the equilibrium soil solution (Salmon, 1964a).
4. Measurement of supply by root interception, mass flow, and diffusion (Al-Abbas and Barber, 1964).
5. Measurement of predicted magnesium uptake with the mechanistic nutrient uptake model (Barber and Cushman, 1981).

Magnesium in the soil that can move rapidly to the root and be absorbed includes both exchangeable magnesium on the cation-exchange sites and the smaller quantity in solution. Exchangeable magnesium is assumed to go into solution before movement to, and absorption by, the root. The quantity in solution is usually from 1 to 10% of the amount of exchangeable magnesium. The measure of exchangeable magnesium plus that in solution, C_{si}, reflects the total available for movement to the root. When magnesium uptake by plants is used to evaluate availability of increasing rates of magnesium added to one soil, this value would be closely correlated with magnesium availability expressed as either exchangeable magnesium, percent saturation, or calcium/magnesium ratio.

When soils varying in cation-exchange capacity are compared, percent saturation is suggested as a factor influencing magnesium availability, with the belief that the supply to the root will be less when soils of higher cation-exchange capacity hold the magnesium more tightly. In addition, higher amounts of calcium will likely be present, so there will be less magnesium per unit amount of calcium at the root surface. McLean and Carbonell (1972) compared two soils at five levels of exchangeable magnesium and two levels of potassium. The soils had exchange capacities of 8.85 and 18.7 cmol(p⁺)/kg. One crop of German millet (*Setaria italica*) and then five cuttings of alfalfa (*Medicago sativa*) were grown. Adding magnesium did not affect yield, but it did affect magnesium concentration in the shoot, which was used as a measure of plant-available magnesium. Plant magnesium concentration was related to percent of magnesium saturation of cation-exchange sites as well as the amount of magnesium. Magnesium concentration of millet was related to percent of magnesium saturation of the two soils. However, when alfalfa was grown, the difference between the soils in magnesium concentration of the shoot was greater than could be accounted for by differences in percent saturation. Results for millet supported the use of exchangeable magnesium as an index; results for alfalfa did not support use of either exchangeable or percent saturation.

Alston (1972) evaluated magnesium uptake by ryegrass (*Lolium perenne* L.) from 25 soils from Northern Ireland. Magnesium concentration in the plants was correlated with exchangeable magnesium ($r = 0.67$), exchangeable magnesium expressed as percent saturation ($r = 0.73$), exchangeable magnesium expressed as a percent of exchangeable bases ($r = 0.81$), and calcium/magnesium activity ratio ($r = 0.75$). The highest correlation was with the activity ratio adjusted for potassium level ($r = 0.88$). Salmon (1964b) used 40 British soils in an investigation of magnesium availability; he obtained a correlation between magnesium concentration in ryegrass and exchangeable soil magnesium ($r = 0.51$). Al-Abbas and Barber (1964) used three soils each having five calcium or magnesium treatments and three soils untreated with calcium or magnesium; the soils were from Indiana and South Carolina. Magnesium uptake by soybeans was correlated with exchangeable magnesium ($r = 0.87$) and soil solution magnesium ($r = 0.74$). McNaught et al. (1973) found that soil solution magnesium accounted for only 46 to 61% of the variability in forage magnesium levels.

Salmon (1964b) found that the calcium/magnesium activity ratio in solution was a closer measurement of magnesium availability than exchangeable magnesium; the effectiveness was increased when the variability of potassium in the soil solution was accounted for. However, since the factor accounting for the effect of potassium on magnesium uptake varied from soil to soil, and was determined using both plant and soil analysis, it would not be a practical measure for soil magnesium availability. With adjustments for both potassium level and soil pH, Salmon (1964b) was able to obtain $r = 0.95$ for the first ryegrass crop, but it dropped to 0.74 for the fourth harvest and to 0.77 for the sixth harvest.

Calcium/Magnesium Ratio in the Soil and Magnesium Uptake

The effect of calcium/magnesium ratio of exchangeable cations on the calcium/magnesium ratio of uptake by corn and alfalfa was investigated in field experiments by Simson et al. (1979). They obtained a correlation between the exchangeable calcium/magnesium ratio and the ratio in the plant (*r* values varied from 0.81 to 0.96 for alfalfa (*Medicago sativia*) and corn grown in 1974 and 1975). The relationship is shown in Figure 12.7. The ratio in the

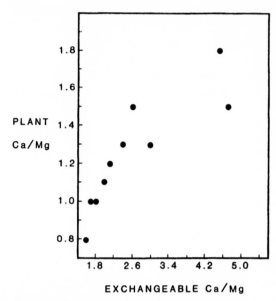

FIGURE 12.7   Relation between calcium/magnesium ratio in alfalfa and ratio exchange calcium/magnesium. Drawn from data of Simson et al. (1979) and reprinted by permission of Marcel Dekker.

plant was 1/3 that on the exchange sites; previously, it was shown that the calcium/magnesium ratio in solution was about one-half that on the exchange sites. In this experiment, the authors calculated that supply to the root by mass flow and root interception was two to four times the rate of uptake, so that these cations would accumulate at the root and uptake would be from a solution of high concentration. The selectivity for magnesium uptake from the soil by plants occurs partly between the exchange sites and solution and partly between the solution and plant uptake. Lazaroff and Pitman (1966) have shown the latter selectivity when plants were grown in a solution of calcium plus magnesium concentration of 40 mmol/L.

The selectivity of the plant, however, can also exert a considerable effect. Schulte et al. reported the results for table beets, sweet corn, and peas shown in Table 12.3. The calcium/magnesium ratio for uptake varied greatly among species growing in the same soil.

Mechanisms of Soil Magnesium Supply to the Root

Supply of magnesium to the root depends on root interception, mass flow, and diffusion. Mass flow is frequently a major supply mechanism for magnesium (Oliver and Barber, 1966). If plants contain from 0.2 to 0.6% magnesium and transpire 300 g of water per g of plant dry matter, the magnesium concentration needed in the soil solution to supply all of the magnesium

TABLE 12.3   Effect of Plant Species on Calcium/Magnesium-Uptake Ratio for a Soil Having Several Ratios of Exchangeable Calcium/Magnesium

Soil pH	Soil Ca/Mg	Ca/Mg in plant shoots		
		Table beet	Sweet corn	Peas
4.7	4.8	0.78	1.4	3.6
5.3	4.0	0.78	1.2	3.6
5.8	4.0	0.77	1.1	3.6
6.3	3.7	0.75	1.1	3.3
7.2	2.8	0.79	1.0	2.9

Source: Schulte et al. (1980). Courtesy of Solutions Magazine.

needed by mass flow alone can be calculated. The calculation for plants containing 0.6% magnesium is

$$\left(\frac{0.6}{300}\right) \times 10\ 000\ =\ 20\ \text{mg}/\text{L, or }0.83\ \text{mmol}/\text{L} \qquad (12.1)$$

If the plant contained 0.2% magnesium, the required concentration would be 6.67 mg/L, or 0.28 mmol/L. Table 12.2 showed that the most common magnesium concentration in the soil solution is about 1 mmol/L. Hence, the most common concentration would supply more magnesium to the root by mass flow than the plant required.

Al Abbas and Barber (1964) measured the magnesium supplied by mass flow to the root and added to it an estimation of magnesium intercepted by the root as it grew through the soil. They obtained the relation shown in Figure 12.8, where correlation between the values gave $r^2 = 0.84$. Magnesium uptake was similar to magnesium supply by mass flow and root interception. At low levels of magnesium, magnesium uptake was more than double the supply to the root. Hence, diffusion of magnesium to the root must also have occurred.

MODELING MAGNESIUM UPTAKE

The mechanistic nutrient uptake of Barber and Cushman, described in Chapter 5, has been used by Rengel and Robinson (1990a,b) for pot studies of magnesium uptake by ryegrass and by Kelly et al. (1992) for pot studies with loblolly pine. Rengel and Robinson (1990a) found predicted magnesium uptake was greater than observed uptake by factors ranging from 1.1 to 1.7 depending on the magnesium application rate applied to the soil. They believed I_{max} was the main reason for the disparity and concluded the major obstacle to using the Barber–Cushman model for predicting magnesium

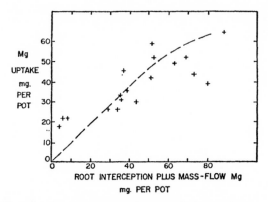

FIGURE   12.8   Relation between magnesium uptake by soybeans and sum of magnesium supplied by mass flow and magnesium intercepted by the roots. From Al Abbas and Barber (1964) by permission of the Williams & Wilkins Co.

uptake is the lack of information about kinetic parameters for plant uptake of magnesium from soil solution.

Kelly et al. (1992) used the Barber–Cushman model to study magnesium uptake by loblolly pine seedlings. They found the model only predicted 38% of observed uptake for growth in a pot study where seedlings were grown for 180 days. The model showed that C_{l0} increased with time since more magnesium reached the root than was absorbed by the root. A pattern similar to curve C of Figure 5.9 was predicted. The C_{l0} value predicted at equilibrium was approximately 100 times larger than the initial solution value used when using the magnesium depletion procedure for obtaining I_{max}, K_m, and C_{min}. Influx would always be at I_{max} and if I_{max} increased with an increase in C_{l0} this would explain the underprediction. The same experiment gave predicted potassium and phosphorus uptake of 94 and 89%, respectively, of observed uptake. Since phosphorus and potassium mainly reached the root by diffusion in this system, there would be no accumulation of these nutrients at the root surface but a decrease as shown for potassium in Figure 10.8. Kelly et al. (1992) produced sensitivity diagrams where I_{max} was increased, which showed a large effect of I_{max} on predicted magnesium uptake by loblolly pine. Just as for the research of Rengel and Robinson (1990b), the value for I_{max} appeared to be the reason for lack of close agreement between observed and predicted magnesium uptake. In both cases discussed here the C_{li} level was high enough that the magnesium supply to the root was greater than the rate of magnesium uptake. In soils that are low in magnesium and have low C_{li} values the supply to the root may have an effect on influx since the C_{l0} value would decrease with time. However, since the b value for magnesium is usually small, the diffusion coefficient will be larger than for phosphorus and potassium and the I_{max} value will likely affect uptake.

The results of these two experiments illustrate how use of the mechanistic model can determine the factor having the greatest effect on magnesium uptake. The distribution of magnesium around the root will be altered by the balance between rate of magnesium uptake and rate of supply from the soil to the root, as described in Chapter 5. Mass flow supplies more magnesium than plant uptake on well-supplied soils and less where soil magnesium is low. Hence, magnesium distributions perpendicular to the root will be similar to the range of possibilities shown in Figure 5.9.

In nature, magnesium may be accumulating at the root surface, and calcium may accumulate even faster than magnesium, while potassium level at the root will decrease with time, since this nutrient is usually supplied by diffusion. Since magnesium uptake depends on levels of calcium and potassium in solution as well as the magnesium level, we need information on all three nutrients so that their simultaneous flux to the root can be modeled mathematically.

Bouldin (1989) developed an uptake model that predicts uptake of calcium, magnesium, and potassium simultaneously, providing the soil supply interactions of these three nutrients are known. We then can run the uptake patterns for calcium, magnesium, and potassium simultaneously, using knowledge of how uptake of each cation depends on the levels of the other two. By combining this information on soil supply with the kinetics of magnesium uptake, we should then be in a position to describe magnesium uptake, including its dependence on various factors involved.

Unlike potassium, plant uptake of magnesium is usually a small portion of the total exchangeable magnesium present in the soil. Hence, depletion of magnesium from the soil by plant uptake is a minor factor in evaluating uptake mechanisms and $C_l v_0$ becomes a major factor since it supplies the root by mass flow.

Magnesium availability in the soil is related to C_{li}, b, and D_e. Deficiencies in magnesium that reduce plant growth usually occur on sandy soils, low in pH and high in potassium and/or ammonium. Under these conditions, the exchangeable-magnesium level is low, and, hence, C_{li} and b are low. If C_{li} levels are adequate for initial growth, b will be too low to maintain supply. Measuring C_{si} has been found to indicate sufficiency. Levels above 50 mg/kg are usually sufficient. Effects of competition from exchangeable calcium will probably depend on the relative bonding strength of the calcium.

Livestock magnesium deficiencies (hypomagnesemia) occur when livestock eat forage produced on soils having low magnesium bioavailability; this can occur even at magnesium levels high enough to maximize crop yield. Environmental factors in addition to soil factors influence magnesium concentrations in the shoot (Hannaway et al., 1980).

Excess magnesium does not affect yield of most crops so long as there is more exchangeable calcium than magnesium present. Hence, the ratio of calcium to magnesium can vary over the a range of 1 to 20, and magnesium availability may still be sufficient as long as the total amount of magnesium is high enough (Barber, 1969). The relative calcium and magnesium content of the plant will vary, but this is of little consequence when there is not a livestock nutritional problem. When only grain is harvested, the calcium/magnesium ratio is probably even less important. Plants take up similar quantities of calcium and magnesium. Because calcium is held more tightly by soil-exchange sites, however, larger amounts of calcium than magnesium are needed on exchange sites to give equal rates of supply by mass flow and diffusion to the root surface. Ideally, the exchangeable calcium/magnesium ratio should be between two and seven. However, this may also change due to differences between soils in relative strengths of bonding of these cations to cation-exchange sites present in the soil. Magnesium deficiency has been considered as one reason for forest decline. Magnesium deficiencies are more common in horticultural crops than in forage or grain crops (Wilkinson et al., 1990).

REFERENCES

Al Abbas, H., and S. A. Barber. 1964. The effect of root growth and mass flow on the availability of soil calcium and magnesium to soybeans in a greenhouse experiment. *Soil Sci.* **97**:103–107.

Alston, A. M. 1972. Availability of magnesium in soils. *J. Agric. Sci.* **79**:197–204.

Arnold, P. W. 1967. Magnesium and potassium supplying power of soils. In W. Dermott and D. J. Eagle, Eds. *Soil Potassium and Magnesium. Gr. Br. Min. Agr. Food Tech. Bull.* **14**:39–48.

Baker, D. E. 1972. Soil chemistry of magnesium. In J. B. Jones, M. C. Blount, and S. R. Wilkenson, Eds. *Magnesium in the Environment; Soils, Crops, Animals, and Man.* Taylor County Printing Co., Reynolds, GA. Pp. 1–39.

Barber, S. A. 1969. Dolomitic or calcitic lime. Publication AY-155. Agronomy Dept., Purdue University, West Lafayette, Indiana.

Barber, S. A., J. M. Walker, and E. H. Vasey. 1962. Principles of ion movement through the soil to the plant root. *Proc. Int. Soil Conf.,* New Zealand 121–124.

Barber, S. A., and J. H. Cushman. 1981. Nitrogen uptake model for agronomic crops. In J. K. Iskandar, ed., *Modeling Waste Water Renovation—Land Treatment.* John Wiley, New York.

Bouldin, D. R. 1989. A multiple ion uptake model. *J. Soil Sci.* **40**:309–319.

Claassen, M. E., and G. E. Wilcox. 1974. Comparative reduction of calcium and magnesium composition of corn tissue by NH_4-N and K fertilization. *Agron. J.* **66**:521–522.

Claassen, N., and S. A. Barber. 1977. Potassium influx characteristics of corn roots and interaction with N, P, Ca, and Mg influx. *Agron. J.* **69**:860–864.

Chan, K. Y., B. G. Davey, and H. R. Geering. 1979. Adsorption of magnesium and calcium by a soil with variable charge. *Soil Sci. Soc. Am. J.* **43**:301–304.

Christenson, D. R., and E. C. Doll. 1973. Release of magnesium from soil, clay, and silt fractions during cropping. *Soil Sci.* **116**:59–63.

Cox, W. J., and H. M. Reisenauer. 1973. Growth and ion uptake by wheat supplied nitrogen as nitrate, or ammonium, or both. *Plant Soil* **38**:363–380.

Ferguson, I. B., and D. T. Clarkson. 1976. Simultaneous uptake and translocation of Mg and Ca in barley roots. *Planta* **128**:167–169.

Grunes, D. L., J. F. Thompson, J. Kubata, and V. S. Laxar. 1968. Effect of Mg, K, and temperature on growth and composition of *Lolium perenne*. *Int. Cong. Soil Sci. Trans. Ninth,* Adelaide, Australia **2**:597–603.

Joseph, R. A., and T. van Hai. 1976. Kinetics of potassium and magnesium uptake by intact soybean roots. *Physiol. Plant.* **36**:233–235.

Kelly, J. M., S. A. Barber, and G. S. Edwards. 1992. Modeling magnesium, phosphorus and potassium uptake by loblolly pine seedlings using a Barber-Cushman approach. *Plant Soil* **139**:209–218.

Kidson, E. B., F. A. Hole, and A. J. Metson. 1975. Magnesium in New Zealand soils. III. Availability of non-exchangeable magnesium to white clover during exhaustive cropping in a pot trial. *N.Z. J. Agric. Res.* **18**:337–349.

Lazaroff, N., and M. G. Pitman. 1966. Calcium and magnesium uptake by barley seedlings. *Aust. J. Biol. Sci.* **19**:991–1005.

Leggett, J. E., and W. A. Bilbert. 1969. Magnesium uptake by soybeans. *Plant Physiol.* **44**:1182–1186.

Lindsay, W. L. 1979. *Chemical Equilibria in Soils.* Wiley-Interscience, New York.

Maas, E. V., and G. Ogato. 1971. Absorption of magnesium and chloride by excised corn roots. *Plant Physiol.* **47**:357–360.

Mayland, H. F., and D. L. Grunes. 1979. Soil-climate-plant relationships in the etiology of grass tetany. In V. V. Rendig and D. L. Grunes, Eds. *Grass Tetany.* Spec. Pub. No. 35, American Society of Agronomy, Madison, WI. Pp. 123–175.

McLean, E. O., and M. D. Carbonell. 1972. Calcium, magnesium, and potassium saturation ratios in two soils and their effects upon yield and nutrient contents of German millet and alfalfa. *Soil Sci. Soc. Am. Proc.* **36**:927–930.

McNaught, K. J., F. D. Dorofaeff, T. E. Ludecke, and K. Cottier. 1973. Effect of potassium fertilizer, soil magnesium status, and soil type on uptake of magnesium by pasture plants from magnesium fertilizers. *N.Z. J. Exp. Agric.* **2**:277–319.

Metson, A. J. 1974. Magnesium in New Zealand soils. I. Some factors governing the availability of soil magnesium: A review. *N.Z. J. Exp. Agric.* **2**:277–319.

Mokwunye, A. V., and S. W. Metsted. 1972. Magnesium forms in selected temperate and tropical soils. *Soil Sci. Soc. Am. Proc.* **72**:762–764.

Moore, D. P., R. Overstreet, and L. Jacobsen. 1961. Uptake of magnesium and its interaction with calcium in excised barley roots. *Plant Physiol.* **36**:290–295.

Nissen, P. 1977. Ion uptake in higher plants and KCl stimulation of plasmalemma adenosine triphosphatase: Comparison of model. *Physiol. Plant* **40**:205–214.

Oliver, S., and S. A. Barber. 1966. An evaluation of the mechanisms governing the supply of Ca, Mg, K, and Na to soybean roots (*Glycine max*). *Soil Sci. Soc. Am. Proc.* **30**:82–86.

Reisenauer, H. M. 1964. Mineral nutrients in soil solution. In P. L. Altman and D. S. Dittmer, Eds. *Environmental Biology.* Fed. Amer. Soc. Exp. Biol., Bethesda, MD. Pp. 507–508

Rengel, Z., and D. L. Robinson. 1990a. Modeling magnesium uptake from an acid soil. I. Nutrient relationships at the soil-root interface. *Soil Sci. Soc. Am. J.* **54**:785–791.

Rengel, Z., and D. L. Robinson. 1990b. Modeling magnesium uptake from an acid soil. II. Barber-Cushman model. *Soil Sci. Soc. Am. J.* **54**:791–795.

Rice, M. A., and E. J. Kamprath. 1968. Availability of exchangeable and nonexchangeable Mg in sandy Coastal Plain soils. *Soil Sci. Soc. Am. Proc.* **32**:386–388.

Roberts, S., and W. H. Weaver. 1974. Magnesium accumulation and mineral balance in sudan grass as influenced by potassium, calcium, and sodium. *Commun. Soil Sci. Plant Anal.* **5**:303–312.

Salmon, R. C. 1963. Magnesium relationships in soils and plants. *J. Sci. Fd. Agric.* **14**:605–610.

Salmon, R. C. 1964a. Cation exchange reactions. *J. Soil Sci.* **15**:273–283.

Salmon, R. C. 1964b. Cation-activity ratios in equilibrium soil solutions and availability of magnesium. *Soil Sci.* **98**:213–221.

Salmon, R. C., and P. W. Arnold. 1963. The uptake of magnesium under exhaustive cropping. *J. Agr. Sci.* **61**:421–425.

Schulte, E. E., K. A. Kelling, and C. R. Simson. 1980. Too much magnesium in soil? *Solutions* **24**:106–116.

Simson, C. R., R. B. Corey, and M. E. Summer. 1979. Effect of varying Ca:Mg ratios on yield and composition of corn (*Zea mays*) and alfalfa (*Medicago sativa*). *Commun. Soil Sci. Plant Anal.* **10**:153–162.

Sumner, M. E., P. M. W. Farina, and V. J. Hurst. 1978. Magnesium fixation—a possible cause of negative yield responses to lime applications. *Commun. Soil Sci. Plant Anal.* **9**:995–1008.

Udo, E. J. 1978. Thermodynamics of potassium-calcium and magnesium-calcium exchange reactions in a kaolinite clay soil. *Soil Sci. Soc. Am. J.* **42**:556–560.

Warncke, D. D., and S. A. Barber. 1972. Diffusion of zinc in soils. I. The influence of soil moisture. *Soil Sci. Soc. Am. Proc.* **36**:39–42.

Wilkinson, S. R., R. M. Welch, H. F. Mayland, and D. L. Grunes. 1990. Magnesium in plant uptake, distribution, function, and utilization by man and animals. In H. Sigel and A. Sigel, Eds. *Compendium on Magnesium and Its Role in Biology, Nutrition, and Physiology.* Marcel Dekker, New York.

CHAPTER **13**

Sulfur

The earth's crust contains about 0.06% sulfur, but the concentration in soils varies widely. Sulfur in soils may occur as sulfate in solution, sulfate adsorbed on minerals, sulfate minerals, sulfur-bearing minerals, and in organic matter. The original rock often contained metal sulfides. With exposure to air, sulfide oxidizes to sulfate and hence goes into solution. In weathered soils, only small amounts of mineral sulfur remain. In many soils, the major portion of sulfur is combined organically in the soil's organic matter.

For 64 Iowa soils, Tabatabai and Bremner (1972) found that total sulfur ranged from 57 to 618 mg/kg, averaging 294 mg/kg. Spencer and Freney (1960) reported a range of 30 to 545 mg/kg, averaging 188 mg/kg, for 24 Australian soils. Rehm and Caldwell (1968) found an average of 501 mg/kg for 18 selected Minnesota soils, with a range of 131 to 940 mg/kg. Differences in sulfur content are closely related to differences in organic matter level of the soil. Tabatabai and Bremner (1972) found 95 to 98% of the sulfur in Iowa soils to occur in the organic matter. The cycling of sulfur among the various phases in the soil has been discussed by Stevenson (1986). Greater detail can be found in his book.

SOIL INORGANIC SULFUR

Inorganic sulfur in the soil can be divided into soil solution sulfate, adsorbed sulfate, and mineral sulfur.

Soil Solution Sulfate

Sulfate in the soil solution is immediately available for uptake. Reisenauer (1964) summarized data on sulfate concentrations for solutions extracted from soils at soil moisture contents approximately equal to or less than field capacity. He included 693 samples in his survey, with the distribution of samples according to range of sulfate-S concentration (elemental basis) shown in Table 13.1; 40% of the samples were between 25 and 50 mg/L. Many of the analyses were from soils in arid regions, which tend to be higher in soluble salts than soils that have been more highly leached. Fox (1980) found sulfur levels in soil solutions from tropical soils to range from 0.4 to 16 mg/L. David (1990) found the average soil solution, C_{li}, sulfate-S concentration in the 0- to 10-cm layer of New Zealand beech forest soils to be 16 µmol/L. He also sampled German beech and spruce forest soils and found the sulfur C_{li} value in the 0- to 10-cm layer to average 265 µmol/L. Samples of deeper layers reflect those in the 0- to 10-cm layer. The New Zealand forests receive little sulfur from precipitation while those in Germany received greater amounts, hence a wide range in C_{li} values for sulfur can occur.

The K_{sp} for $CaSO_4$ is 1.95×10^{-4} at 10°C. If the soil solution has a calcium level of 3 mmol/L, and the sulfate level is regulated by the solubility of $CaSO_4$, then the sulfate level in solution would be 65 mmol/L. As calcium level increases, sulfate level will decrease, and they will be equal at 14

TABLE 13.1   Distribution of Soil Solution Sulfate Concentrations in 693 Soil Samples Reported in the Literature

Concentration range as elemental (mg/L)	Fraction of samples (%)
0–25	16.5
26–50	40.1
51–100	38.1
101–200	3.2
201–400	1.3
>400	0.8

Source: Reproduced from Reisenauer (1964) by permission of the Federation of Amer. Soc. Exp. Biol.

mmol/L. Where sodium and magnesium are the dominant cations, sulfate levels may be much higher than 65 mmol/L.

Adsorbed Sulfate

Harward and Reisenauer (1966) summarized the sulfate-adsorption literature and found that most soils have some capacity to adsorb sulfate. Lower soil horizons usually adsorb more sulfate than upper soil horizons; kaolinite adsorbs more than montmorillonite; soils containing aluminum and iron oxides adsorb more sulfate, and the amount of adsorption decreases with increasing soil pH. Barrow (1967) found sulfate adsorption to be positively correlated with extractable aluminum and negatively correlated with total soil nitrogen. Hence, while many soils adsorb little sulfate, soils containing allophane or amorphous aluminum and iron oxides adsorb sulfate on their surfaces; this adsorption is greatest at low pH. Increasing soil pH, furthermore, releases adsorbed sulfate into solution (Elkins and Ensminger, 1971). Adsorbed sulfate can be desorbed with phosphate, though sulfate does not desorb appreciable phosphate.

Ajwa and Tabatabai (1993) determined sulfate extracted by 0.1 mol/L LiCl in 11 Iowa soils and obtained an average of 5.7 mg S/kg soil with a range of 1.9 to 10.3. Six soils were in the range 5.0 to 6.7 mg/kg.

For soils where sulfate is adsorbed, the amount adsorbed increases as solution-sulfate concentration is increased. The adsorption usually approximates a Langmuir adsorption curve. The $\Delta C_s/\Delta C_l$ or buffer power of the adsorbed sulfate for that in solution is often between 2 and 10 and tends to be higher for tropical soils. Data in Figure 13.1 illustrate the relation between adsorbed sulfate and solution sulfate concentration for acid tropical soils where sulfate adsorption was significant. The most highly leached soils had the greatest sulfate-adsorption capacities. Adsorbed sulfate desorbs into solution as dissolved sulfate is removed.

Sulfur Minerals

Sulfur may occur as sulfates or as sulfides. While sulfate may precipitate as calcium, magnesium, or sodium sulfate, only calcium sulfate is of sufficiently low solubility that solid-phase sulfates ordinarily occur. Sulfide metals such as FeS_2, pyrite, and ZnS_2, sphalerite, are common in soil-forming materials. These minerals occur primarily under anaerobic conditions. Upon exposure to air, sulfide oxidizes to sulfuric acid and increases the acidity of the medium. Acidity due to sulfides is sometimes a problem with reclaimed mine spoils.

SOIL ORGANIC SULFUR

Inorganic sulfur is usually only 5 to 10% of total sulfur in the soil (Neptune et al., 1975). Most soil sulfur is present in the organic fraction; soil organic

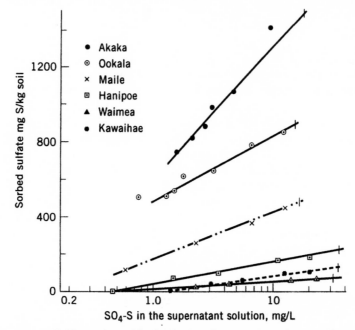

FIGURE   13.1   Sulfate-adsorption isotherms for a series of six soils developed from volcanic ash, Island of Hawaii. The soils are listed in order of decreasing rainfall. Reproduced from Hasan et al. (1970) by permission of Soil Science Society of America.

matter contains approximately 0.5% sulfur. Organic sulfur has been fractionated (Anderson, 1975; Neptune et al., 1975) into reduceable sulfur, estersulfate sulfur, carbon-bonded sulfur and unidentified organic sulfur. Unidentified sulfur is believed to be additional carbon-bonded sulfur not detected in the fractionation procedure.

Sulfur-containing compounds in soil organic matter include the amino acids cystine and methionine and related compounds. The vitamins thiamine and biotin also contain sulfur in their ring structures.

Release of Organic Sulfur

Other than sulfur added in rainfall or fertilizers, release of sulfur from organic matter usually constitutes the largest supply of sulfur to soil solution. Release of sulfur is related to rate of organic-matter decomposition. In soil organic matter, the ratio of carbon to sulfur is approximately 100, though values ranging from 80 to 200 have been obtained. There is approximately one-tenth as much sulfur as nitrogen in soil organic matter. During organic matter decomposition, the relative release of sulfur is slightly more rapid than the release of nitrogen when compared on the basis of amounts present. Sulfur released from organic matter is oxidized to sulfate by sulfur-oxidizing microorganisms. The most common species is *Thiobacillus thiooxidans.*

The nitrogen/sulfur ratio in the soil varies from 10 to 6.7. If the sulfur level in the soil is 300 mg/kg and 95% is in the organic fraction, we can estimate the amount of sulfur that will be released by organic-matter decomposition. Organic matter in an Indiana soil having an organic-matter content of 3.0% was mineralized at a rate of 2.4% per year (Barber, 1979). Assuming that the mineralized organic matter contained 0.5% sulfur, this would release about 3.5 mg sulfur/kg soil, or approximately 7 kg/ha of sulfur annually. This quantity would continue in the sulfur cycle of the soil and be used for plant growth and subsequent formation of new sulfur-containing organic matter to replace that which had been decomposed. Some sulfur could also be lost by leaching.

ATMOSPHERIC SULFUR

There are variable amounts of sulfur gases, primarily SO_2, and soluble sulfur compounds in the atmosphere, which may be added to the soil through rainfall and dry fallout. Amounts added per year vary from as little as 2 to as much as 200 kg sulfur/ha. Values in midwestern United States average approximately 10 to 20 kg/ha · year. Sulfur dioxide and sulfur compounds are emitted by power and industrial plants as well as from such natural sources as organic matter decomposition, volcanic activity, and dust particles from gypsiferous areas (Reisenauer et al., 1973). These sulfur additions often oxidize further to sulfate in the soil and frequently contribute significantly to the eventual sulfate concentration of the soil solution. Assuming 20 kg/ha brought down annually in rainfall in Indiana, the mean sulfur concentration of the rainfall would be about 2 mg/L. In addition, plants can absorb atmospheric SO_2 directly through their leaves.

SULFUR-UPTAKE KINETICS

Sulfur as sulfate is mainly absorbed metabolically (Leggett and Epstein, 1956) by plant roots. The influx relationship can be described by I_{max}, the maximal influx rate, and K_m, the concentration where net influx, In, minus C_{min}, is $1/2$ I_{max}. No information could be found about levels of C_{min}, the concentration when In is zero. Values obtained for different crops by various observers are shown in Table 13.2. Cacco et al. (1977) measured the sulfate-uptake kinetics of corn roots that had been grown to lengths varying from 2 to 14 cm. Uptake followed Michaelis–Menten kinetics, with I_{max} increasing with root length to a maximum at 9 to 10 cm. Values of K_m varied from 10 to 20 μmol/L. The lowest value was for 7- to 8-cm roots and the highest for

TABLE 13.2   Kinetic Factors Describing Sulfate Absorption by Plant Roots

Plant	Phase	I_{max} (μmol/ $g_{D.W.} \cdot$ hr)	K_m (mmol/L)	Inflection point	Source
Barley (excised roots)		0.1[a]	0.010		Leggett and Epstein (1956)
Corn (excised roots)	1	0.61	0.039		Berlier et al. (1969; calculated by Nissen, 1973)
	2	6.3	2.8		
	3	17.4	11.0		
Corn hybrid 79A		0.77	0.179		Ferrari and Renosto (1972)
Corn hybrid 75		0.835	0.166		
Corn hybrid 79A + 75		1.43	0.055		
Barley	1	0.372	0.015	0.040	Nissen (1973)
	2	0.538	0.039	0.10	
	3	0.752	0.094	0.40	
	4	1.65	0.67	1.6	
	5	3.33	7.8	4.0	
	6	7.69	12.0	10.0	
	7	22.00	53.0	40	

[a]Per $g_{F.W.}$ for this value only.

3- to 4-cm roots. Nissen (1971) has shown that sulfate uptake is multiphasic, because there is continuing increase in I_{max} and K_m as the sulfate concentration in solution increases. Hence, at high levels of sulfate in the soil solution, sulfate influx can be high. However, this obviously occurs for only part of a root or for a short time period; otherwise, sulfur content of the plant would become excessive.

Factors Affecting Sulfate Influx

Sulfate influx was most rapid at pH 4.0 and decreased with increased pH above 4.0 (Leggett and Epstein, 1956). Phosphate, nitrate, and chloride concentrations had no measureable effect on sulfate uptake (Leggett and Epstein, 1956), while selenite competitively interfered with sulfate uptake. Higher-temperature increased sulfate uptake when temperatures of 15, 25, and 35°C were compared (Rajan, 1966).

Cacco et al. (1977) found a close similarity between sulfate-uptake capacity and ATP-sulfurylase activity in plant roots. Sulfate uptake increased with increased root length, reaching a maximum for maize with 10-cm roots.

Bowen and Rovira (1971) found that sulfate uptake by wheat roots was greatest in the first 5 cm from the root apex; beyond this point, lateral roots provided much of the sulfate uptake. Sulfate was different in this respect from phosphate and chloride, where uptake rate remained relatively uniform along the length of the root.

Cacco et al. (1976) measured differences in sulfate influx with different ploidy levels of the plants involved. Experiments were conducted with wheat, sugar beet (*Beta vulgaris* L.), and tomato (*Lycopersicon esculentum* L.), and uptake efficiency was measured in terms of K_m and I_{max}. Values shown in Table 13.3 indicate that increased ploidy reduced the K_m value of sugar beet and tomato but had no effect on wheat. Increased ploidy increased I_{max} for wheat and sugar beet but reduced the I_{max} value for tomato.

Rehm and Caldwell (1968) found that sulfate uptake was influenced by ammonium and nitrate. They added either ammonium or nitrate nitrogen along with sulfate in the row fertilizer and found that sulfate uptake in the presence of nitrate was similar to that where no nitrogen was added but sulfate uptake was approximately 50% greater when ammonium nitrogen was used. The authors attributed the effect to physiological processes within the plant root, rather than to reactions in the soil.

Soil Sulfate Supply

When plant roots grow in soils, the rate of sulfate uptake depends on the sulfate concentration at the root surface, which, in turn, depends on the rate of supply of sulfate to the root by mass flow and diffusion. The level of sulfate in the soil solution is in the range of 0.1 to 1.0 mmol/L, or 3.2 to 32

TABLE 13.3   Effect of Plant Ploidy on Sulfate-Influx Kinetics of Wheat, Sugar Beet, and Tomato

| | | SO_4^{2-} Influx kinetics | |
Plant species	Ploidy	K_m (mmol/L)	I_{max} (μmol/ $g_{D.M.} \cdot$ hr)
Wheat	2n	0.06	2.5
	4n	0.06	3.0
	6n	0.06	4.1
Sugar beet	2n	0.22	2.8
	3n	0.19	3.6
	4n	0.12	4.5
Tomato	2n	0.12	2.4
	4n	0.06	1.0

Source: Reprinted from Cacco et al. (1976) by permission of Cambridge University Press.

mg/L. The sulfur concentration in a corn plant is approximately 0.2%, or 2000 mg/kg. If the plant transpires 400 g of water to produce 1 g of dry matter, an average sulfur level in the soil solution of 5 mg/kg would satisfy the plant's sulfur requirement by supply to the root via mass flow. However, there is little buffer power for sulfur in many soils, so that the only solution replenishment is by rainfall and the decomposition of organic matter. This makes mathematical description of sulfate uptake difficult to assess unless values for the rates of addition are known. For many soils, there is enough sulfate present to cause accumulation at the root as more is supplied by mass flow than is absorbed by the root. When this occurs and both calcium and sulfate accumulate, $CaSO_4 \cdot 2H_2O$ can precipitate at the surface. Barber et al. (1963) showed accumulation of sulfate around corn roots by using autoradiographs; the pattern of accumulation suggest that sulfate had precipitated. Using both ^{45}Ca and ^{35}S, Malzer and Barber (1975) later demonstrated that calcium sulfate precipitated around roots of young corn plants on soils where solution sulfate was 10 mmol/L but not where the soil solution sulfate concentration was 0.55 mmol/L. A saturated solution of $CaSO_4 \cdot 2H_2O$ is approximately 14 mmol/L. The level of sulfate in solution where calcium and sulfate are being supplied faster by mass flow than uptake can occur often results in accumulation to at least this level. Levels of sulfate in the soil solution can be even higher than 14 mmol/L if calcium concentrations are lower than 14 mmol/L.

Sulfur deficiencies may occur in areas where natural supplies of sulfur are low and when little sulfur is being added by rainfall. Leached sandy soils may lose part of the sulfur added, causing sulfur deficiencies.

MEASURING BIOAVAILABLE SULFATE

Reisenauer et al. (1973) reported that $Ca(H_2PO_4)_2$ and water were the extractants most frequently used by soil-testing laboratories for measuring available sulfate. Spencer and Freney (1960) found that KH_2PO_4 extraction and growth of *Aspergillus niger* were both suitable procedures for assessing available sulfate.

The level of sulfate in the soil solution is significant for determining initial sulfate uptake and supply to the root by mass flow. Supply by diffusion depends on C_{li}, b, and D_e. Since D_e is related to b, measurement of soil buffer power for sulfate is important. The difference between the value for the water extract, which gives a measurement for C_{li} and the value for phosphate-desorbable sulfate, is that the latter removes both dissolved and adsorbed sulfate.

The mechanistic nutrient uptake model has not (to my knowledge) been used to verify the model for sulfur uptake. Parameters for influx and sulfur supply from the soil have been measured in various experiments. Values for C_{li} (Davis, 1990; Fox, 1980) have been shown to vary from 0.4 to 265

μmol/L in leached soils and can be as high as 65 mmol/L in soils solutions saturated with $CaSO_4$. Hence, a very wide range can occur.

Table 13.2 gives values for I_{max} that show variation with C_{li}, sulfate level, however, they are all based on uptake per gram dry weight.

The soil nutrient supply to the root by mass flow is $C_{li} \times v_0$. Values of v_0 are usually about 1×10^{-6} cm³/cm · s. Hence since $C_{li} \times v_0$ vs. I_{max} indicates whether mass flow can provide all the needed sulfate to have soil supply rate equal or exceed I_{max} a value of $C_{li} = I_{max}/v_0$ can be used to obtain the C_{li} level needed.

There is little reserve sulfate in leached soils, and if it were not for additions by rainfall and organic-matter decomposition, crop sulfur deficiencies would occur after harvesting only one or two crops.

REFERENCES

Ajwa, H. A., and M. A. Tabatabai. 1993. Comparison of some methods for determination of sulfate in soils. *Commun. Soil Sci. Plant Anal.* **24**:1817–1832.

Anderson, G. 1975. Sulfur in soil organic substances. In J. Gieseking, Ed. *Soil Components I. Organic Components.* Springer-Verlag, New York. Pp. 333–341.

Barber, S. A. 1979. Corn residue management and soil organic matter. *Agron. J.* **71**:625–628.

Barber, S. A., J. M. Walker, and E. H. Vasey. 1963. Mechanics for the movement of plant nutrients from the soil and fertilizer to the plant root. *J. Agr. Food Chem.* **11**:204–207.

Barrow, N. J. 1967. Studies on adsorption of sulfate by soils. *Soil Sci.* **104**:342–349.

Barrow, N. J. 1969. Effects of adsorption of sulfate by soils on the amount of sulfate present and its availability to plants. *Soil Sci.* **108**:193–201.

Berlier, Y., G. Gurraud, and Y. Sauvaire. 1969. Etude avec l'azote 15 de l'absorption et du metabolisme de l'ammonium fourni a concentration croissante a des racines excisees de mais. *Agrochimica* **13**:250–260.

Bowen, G. D., and A. D. Rovira. 1971. Relationship between root morphology and nutrient uptake. *Recent Adv. Plant Nutr.* **1**:293–305.

Cacco, G., G. Ferrari, and G. C. Lucci. 1976. Uptake efficiency of roots in plants at different ploidy levels. *J. Agric. Sci.* **87**:585–589.

Cacco, G., M. Saccomani, and G. Ferrari. 1977. Development of sulfate uptake capacity and ATP-sulfurylase activity during root elongation in maize. *Plant Physiol.* **60**:582–584.

Davis, M. R. 1990. Chemical composition of soil solutions extracted from New Zealand beech forests and West German beech and spruce forests. *Plant Soil* **126**:237–246.

Elkins, D. M., and L. E. Ensminger. 1971. Effect of soil pH on the availability of adsorbed sulfate. *Soil Sci. Soc. Am. Proc.* **35**:931–934.

Ferrari, G., and F. Renosto. 1972. Comparative studies on the active transport by excised roots of inbred and hybrid maize. *J. Agric. Sci.* **79**:105–108.

Fox, R. L. 1980. Response to sulphur by crops growing in highly weathered soils. *Sulfur Agric.* **4**:16–22.

Harward, M. E., and H. M. Reisenauer. 1966. Movement and reactions of inorganic soil sulfur. *Soil Sci.* **101**:326–335.

Hasan, S. M., R. L. Fox, and C. C. Boyd. 1970. Solubility and availability of sorbed sulfate in Hawaiian soils. *Soil Sci. Soc. Am. Proc.* **34**:897–901.

Leggett. J. E., and E. Epstein. 1956. Kinetics of sulfate absorption by barley roots. *Plant Physiol.* **31**:222–226.

Malzer, G. L., and S. A. Barber. 1975. Precipitation of calcium and strontium sulfates around plant roots and its evaluation. *Soil Sci. Soc. Am. Proc.* **39**:492–495.

Neptune, A. M. L., M. A. Tabatabai, and J. J. Hanway. 1975. Sulfur fractions and carbon-nitrogen-phosphorus-sulfur relationships in some Brazilian and Iowa soils. *Soil Sci. Soc. Am. Proc.* **39**:51–55.

Nissen, P. 1973. Multiphasic uptake in plants. I. Phosphate and sulfate. *Physiol. Plant* **28**:304–316.

Rehm, G. W., and A. C. Caldwell. 1968. Sulfur supplying capacity of soils and the relationship to soil type. *Soil Sci.* **105**:355–361.

Reisenauer, H. M. 1964. Mineral nutrients in soil solution. In P. L. Altman and D. S. Dittmer, Eds. *Environmental Biology.* Fed. Am. Soc. Exp. Biol., Bethesda, MD. Pp. 507–508.

Reisenauer, H. M., L. M. Walsh, and R. G. Hoeft. 1973. Testing soils for sulfur, boron, molybdenum, and chlorine. In L. M. Walsh and J. D. Beaton, Eds. *Soil Testing and Plant Analysis.* Soil Science Society of America, Madison, WI. Pp. 173–200.

Rajan, A. K. 1966. The effect of root temperature on water and sulfate absorption in intact sunflower plants. *J. Exp. Bot.* **17**:1–19.

Spencer, K., and J. R. Freney. 1960. A comparison of several procedures for estimating the sulfur status of soils. *Aust. J. Agr. Res.* **11**:948–959.

Stevenson, F. J. 1986. *Cycles of Soil Carbon, Nitrogen, Phosphorus, Sulfur, Micronutrients.* John Wiley, New York.

Tabatabai, M. A., and J. M. Bremner. 1972. Distribution of total and available sulfur in selected soils and soil profiles. *Agron. J.* **64**:40–44.

CHAPTER **14**

Boron

Total boron concentration in the earth's crust is about 50 mg/kg (Aubert and Pinta, 1977). Concentration varies with the nature of the rock. Basic rocks have boron contents of 1 to 5 mg/kg, acidic rocks have 3 to 10 mg/kg, metamorphic rocks have 5 to 12 mg/kg, and sedimentary rocks of marine origin have 500 or more mg/kg. Total boron concentration in soil varies with its parent material and degree of weathering; values may range from 1 to 270 mg/kg, with an average of 20 to 50 mg/kg. Sandy soils tend to be low in boron, while clay and high organic soils tend to have higher boron contents. Less weathered soils from arid regions are usually higher in boron. Sea water contains 4.6 mg/L of boron, which is an indication that soluble boron minerals have been removed from the earth's land surface by weathering.

FORMS OF SOIL BORON

Soil boron may be subdivided into soil solution boron, adsorbed boron, and mineral boron. The most common boron mineral is tourmaline, a complex borosilicate. Boron in solution usually occurs as the undissociated acid

H_3BO_3. Curve A in Figure 14.1 gives the relation between solution pH and the form of boron in solution. Since most soils are in the pH range 5 to 8, the form of boron in solution is H_3BO_3.

Little is known about the dissolution of boron minerals in the soil. Solution boron is assumed to be in equilibrium with boron adsorbed on the soil's mineral surfaces.

Adsorbed Boron

When boron is added to soil, part of the element is adsorbed on the solid phase of the soil and part remains in solution. Biggar and Fireman (1960) found that adsorption of added boron by the soil solid phase followed the Langmuir adsorption equation. Values for maximum adsorption by four soils varied from 7.3 to 21.3 mg/kg. Adsorption varies with the type of clay mineral and the presence of iron and aluminum oxides. Sims and Bingham (1968) found oxide coatings of clays to be more important than the type of clay in determining the amount of adsorption. They obtained a linear relation between the amount of boron adsorbed and the amount of iron oxide or aluminum oxide present.

Boron adsorption increases as pH is increased above pH 4, reaching a maximum at pH 8 to 9 and then decreasing once more at higher pH values (Sims and Bingham, 1968). Soil texture also affects boron adsorption; Singh (1964) measured boron adsorption by light- and heavy-textured soils and found that the relationship between the amount of boron added and boron adsorbed followed the Langmuir adsorption equation. Maximum adsorption

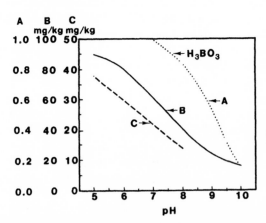

FIGURE   14.1   Relation between A, solution pH and the fraction of boron present as H_3BO_3; B, solution pH and the rate of boron uptake by excised barley (*Hordeum vulgare*) roots (reproduced from Bingham et al., 1970, by permission of Soil Science Society of America); and C, rhizocylinder pH and the boron concentration in soybean shoots (Barber, 1971).

values calculated from the equation were 7.76 and 22.8 mg/kg for the light- and heavy-textured soils, respectively. The buffer-power value decreased with an increase in boron added.

Some soils contain toxic levels of boron, so boron removal by desorption and leaching becomes important. Griffin and Burau (1974) used mannitol to desorb boron, because mannitol forms a high-affinity, soluble chelate with boric acid. This reduces the boric acid concentration in solution to a low level, which favors movement of adsorbed boron into solution. Boron desorption from soil was measured over a 12-h period; it followed two pseudo-first-order reactions and one very slow reaction. The first-order reactions were postulated to be due to two independent retention sites on hydroxy iron, magnesium, and aluminum materials. The desorption followed a two-site analog of the one-site Langmuir adsorption equation; hence, desorption followed the same kinetics as adsorption.

Soil Solution Boron

Only a few measurements have been made of boron levels in displaced soil solutions. Oliver and Barber (1966) measured a level of 15.7 μmol/L in solution from a Sidell silt loam (Typic Argiudoll) subsoil. Rhoades et al. (1970) reported values from arid soils with high boron levels of 0.14 to 3.5 mmol/L. Sulaiman and Kay (1972) measured solution-boron values of 1.1 mg/L or 0.1 mmol/L of boron in a Woburn loam from Ontario, Canada.

The amount of boron that would have to be maintained in soil solution in order for mass flow to supply the crop's total boron requirement can be approximated for a legume crop having a boron level of 20 mg/kg. Assuming water use to be 500 ml/g of shoot dry matter produced, the boron concentration in the water would have to be 20/500, or 0.04 mg/L, which is 3.7 μmol/L. Hence, low concentrations maintained in the soil solution would supply the plant's boron requirement. It appears that mass flow is probably an important mechanism for supplying boron to plant roots. However, mass flow obviously does not supply enough of the element in boron-deficient soils so that diffusive supply also becomes important. The preceding calculations indicate that boron-deficient soils probably have an average boron concentrations in their soil solutions of less than 5 μmol/L during crop growth.

Buffer Power

The relation between solution boron and adsorbed boron is termed the buffer power of boron in solution. Values from adsorption curves indicate that values for b are usually less than 3. Sulaiman and Kay (1972) reported values of 1.7 and 1.35 for a loam soil.

Diffusion Coefficient

Sulaiman and Kay (1972) measured boron diffusion coefficients in a loam soil and obtained values for the apparent diffusion coefficient of 0.94×10^{-6} cm^2/s where no boron had been added. Scott et al. (1975) studied boron diffusion in two Arkansas soils, Toloka silt loam (Mollic Albaqualf) and Calloway silt loam (Glossic Fragiudolf). Diffusion coefficients were measured after equilibration of the soil with 37.5 μg boron/kg of soil. Diffusion coefficients averaged over various lime and moisture treatments were 2.68 and 4.63×10^{-6} cm^2/s for Toloka and Calloway soils, respectively. No measurement of soil solution boron was made. When the soils were limed, the value of D_e decreased; when volumetric moisture contents decreased, D_e decreased.

BORON-UPTAKE KINETICS

Since boron is present in solution as undissociated H_3BO_3, its absorption by plant roots may differ from absorption of ionized species. Bingham et al. (1970) investigated boron uptake by excised barley roots over a 4-h period using solutions ranging in boron concentration from 0.2 to 80 mg B/L (0.18 to 7.39 mmol/L). They obtained a linear relation between boron uptake and boron concentration in solution. Reduced temperature or addition of respiration inhibitors had little effect on boron uptake; this evidence indicates that the uncharged boric acid molecule is passively absorbed. After absorption of boron from solution, these researchers transferred the roots to boron-free solution and measured the diffusion of boron from the roots back into solution. Bingham et al. (1970) concluded that the process was reversible. Apparently, boron in the soil solution diffuses into the root or moves in with the water flow until levels inside the root are in equilibrium with levels in the soil solution. Because of this passive uptake, toxic levels of boron are absorbed into the plant when levels in the soil solution are high.

Boron only moves up the xylem, and not down the phloem; hence, boron does not move into the roots from the shoots. Boron is also not mobile within the plant once it has been stabilized. Boron deficiencies occur on the youngest growth, with boron not translocated from older plant parts to the most recent growth.

Factors Affecting Boron Uptake

Solution pH and Boron Uptake

Bingham et al. (1970) measured the effect of solution pH on boron uptake by barley. They found a gradual decrease in boron uptake as pH was raised from pH 5 to 11; their data are plotted in curve *B* in Figure 14.1. Boron uptake decreased rapidly as pH was increased from 6 to 9; the change in solution boron form is also shown in Figure 14.1. All of the boron is, as H_3BO_3, below pH 7.0 and is gradually dissociating as pH is increased above 7.0. The decrease in H_3BO_3 in solution does not exactly correspond with the soil pH range where boron uptake decreases; hence, this cannot be the sole reason for the decrease. Decreases in boron uptake with increases in pH above 7.0 may be attributed to changes in boron form, though not below pH 7.0. Bingham et al. (1970) also found that calcium and potassium levels in the solution had little effect on boron uptake.

Soil pH and Boron Uptake

Increasing soil pH reduces the level of available boron in the soil. Scott et al. (1975) showed that increasing soil pH decreased the apparent diffusion coefficient for soil boron and increased the amount of added boron that was adsorbed by the soil.

Several investigators have found that increasing soil pH by liming reduces the boron content in plants grown on the limed soil. Gupta (1972) found an interaction between pH and boron concentration in barley plants: Increasing soil pH reduced boron uptake. Barber (1971) reported a relation between the pH of the rhizocylinder (root plus rhizosphere soil) and boron concentrations in soybean plants. Soybeans were grown on Chalmers silt loam (Typic Argiaquoll) in the growth chamber; data are plotted in curve *C* in Figure 14.1. As pH increased, boron concentration in the shoot decreased. The slope of the line was similar to that for boron uptake from solution, as reported by Bingham et al. (1970). It would appear that the pH effect was primarily related to the pH effect on boron-absorption kinetics of the root as well as changes in boron concentration in the soil solution due to adsorption by the soil. Peterson and Newman (1976) increased the pH of Plano silt loam (Typic Argiudoll) stepwise from 4.7 to 7.4 by liming. Increasing the pH to 6.3 did not affect boron concentration in tall fescue (*Festuca arundinacea* Schreb.), but increasing the pH further (to 7.4) reduced boron concentration to approximately half the level obtained when no boron was added. Even greater relative reductions occurred when boron was added.

Reduction in boron concentrations of soil-grown plants with increasing pH apparently results from both a decrease in the rate of boron absorption by the root and a decrease in the rate of boron supply to the root by the soil. The latter may result from higher boron adsorption by soil, with resultant lower soil solution boron levels.

Boron-Influx Rates

Scott et al. (1975) determined boron flux into soybean roots; they found influx decreased as plants increased in age from 30 to 51 days. Boron influx then increased from 51 to 65 days. This increase after 51 days coincided with seed development. Influx values varied from 43 to 10 pmol/m² · s.

Mengel and Barber (1974) measured boron influx for corn growing in the field and found that mean influx decreased with plant age. Values varied from 0.98 pmol/m^2 · s to no uptake or even loss from the roots at 70 days.

MEASURING BIOAVAILABLE BORON

Supply of boron to the root depends on C_{li}, b and D_e. For boron, values for C_{li} and b are usually small, but values for D_e are large because of the small b values. Values for C_{li} are probably the most important in determining boron availability. Levels of diffusible boron are difficult to measure; the most common available boron measurement is extraction with boiling water. It is accomplished by boiling a 1:2 soil/H$_2$O suspension for 5 min, then analyzing the filtrate for boron. Critical levels reported in the literature range from 0.15 to 0.75 mg/L. Variation may be due to crop and types of soils evaluated.

Since b is usually small, D_e values for boron will be large. Hence, boron can move relatively long distances by mass flow and diffusion in order to reach the root. Much of the diffusible boron in the soil is able to reach the root during the period of plant growth.

Boron deficiency in crops commonly occurs during drought periods, possibly because water flow to the root is reduced, which reduces mass flow supply of boron to the root. The reduced level of water in the soil also causes a proportionate decrease in the rate of boron diffusion to the root.

REFERENCES

Aubert, H., and M. Pinta. 1977. *Trace Elements in Soils.* Elsevier Scientific Publishing Co., Amsterdam. Pp. 5–11.

Barber, S. A. 1971. The influence of the plant root system in the evaluation of soil fertility. *Proc. Int. Symp. Soil Fert. Evaln.,* New Delhi 1:249–256.

Biggar, J. W., and M. Fireman. 1960. Boron adsorption and release by soils. *Soil Sci. Soc. Am. Proc.* **24**:115–120.

Bingham, F. T., A. Elseewi, and J. J. Oertli. 1970. Characteristics of boron absorption by excised barley roots. *Soil Sci. Soc. Am. Proc.* **34**:613–617.

Griffin, R. A., and R. G. Burau. 1974. Kinetic and equilibrium studies of boron desorption from soil. *Soil Sci. Soc. Am. Proc.* **38**:892–897.

Gupta, U. C. 1972. Interaction effects of boron and lime on barley. *Soil Sci. Soc. Am. Proc.* **36**:332–334.

Mengel, D. B., and S. A. Barber. 1974. Rate of nutrient uptake per unit of corn root under field conditions. *Agron. J.* **66**:399–402.

Oliver, S., and S. A. Barber. 1966. Mechanisms for the movement of Mn, Fe, B, Cu, Zn, Al, and Sr from one soil to the surface of soybean roots (*Glycine max*). *Soil Sci. Soc. Am. Proc.* **30**:468–470.

Peterson, L. A., and R. C. Newman. 1976. Influence of soil pH on the availability of added boron. *Soil Sci. Soc. Am. J.* **40**:280–282.

Rhoades, J. D., R. D. Ingvalson, and J. T. Hatcher. 1970. Laboratory determination of leachable soil boron. *Soil Sci. Soc. Am. Proc.* **34**:871–875.

Scott, H. D., S. D. Beasley, and L. F. Thompson. 1975. Effect of lime on boron transport to and uptake by cotton. *Soil Sci. Soc. Am. Proc.* **39**:1116–1121.

Sims, J. R., and F. T. Bingham. 1968. Retention of boron by layer silicates, sesquioxides, and soil materials: III. Iron- and aluminum-coated layer silicates and soil materials. *Soil Sci. Soc. Am. Proc.* **32**:369–373.

Singh, S. S. 1964. Boron adsorption equilibrium in soils. *Soil Sci.* **98**:383–387.

Sulaiman, W., and B. D. Kay. 1972. Measurement of the diffusion coefficient of boron in soil using a single cell technique. *Soil Sci. Soc. Am. Proc.* **36**:746–752.

CHAPTER **15**

Copper

The average copper concentration in the earth's crust is approximately 70 mg/kg (Hodgson, 1963). Copper concentration varies with the type of rock; basalt may contain as much as 100 mg/kg, while granite contains only approximately 10 mg/kg. Sedimentary rocks also vary in copper concentration. Limestone, sandstone, and shale average approximately 4, 30, and 45 mg/kg of copper, respectively. The most common mineral form of copper is the sulfide, where sulfur has combined with Cu^{2+} in minerals such as chalcopyrite, $CuFeS_2$. Under oxidizing conditions, the copper is oxidized to the divalent form as the mineral is dissolved by weathering (McBride, 1981).

Copper can also be substituted isomorphously for manganese, iron, and magnesium in various minerals. Copper oxides, carbonates, and sulfates are not present in most soils, because the copper concentration found in soil solution is much lower than would be maintained by the solubilities of these minerals (Lindsay, 1979). Detailed information on copper in soils and plants is available in the proceedings of a symposium on copper (Loneragan et al., 1981).

SOIL COPPER

Soils contain a total of 1 to 50 mg of copper/kg. This relatively small amount could all be present as substitutions in noncupric minerals or adsorbed copper on mineral surfaces and in organic matter. Copper in soil can occur in soil solution, both ionic and complexed; as an exchangeable cation on the exchange complex; as a specifically adsorbed (nonexchangeable) ion; in organic matter; in occluded oxides; and in minerals.

McLaren and Crawford (1973a) attempted to fractionate the copper in 24 British soils into the forms described by using a series of extraction procedures. They used 0.05 mol/L $CaCl_2$ to extract soil solution and exchangeable copper; 2.5% acetic acid to extract specifically adsorbed copper; 1.0 mol/L potassium pyrophosphate to extract organic copper; acid oxalate to obtain occluded copper; and HF to obtain residual copper. They did not measure solution copper. Exchangeable plus solution copper amounted to less than 0.2% of total for all but one soil, where it was 2.3%. Total copper varied from 4.4 to 63.5 mg/kg. Exchangeable plus solution copper was less than 0.06 mg/kg for all but one soil. Hence, values for readily extractable copper are relatively low. Copper specifically adsorbed by inorganic sites was less than 1.9%, while that specifically adsorbed by organic sites varied from 13.1 to 46.9% of the total. Copper combined with oxides ranged from 0 to 28.9% of total and residual copper varied from 23.7 to 77.2%. The residual copper is probably too insoluble to have much significance for plant uptake. On the average, over 50% of the copper remained in the residue, about 30% was extracted by pyrophosphate (copper largely bound by organic sites), and about 15% was extracted by oxalate (copper occluded in free oxides). Little is known of how such values vary for soils differing widely in parent materials and degree of weathering.

Soil Solution Copper

Copper concentrations in the soil solution are low, with only a few measurements having been made. Hodgson et al. (1966) measured copper in solution for 20 Colorado soils, which had pH levels ranging from 6.6 to 7.9. Total copper in solution ranged from 3.5 to 39.2 µg/L, with an average of 10.8 µg/L. However, a large part of the copper in solution was found to be in association with soluble organic compounds, so that free Cu^{2+} in solution ranged from 0.0005 to 0.038 µg/L, with an average of 0.009 µg/L. These values for Colorado soils are in contrast with those found for New York soils (Hodgson et al. 1965), where total copper in solution averaged 9.4 µg/L but free Cu^{2+} averaged 0.31 µg/L. Much less was complexed in the solutions from New York soils. Cavallaro and McBride (1980) have shown that pH affects the proportion of copper remaining in ionic form (Figure 15.1). Cavallaro and McBride (1978) also found that the proportion of copper present as Cu^{2+} depends on the total copper level in solution (Figure

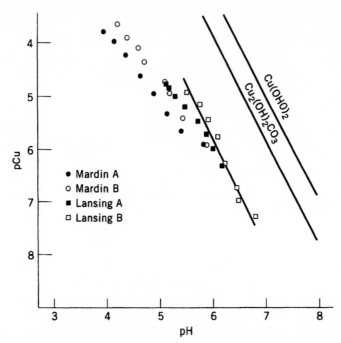

FIGURE   15.1   Solubility diagrams for Cu²⁺–soil equilibrations at varying pH. The malachite solubility line assumes an atmospheric level of CO₂. Reproduced from Cavallaro and McBride (1980) by permission of Soil Science Society of America.

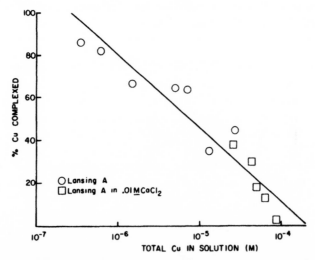

FIGURE   15.2   Relation between total soluble Cu²⁺ and percentage copper complexed in solution. Reproduced from Cavallaro and McBride (1978) by permission of Soil Science Society of America.

15.2). As solution copper increased, the proportion of Cu^{2+} increased, indicating that there was only a limited quantity of suitable organic complexes in solution. Complexing of copper in solution occurs at solution pH levels above 6.5.

Sanders and Bloomfield (1980) measured the ability of soluble organic matter to complex copper. They found increased complex formation with increased pH, decreased ionic strength, and increased organic matter/copper ratios. In addition to organic complexes in solution, inorganic complexes of the hydroxy and carbonate forms of copper may exist at soil solution pH values above 7.0 (McBride, 1981).

The levels of Cu^{2+} in soil solution indicate undersaturation with respect to the solubility of copper minerals that might be present in soils, so that mineral solubility probably does not control solution copper concentrations. They are more likely controlled by adsorption reactions involving both inorganic and organic components of the soil.

Exchangeable Copper

Copper can also be present as a cation held on soil-exchange sites. This form of soil copper can be displaced by an excess of a cation, such as calcium. The quantity of exchangeable copper measured by McLaren and Crawford (1973a) was <0.01 to 0.07 mg/kg, except for one value of 0.50 mg/kg. These values included solution copper, but solution copper would normally be a relatively small portion of the amount measured. Misra et al. (1973) measured copper extracted by ammonium acetate pH 7.0 from 30 Indian soils; the amounts extracted varied from undetectable amounts to 0.6 mg/kg. The authors used 15 black soils and 15 red soils from various locations in India. Kline and Rust (1966) found values for ammonium acetate exchangeable copper ranging from 0.2 to 0.5 mg/kg for six Minnesota soils. These soils had total copper contents that ranged from 2.26 to 17.88 mg/kg. Shuman (1979) measured exchangeable copper values of 0.03 to 0.26 mg/kg for eight southeast U.S. soils. Only limited information is available on amounts of exchangeable copper present in various soils, because the difficulty of measuring low amounts of copper discourages use of exchangeable copper for predicting copper availability to plants.

Specifically Adsorbed Copper

This copper fraction is not displaced with normal cation-exchange reagents, such as calcium chloride or ammonium acetate. McLaren and Crawford (1973a) used 2.5% acetic acid to displace inorganically bound copper and 1.0 mol/L of potassium pyrophosphate to displace organically bound copper. About 1% of the total copper, 0.030 to 0.595 mg/kg, was extracted as copper specifically absorbed by the inorganic soil fraction. An average of about 30% of the total soil copper was extracted with pyrophosphate (1.1 to 15.8 mg/kg). This copper was regarded as specifically bound by the soil organic fraction. These quantities of copper are much larger than those found in solution or on cation-exchange sites.

McLaren and Crawford (1974) used isotopic-exchange procedures to evaluate the amount of copper in the soil that will equilibrate with the copper in soil solution. In addition to estimating the amount of copper in the solid phase, they also estimated the rate of exchange and the soil constituent associated with most of the exchangeable copper. They found that the amounts of isotopically exchangeable copper in several British soils examined varied from 0.19 to 12.24 mg/kg. This represented a range of 2 to 21% of the total soil copper. Furthermore, isotopic exchange for most soils was virtually complete within 24 h. The amount of isotopically exchangeable copper was correlated ($r = 0.88$) with copper extracted by pyrophosphate. Further experiments showed that most of the ^{64}Cu was found in the pyrophosphate extract when ^{64}Cu was equilibrated with the soil prior to copper fractionation; the copper adsorbed on the inorganic fraction contributed much less to the total copper extracted. Specific activity measurements indicated that 25 to 70% of the copper regarded as specifically adsorbed on inorganic soil constituents was isotopically exchangeable with copper in solution. For copper specifically absorbed on organic soil materials, 19 to 45% of that extracted with pyrophosphate was isotopically exchangeable with copper in solution. Hence, from their data, it is apparent that there is a large amount of specifically adsorbed copper that will equilibriate with copper in the soil solution. A large share of this specifically adsorbed copper is adsorbed by soil organic matter. Copper differs in this respect from calcium and magnesium where exchangeable cations represent the ions in the solid phase that most readily buffer those in solution.

Kline and Rust (1966), working with six Minnesota soils, found that 35 to 70% of total soil copper equilibrated with copper in the soil solution. These soils were relatively low in total copper content (2.26 to 17.88 mg/kg) and had organic matter levels ranging from 1.2 to 7.9%. The authors found that treating the soil with NaOH released copper that had equilibrated with copper in solution, as detected with radioactive isotopes. Hence, they concluded that much of the soil copper that is diffusible and will equilibrate with copper in solution is associated with the soil organic matter. The specifically adsorbed copper equilibrated rapidly with copper in solution, with equilibrium essentially reached in one hour. Results obtained by Kline and Rust (1966) thus agree with those of McLaren and Crawford.

Specifically adsorbed copper appears to constitute the major copper reserve that buffers the copper concentration in the soil solution. Hence, buffer power is most appropriately measured using isotopic equilibration techniques.

Residual Copper

Copper not present in solution or as exchangeable copper or specifically adsorbed has been termed residual copper; it is measured as the difference between the aforementioned forms and total copper. Residual copper usually constitutes about 50% of the total copper. This is a smaller proportion than for some other ions, such as zinc, phosphorus, and potassium, where a major portion of the nutrient may be in the residual fraction. Little, if any, information is available on the rate of release of copper from residual forms. However, since the residual fraction shows little equilibration with solution copper over 24 h, it is doubtful that release of residual copper is rapid enough to be of much consequence for plant nutrition.

Copper Adsorption by Soil

Much of the copper adsorbed by soil is specifically adsorbed with soil pH having a large effect on specific adsorption. McLaren and Crawford (1973b) investigated the effect of pH on copper absorption by various soil components. One-gram samples of soil or soil materials were equilibrated with 200 mL of 0.01 mol/L $CaCl_2$ containing 5 mg Cu/L; these samples were adjusted to a range of pH levels. After equilibration on a shaker for 24 h, solution copper was measured and copper adsorption calculated; the results are shown in Figure 15.3. As pH increased, more copper was adsorbed. Organic matter (a peat soil) adsorbed copper at much lower pH values than did mont-

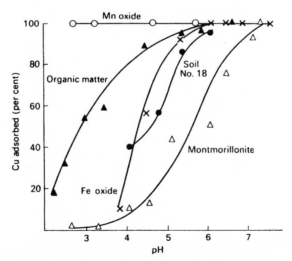

FIGURE 15.3   Effect of pH on the specific adsorption of copper from a solution initially containing 5 mg copper/L. Reprinted from McLaren and Crawford (1973b) by permission of Blackwell Scientific Publications Ltd.

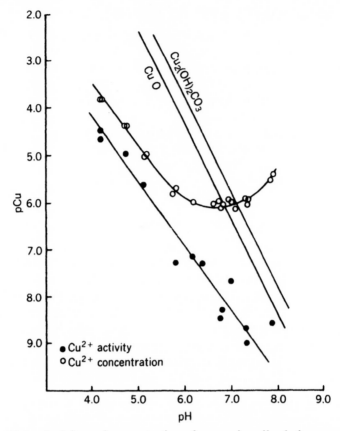

FIGURE   15.4   Activity and concentration of copper in soil solution as a function of pH for a system initially containing 40 mg/L copper. Solubility lines assume $PCO_2 = -3.5$. Ten grams of Mardin silt loam was equilibrated with 80 mL of 0.015 mol/L Cu^{2+} solution. Reproduced from McBride and Blasiak (1979) by permission of Soil Science Society of America.

morillonite; data agree with the concept that organic matter is the primary constituent specifically adsorbing copper. Calcium chloride was added to minimize any exchangeable copper adsorption. McBride and Blasiak (1979) measured the effect of soil pH on the amount of copper in soil solution; their results are reported in Figure 15.4. Both free Cu^{2+} and complexed copper were measured; as indicated earlier, complexed copper as a proportion of the copper in solution increased as pH increased. Total copper in solution reached a minimum at pH 7.0 on the Mardin soil, Typic Haplaquolls. Comparison of the values found with calculated solubility lines for CuO and $Cu_2(OH)_2CO_3$ indicated that the solution was undersaturated with respect to either of these minerals. Hence, the effect of pH appeared to be on adsorption rather than precipitation.

Copper Adsorption Curves

Adsorption curves for copper at pH 5.5 were obtained by McLaren and Crawford (1973b), using 24 British soils; results for four of the soils are shown in Figure 15.5. Maximum adsorption was correlated with organic matter level and free manganese oxides ($r = 0.93$). Adsorption maxima varied from 340 to 3000 mg/kg. Since this is many times greater than the content of adsorbed copper for most soils, soils will usually have adsorbed copper values on the steep portion of the adsorption curves; adsorption sites will be only partially filled with copper.

COPPER-UPTAKE KINETICS

Copper uptake is an active process, since it is reduced by metabolic inhibitors (Dokiya et al. 1964). Copper uptake influenced by copper concentration in a stirred solution appears to follow Michaelis–Menten kinetics. Nielsen (1976a) studied the kinetics of copper uptake by barley; uptake rates were measured for solutions varying in copper concentration from 0.08 to 3.59 μmol/L. He obtained a mean K_m value of 0.11 μmol/L and a mean C_{min} value of 0.045 μmol/L. No other measurements of copper uptake kinetics have been found in the literature. Soil solution copper concentrations found in the literature were approximately 10 μg/L, which is about 0.15 μmol/L. However, only part of the copper in solution was free Cu^{2+}, and if we look at solution Cu^{2+} values, the concentrations reported

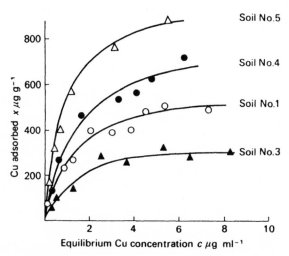

FIGURE 15.5 Variation among soils in the specific adsorption of copper. Reprinted from McLaren and Crawford (1973b) by permission of Blackwell Scientific Publications Ltd.

range from 0.009 to 0.31 µg/L, or 0.0001 to 0.005 µmol/L, which is much lower than the C_{min} value of 0.045 µmol/L reported by Nielsen. Since plants are able to obtain sufficient copper under these conditions, it appears that complexed copper is involved in copper uptake from solution. No information was found on this problem or on the effect of pH on copper uptake. Studies of uptake from soil solution versus pH might shed some light on the situation, since the fraction of copper complexed varies with pH.

SOIL COPPER SUPPLY

Nielsen (1976b) measured the copper concentration of the soil solution from a calcareous peat soil during growth of barley in pots. Copper concentration of the solution changed during barley growth, particularly where higher levels of copper were present in the soil. Increases in levels of copper in solution corresponded with increases in organic solute content of the solution. Where no plants were grown, the copper concentration in the soil solution remained relatively stable once it had reached a minimum value. For this soil, the copper concentration in the soil solution from untreated soil was 0.09 µmol/L, which is below the K_m value for copper uptake.

Experimental results indicate that copper uptake is not solely from free Cu^{2+} in solution, since such levels are too low for soil solutions of pH 7.0 and above to supply the plant with appreciable amounts of copper. To satisfy copper uptake rates, it is necessary to assume that complexed as well as ionic copper is absorbed by the plant root or at least that the complex supplies copper to the plasma membrane where it then dissociates before absorption. Sinclair et al. (1990) found the mobilization of copper by chelates increased copper uptake.

Copper in the soil solution equilibrates with both exchangeable and specifically adsorbed copper associated with the solid phase. The equilibration between specifically adsorbed and solution copper is rapid (Kline and Rust, 1966). In addition, a large portion (50% or more) of total soil copper is labile. However, even though it would appear that copper should be readily weathered from soils, most have sufficient copper for adequate supply to the plant. Those soils where copper deficiency occurs are usually organic soils or sandy soils, each having low total amounts of copper. Laboratory measurements of free Cu^{2+} in solution versus soil pH suggest that soil pH should have a considerable effect on supplying copper to the root. A few experiments where copper uptake has been compared at several soil pH levels indicate that soil pH does not greatly affect copper uptake in practice (Blevins and Massey, 1959; Lucas and Knezek, 1972). This may be partly due to the soil being able to maintain an adequate level of copper even when free Cu^{2+} is reduced by increasing soil pH.

Measurement of copper in solution, copper buffering power, and the effective-diffusion coefficient for copper is complicated by the presence of more than one fraction both in solution and adsorbed by the soil. No detailed investigations of buffering power or effective diffusion coefficients have been made.

Copper-uptake rates are lower than for most other micronutrients. In a nutrient-uptake study using field-grown corn, Mengel and Barber (1974) found that the uptake rate of copper when the corn plant was 40 days old was 0.23 that of boron, 0.17 that of manganese, 0.35 that of zinc, and 0.03 that of iron. The low requirement of many plants for copper is probably the reason why copper deficiencies occur in relatively few soils. Little information is available on differences between species with respect to their copper requirements; most plants contain less than 10 mg of copper/kg.

As discussed in Chapter 7, Li et al. (1991) studied the effect of mycorrhiza hyphae on copper uptake. They found that 53 to 62% of copper uptake by white clover was due to uptake by mycorrhizal hyphae.

A wide variety of extracting procedures have been proposed for measuring soil copper availability (Cox and Kamprath, 1972). Those measurements of available copper that extract some portion of the specifically adsorbed copper suggest that a value of 2 mg copper/kg of soil is the level below which response to added copper may occur.

Pettersson (1976) found that 10 μmol/L copper in solution culture depressed plant growth. The copper concentration used in many nutrient solutions is 0.5 μmol/L. Hence, there appears to be a relatively narrow range over which copper concentration is satisfactory for plant growth.

REFERENCES

Blevins, R. L., and H. F. Massey. 1959. Evaluation of two methods of measuring available soil copper and the effects of soil pH and extractable aluminum on copper uptake by plants. *Soil Sci. Soc. Am. Proc.* **23**:296–298.

Cavallaro, N., and M. B. McBride. 1978. Copper and cadmium adsorption characteristics of selected acid and calcareous soils. *Soil Sci. Soc. Am. J.* **42**:550–556.

Cavallaro, N., and M. B. McBride. 1980. Activities of Cu^{2+} and Cd^{2+} soil solutions as affected by pH. *Soil Sci. Soc. Am. J.* **44**:729–732.

Cox, F. R., and E. J. Kamprath. 1972. Micronutrient soil tests. In J. J. Mortvedt, P. M. Giordano. and W. L. Lindsay, Eds. *Micronutrients in Agriculture.* Soil Science Society of America, Madison, WI. Pp. 289–317.

Dokiya, Y., K. Kumazawa, and S. Mitsui. 1964. Nutrient uptake of crop plants. 4. The comparative physiological study on the uptake of iron, manganese, and copper by plants. The uptake of [58]Fe, [54]Mn, and ^{64}Cu by rice and barley seedlings as influenced by metabolic inhibitors. *J. Sci. Soil Manure, Japan* **35**:367–378.

Hodgson, J. F. 1963. Chemistry of the micronutrient elements in soils. *Adv. Agron.* **15**:119–159.

Hodgson, J. F., W. L. Lindsay, and J. F. Trierweiler. 1966. Micronutrient cation complexing in soil solution: II. Complexing of zinc and copper in displaced solution from calcareous soils. *Soil Sci. Soc. Am. Proc.* **30**:723–726.

Hodgson, J. F., H. R. Geering, and W. A. Norvell. 1965. Micronutrient cation complexes in soil solution: Partition between complexed and uncomplexed forms by solvent extraction. *Soil Sci. Soc. Am. Proc.* **29**:665–669.

Kline, J. R., and R. B. Rust. 1966. Fractionation of copper in neutron activated soils. *Soil Sci. Soc. Am. Proc.* **30**:188–192.

Li, X.-L., H. Marschner, and F. George. 1991. Acquisition of phosphorus and copper by VA-mycorrhizal hyphae and root to shoot transport in white clover. *Plant Soil* **136**:49–57.

Lindsay, W. L. 1979. *Chemical Equilibria in Soils.* Wiley-Interscience, New York.

Loneragan J. F., A. D. Robson, and R. D. Graham. Eds. 1981. *Copper in Soils and Plants. Proc. Int Symp. on Copper,* Perth, Australia. Academic Press, New York.

Lucas, R. E., and B. D. Knezek. 1972. Climatic and soil conditions promoting micronutrient deficiencies in plants. In J. J. Mortvedt, P. M. Giordano, and W. L. Lindsay, Eds. *Micronutrients in Agriculture.* Soil Science Society of America, Madison, WI. Pp. 265–288.

McBride, M. B., and J. J. Blasiak. 1979. Zinc and copper solubility as a function of pH in an acid soil. *Soil Sci. Soc. Am. J.* **43**:866–870.

McBride, M. B. 1981. Forms and distribution of copper in solid and solution phases of soil. In J. F. Loneragan, A. D. Robson, and R. D. Graham, Eds. *Copper in Soils and Plants. Proc. Int. Symp. on Copper,* Perth, Australia. Academic Press, New York. Pp. 25–46.

McLaren, R. G., and D. V. Crawford. 1973a. Studies on soil copper. I. The fractionation of copper in soils. *J. Soil Sci.* **24**:172–181.

McLaren, R. G., and D. V. Crawford. 1973b. Studies on soil copper. II. The specific adsorption of copper by soils. *J. Soil Sci.* **24**:443–452.

McLaren, R. G., and D. V. Crawford. 1974. Studies on soil copper. III. Isotopically exchangeable copper in soils. *J. Soil Sci.* **25**:111–119.

Mengel, D. B., and S. A. Barber. 1974. Rate or nutrient uptake per unit of corn root under field conditions. *Agron. J.* **66**:399–402.

Misra, P. C., M. K. Misra, and S. G. Misra. 1973. Note on the evaluation of methods for estimating available copper in soils. *Ind. J. Agric. Res.* **43**:609–610.

Nielsen, N. E. 1976a. A transport kinetic concept for ion uptake by plants. III. Test of a concept by results from water culture and pot experiments. *Plant Soil* **45**:659–677.

Nielsen, N. E. 1976b. The effect of plants on the copper concentration in the soil solution. *Plant Soil* **45**:679–687.

Pettersson, O. 1976. Heavy-metal ion uptake by plants from nutrient solutions with metal ion, plant species, and growth period variations. *Plant Soil* **45**:445–459.

Sanders, J. R., and C. Bloomfield. 1980. The influence of pH, ionic strength, and reactant concentration on copper complexing by humified organic matter. *J. Soil Sci.* **31**:53–63.

Shuman, L. M. 1979. Zinc, manganese, and copper in soil fractions. *Soil Sci.* **127**:10–17.

Sinclair, A. H., L. A. Mackie-Dawson, and D. J. Linehan. 1990. Micronutrient inflow rates and mobilization into soil solution in the root zone of winter wheat (*Triticum aestivum* L.). *Plant Soil* **122**:143–146.

CHAPTER **16**

Iron

Iron is present in soils in higher concentrations than any other nutrient. The lithosphere contains 5.1% iron, which forms compounds with sulfur and oxygen. However, in spite of the large amount of iron in the soil and the low quantities needed for plant growth, iron deficiencies occur because so little of the element is in an available form. Soils vary in total iron content from 0.02 to 10%, depending on their origin.

FORMS OF SOIL IRON

Iron may be present in the soil as oxides, hydroxides, and silicate minerals, amorphous oxides, adsorbed iron, iron complexed by organic materials, and iron in solution. Iron is present in the oxidation states Fe^{2+} and Fe^{3+}.

Mineral Iron

Iron minerals commonly found in soils include goethite, (FeOOH), hematite, (Fe_2O_3), lepidocrocite (FeOOH), maghemite, (Fe_2O_3), and magnetite (Fe_3O_4). Hematite gives a red color to soils, while goethite imparts a yellow color. Amorphous iron as $Fe(OH)_3$ is probably the most significant form in supplying iron for uptake by the plant. Olivene, $Mg_{1.6}Fe^{2+}_{0.4}SiO_4$, is an example of an iron magnesium silicate that may be present in the soil.

Organic-Iron Compounds

Iron forms stable complexes with organic compounds that occur in both the soil's solid phase and soluble organic compounds. Simpler organic compounds are citrate and oxalate. Iron compounds are more stable than combinations with most other nutrients, so that iron can replace them in the chelate except where mass-action displacement occurs due to high concentrations of other ions present. Iron chelates have high stability constants, so that at pH levels below 7.0, iron is frequently the dominant cation in the chelate.

Solution Iron

The amount of iron present as Fe^{2+} and Fe^{3+} in the soil solution depends on the hydroxide forms present in the soil, which, in turn, depends on pH and *pe,* a parameter related to the redox potential. The calculated Fe^{3+} activity in solution maintained by iron oxides and soil iron related to pH is shown in Figure 16.1. The soil iron calculation is based on a log K_0 of -39.3 for the relation $Fe(OH)_3$ (soil) $\leftrightarrows Fe^{3+} + 3OH$. The activity of Fe^{3+} maintained in solution by this equilibrium decreases 1000-fold for each unit increase in pH. While the exact value of Fe^{3+} activity in solution at equilibrium may change due to the forms of iron present, the relation to pH will be similar. Since equilibration is slow, levels of Fe^{3+} activity may be higher than calculated because of recent changes in soil pH or redox status that have not produced a new equilibrium.

Oxidation Reduction

Plants primarily absorb Fe^{2+}, but iron may occur in solution as Fe^{2+} or Fe^{3+}. The relation between Fe^{2+} and Fe^{3+} in solution depends on the oxidizing or reducing status of the soil. The oxidation reduction of iron is described by Equation 16.1:

$$Fe^{3+} + e^- \leftrightarrows Fe^{2+} \quad \log K_0 = 13.04 \tag{16.1}$$

The equilibrium constant K_0 is actually expressed in terms of activities. The activity of Fe^{2+} and Fe^{3+} in solution is also affected by pe, as described by Equation 16.2

$$\log\left(\frac{Fe^{2+}}{Fe^{3+}}\right) = 13.04 - pe \tag{16.2}$$

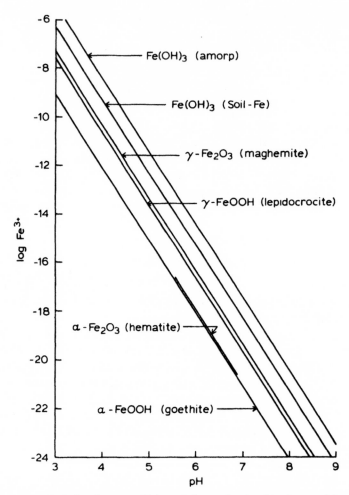

FIGURE   16.1   Relation between Fe^{3+} activity maintained by iron oxides and soil iron and pH. Reprinted from Lindsay (1979) by permission of John Wiley & Sons, Inc.

The value *pe* is the negative log of the electron activity, where electron activity is unity for the standard hydrogen electrode. The *pe* may vary from −4 to 12 in aqueous systems (Lindsay, 1979).

Values for *pe* + pH may vary from zero under the most reduced conditions (an H_2 atmosphere) to 20.78 under the most oxidized conditions of one atmosphere O_2. The relation between Fe^{2+} in solution and soil iron affected by pH is shown in Figure 16.2. This figure was calculated by assuming values of *pe* plus pH from 10 to 20.61; the relation for calculating the lines shown is given in equation 16.3:

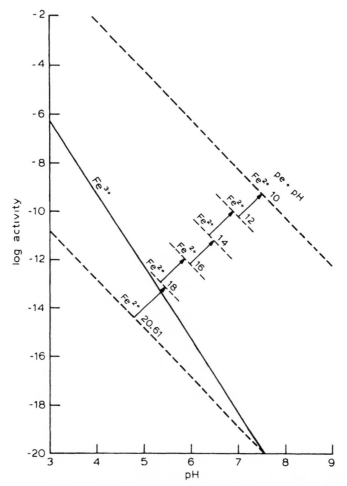

FIGURE   16.2   Effect of redox and pH on the equilibrium of Fe^{2+} with soil iron. Reprinted from Lindsay (1979) by permission of John Wiley & Sons, Inc.

$$\log Fe^{2+} = 15.74 - (pe + pH) - 2\,pH \qquad (16.3)$$

Values of $pe + pH$ in the soil may vary from 2 to 18, depending on pH and the oxidation status of the soil. If we assume a $pe + pH$ value of 12, the Fe^{2+} activity in solution at pH 7 would be 5.5×10^{-11} molar. This is an extremely low value since Fe^{2+} levels must be higher than this in most soils with pH values higher than 7, otherwise iron deficiency in plants growing on these soils would be widespread. Plant roots may greatly affect the $pe + pH$ value of the soil in the immediate vicinity of the root, which, in turn, may affect iron availability. Measured values of iron levels in the soil solution are scarce; O'Connor et al. (1971) obtained a value of 0.02 mg/L for a Thoroughfare loamy sand (Typic Torrifluvents) at a pH of 7.9. The iron present probably included several hydrolytic species of iron as well as soluble organic complexes.

Buffer Power

O'Connor et al. (1971) measured the buffer power of $^{59}Fe^{3+}$ added to a Thoroughfare loamy sand (pH 7.9) and obtained a value of 1000. However, this value may be affected by precipitation of iron oxides, since the value increased with time. Also, the ^{59}Fe added was not carrier-free; hence, precipitation may have occurred after $^{59}Fe^{3+}$ addition. It is difficult to determine the equilibrium between solution and adsorbed iron, since equilibrium can be affected by precipitation, oxidation reduction, and the presence of iron in organic complexes.

Diffusion Coefficient

The only data on iron diffusion found were those of O'Connor et al. (1971), who obtained a D_e value of 3×10^{-10} cm²/s. After adding EDDHAFe to the soil, the value for D_e increased to 1.7×10^{-7} cm²/s. This large a D_e value would indicate very little interchange between chelated iron in solution and iron on the solid phase that buffered the solution; otherwise D_e would have been smaller.

IRON-UPTAKE KINETICS

Iron uptake by plant roots depends on respiration energy (Moore, 1972), since addition of respiration inhibitors inhibits iron uptake. Shim and Vose (1965) measured Fe^{3+} uptake by excised rice (*Oryza sativa* L.) roots, obtaining K_m values varying between 0.77 and 2.6 μmol/L for the first phase and 45 and 69 μmol/L for the second phase of uptake at higher concentrations.

While uptake studies have been conducted with Fe^{3+} in solution culture, research has indicated that Fe^{3+} is reduced to Fe^{2+} before it is absorbed by the plant. Chaney et al. (1972) found that plant roots exude reducing compounds that reduce Fe^{3+} to Fe^{2+} before it is absorbed. Where reductants were prevented from reducing Fe^{3+}, little iron uptake occurred.

Brown (1978) described iron uptake by plant roots growing in soil. The action of plant roots in their rhizosphere may make soil iron available for absorption. The soil solution's iron concentration can be increased by both reducing soil pH and adding reducing agents that reduce Fe^{3+} to Fe^{2+}. Some plant roots release H^+ and reductants when placed under iron stress. This release increases Fe^{2+} levels in the rhizosphere. Plants have been classified as either Fe efficient or Fe inefficient; plants termed Fe efficient respond to iron deficiency by releasing H^+ ions and reductants into the rhizosphere. Plant roots absorb Fe^{2+} and transport it in this form to the junction of the protoxylem and metaxylem, where it is oxidized back into Fe^{3+}. It is then chelated with citrate and moved into the xylem for transport to the shoot.

Romheld and Marshner (1979) investigated iron uptake by sunflower (*Helianthus anneus*), an Fe-efficient plant, and corn, an Fe-inefficient plant. With iron stress, sunflower reduced solution pH and released reductants, whereas corn did not. They found that pH changes and iron reduction occurred exclusively in the 0- to 1.5-cm section of the root tip. Root tips responded to iron stress before symptoms of chlorosis occurred in the plant, and root tips of iron-stressed plants also produced many more root hairs. The change in root morphology and the physiology of the root tip indicated an effective mechanism for fine regulation of iron supply to the plant. Within each species, there are cultivars that are iron efficient and iron inefficient. This has led to investigations into the development of cultivars for use on high-pH soils that have produced iron deficiencies for some species. Because of the effect of the plant root on solubilizing iron, it is difficult to use the usual parameters for describing iron flux by mass now and diffusion to the root surface.

Siderophores and phytosiderophores can enhance iron uptake from sparingly soluble Fe^{3+} compounds. Siderophores are low-molecular-weight iron-chelating agents produced by virtually all bacteria and fungi under iron-limiting conditions (Buyer and Sikora, 1990). Phytosiderophores are iron-chelating compounds that are released by iron-deficient graminaceous species. Many mobilize Fe^{3+} by chelation. Release of phytosiderophores increases 10- to 20-fold when the species are iron deficient (Romheld and Marschner, 1990). But for some it is much less. Release of phytosiderophores is absent in dicots. Chemical nature of phytosiderophores can differ between species. Phytosiderophores are one reason for differences among species in their ability to absorb iron from iron-deficient soils.

Plant roots absorb iron from Fe^{3+} chelates. Chaney et al. (1972) have shown that the plant root releases a reductant that reduces Fe^{3+} in the chelate. After reduction, Fe^{2+} is released into solution and absorbed by the root. It is hypothesized that there is chelated iron in the soil solution due to natural chelates. This iron moves readily to the root by mass flow and/or diffusion and is then reduced, released from the chelate, and absorbed.

Some ions compete with iron for absorption. Shim and Vose (1965) observed that manganese and copper competitively inhibited iron uptake; there also were indications that high levels of calcium inhibit iron uptake, which may be one reason why iron deficiencies for some plants occur on high-calcium soils. High levels of phosphorus have also been shown to reduce iron uptake.

Clarkson and Sanderson (1978) have investigated sites for iron absorption by barley roots and found that the root section between 1 and 4 cm from the tip absorbed and translocated much more iron than the rest of the root. In their studies, iron influx by iron-stressed roots was 7- to 10-fold higher than for unstressed roots. This observation agrees with that of Romheld and Marshner (1979), who found that most of the proton and reductant exudation from the root occurred in the 0- to 1.5-cm region of the root tip.

Uptake of iron from high pH soils is influenced by the release of siderophores and phytosiderophores that chelate Fe^{3+}, bring it into solution, and allow its movement to the root by mass flow and diffusion where Fe^{3+} is reduced to Fe^{2+} and absorbed by the plant root. Because the release of phytosiderophores varies with species and cultivars within species as well as the effect of soil microorganisms, it is difficult to develop a model for the movement of iron to the root by this mechanism. In addition, plant roots may change the pH of the rhizosphere soil (Chapter 6); when the pH is reduced the solubility of iron is increased and this increases the iron source. Hence the influence of the plant root is a big factor in changing the iron supply to the root from what would be calculated from the initial levels in the soil before the plant grows.

There have been several international symposia on iron nutrition. The fourth such symposia was held in Albuquerque, New Mexico in 1987. The papers given at this symposia were published in the *Journal of Plant Nutrition* **11**:604–1621 (1988). The sixth symposium was held in Logan, Utah in 1991 and the papers were published in *Journal of Plant Nutrition* **15**:1487–2313 (1992). The scope of the papers given indicates the complex nature of iron uptake from the soil by plant roots as well as the utilization of iron within the plant. The information in this chapter gives some significant factors related to iron uptake.

REFERENCES

Brown, J. C. 1978. Mechanisms of iron uptake by plants. *Plant, Cell Environ.* **1**:249–257.

Buyer, J. S., and L. J. Sikora. 1990. Rhizosphere interactions and siderophores. *Plant Soil* **129**:101–107.

Chaney R. L., J. C. Brown, and L. O. Tiffin. 1972. Obligatory reduction of ferric chelates in iron uptake by soybean. *Plant Physiol.* **50**:208–213.

Clarkson, D. T., and J. Sanderson. 1978. Sites of absorption and translocation of iron in barley roots. Tracer and microautoradiographic studies. *Plant Physiol.* **61**:731–736.

Lindsay, W. L. 1979. *Chemical Equilibrium in Soils*. Wiley-Interscience. New York.

Moore, D. P. 1972. Mechanisms of micronutrient uptake by plants. In J. J. Mortvedt, P. M. Giordano, and W. L. Lindsay, Eds. *Micronutrients in Agriculture*. Soil Science Society of America, Madison, WI. Pp. 171–198.

Norvell, W. A. 1972. Equilibria of metal chelates in soil solution. In J. J. Mortvedt, P. M. Giordano, and W. L. Lindsay, Eds. *Micronutrients in Agriculture*. Soil Science Society of America, Madison, WI. Pp. 115–138.

O'Connor, G. A., W. L. Lindsay, and S. R. Olsen. 1971. Diffusion of iron and iron chelates in soil. *Soil Sci. Soc. Am. Proc.* **35**:407–410.

Romheld, V., and H. Marshner. 1979. Fine regulation of iron uptake by Fe-efficient plant *Helianthus annus.* In J. L. Harley and R. S. Russell, Eds. *The Soil Root Interface.* Academic Press, London. Pp. 405–417.

Romheld, V., and H. Marschner. (1990). Genotypical differences among graminaceous species in release of phytosiderophores and uptake of iron phytosiderophores. *Plant Soil* **123**:147–153.

Shim, S. C., and P. B. Vose. 1965. Varietal differences in the kinetics of iron uptake by excised rice roots. *J. Exp. Bot.* **16**:216–232.

CHAPTER **17**

Manganese

Manganese is the eleventh most common element in the earth's crust, with an average concentration of 0.09%, or 900 mg/kg. Manganese is present primarily as oxides and sulfides; it often occurs in association with iron. Igneous rocks have a Mn/Fe ratio of 1:60. Soils have manganese concentrations that are usually in the range of 20 to 3000 mg/kg, with an average of 600 mg/kg (Lindsay, 1979). Walker and Barber (1960) analyzed 12 Indiana soils and found total manganese concentrations ranging from 60 mg/kg for a histisol to 1320 mg/kg for a soil developed from residual limestone. Duangpatra et al. (1979) analyzed 15 Kentucky soils and found a range of 640 to 3040 mg/kg. Heintz and Mann (1951) measured total manganese in 18 English soils varying widely in organic matter and obtained a range of 163 to 2320 mg/kg.

Soil manganese exists in three oxidation states—Mn^{2+}, Mn^{3+}, and Mn^{4+}; manganese absorbed by plant roots is primarily as Mn^{2+}. Oxidation–reduction reactions in the soil influence the amount of each oxidation state present. The predominant oxidation states in most soils are Mn^{2+} and Mn^{4+}, with much more as Mn^{4+} than Mn^{2+} in aerated soils.

FORMS OF SOIL MANGANESE

Total soil manganese may be divided into mineral manganese, organically complexed manganese, exchangeable manganese, and solution manganese. Manganese in solution may be either Mn^{2+} or manganese combined with soluble organic compounds. The equilibrium of manganese between these forms is influenced greatly by soil pH and redox conditions.

Mineral Manganese

Manganese oxides, the most common manganese minerals in soil, include pyrolusite (MnO_2), manganite ($MnOOH$), and hausmanite (Mn_3O_4). There is a series of manganese oxides that varies according to substitutions of O^{2-} for OH^-. Manganese oxides may also occur as coatings on other minerals. Some soils contain iron–manganese concretions that range in size from 0.1 to 15 mm in diameter. They have an iron content of 5 to 17% and a manganese content of 0.5 to 8%.

Organically Complexed Manganese

Divalent manganese forms complexes with soil organic compounds, which may be either soluble or insoluble, the nature of these complexes is described in Chapter 2. The amount of complexed manganese can be measured by displacing it with a more strongly adsorbed ion, such as zinc. Walker and Barber (1960) measured the amount of organically complexed manganese in 12 Indiana soils by displacement with copper after removing exchangeable manganese with ammonium acetate. Amounts of complexed manganese varied from essentially none to 24 mg/kg. Liming the soil to increase soil pH greatly reduced the amount of manganese extracted. Since there were similar amounts of each present, complexed manganese was correlated with exchangeable manganese.

Exchangeable Manganese

Exchangeable manganese is held by the soil cation-exchange sites and measured by displacement with ammonium acetate. It exists essentially as Mn^{2+}. A wide range of values can be obtained, with very acid soils having values above 1000 mg/kg, while organic soils of high pH may have values less than 0.1 mg/kg.

Adding Mn^{2+} to the soil may not increase exchangeable manganese appreciably, because it is readily oxidized to Mn^{4+} and precipitated as an oxide. Extracting exchangeable manganese with 1 mol/L ammonium acetate at pH 7.0 or an acid extractant, such as 0.1 mol/L H_3PO_4, has been used to measure the manganese presumed available to plants. The acid extractants probably extract more than exchangeable manganese, however.

Air drying soil increases the amount of exchangeable manganese. Walker and Barber (1960) found, for example, that air drying doubled exchangeable manganese levels. The increase was correlated with the total amount of manganese present in the soil. Exchangeable manganese decreases as soil pH increases.

Soil Solution Manganese

Barber et al. (1967) found that total manganese in saturation extracts from samples of six Indiana surface soils varied from 0.18 to 790 μmol/L. Analysis of A and B horizons from Colorado, New York, South Carolina, and Washington soils varied from <0.18 to 236 μmol/L. (Geering et al., 1969; Sinclair et al., 1990). The total manganese measured in the soil solution included Mn^{2+} and organically complexed manganese. Geering et al. (1969) found that 84 to 99% of the solution manganese was complexed for samples from the soil A horizon. For solution samples from the B horizon, 39 to 73% was complexed.

The Mn^{2+} in solution in equilibrium with pyrolusite and magnatite can be calculated from solubility–product relationships. Solution Mn^{2+} is influenced by pH and the electron activity pe. Using the relation pH + pe = 16.62 (Lindsay, 1979), the relation between pH and Mn^{2+} in solution in equilibrium with manganite can be calculated from the reaction

$$\log Mn^{2+} = 25.27\ (pe + pH) - 2pH$$

$$= 25.27 - 16.62 - 2pH$$

$$= 8.65 - 2pH$$

Thus, the calculated activity would decrease from 4.5×10^{-2} mol/L at pH 5.0 to 4.5×10^{-8} mol/L at pH 8.0. Values predicted at pH 5 are much higher than generally found in soil solutions; hence, more than solubility alone governs Mn^{2+} activity in soil solutions. Values calculated at pH 8.0 are much lower than would be needed in the soil solution to supply manganese to the root.

Linehan et al. (1989) found low C_{li} values for copper, manganese, and zinc in the root zone of barley in late winter for eight Scottish soils; maximum values occurred between May and early July. The ranges of C_{li} values for copper, manganese, and zinc were 0.02 to 0.07, 0.03 to 0.1, and 0.03 to 0.27 μmol/L. These values would include both ionic and chelated forms in solution. The C_{si} values obtained by EDTA extraction were 0.7 to 14.0, 20 to 287, and 1.6 to 21.5 for copper, manganese, and zinc. Sinclair et al. (1990) indicated that the size of C_{li} appears to depend on biologically produced chelating agents in solution.

Buffer Power of Soil for Manganese

Few investigations of the soil buffer power for soil solution manganese have been made. Values for both exchangeable manganese and solution manganese are available and can be used for calculating buffer power. Since solution manganese contains both complexed and ionic forms, however, the appropriateness of values calculated from these data may be questioned. Halstead et al. (1968) found exchangeable manganese/solution manganese values ranging from 65 to 0.4. Barber et al. (1967) found values ranging from 527 to 1.1. High values are for soils low in soluble manganese, while low values are from acid soils where manganese levels are high. The range in b is greater than for most nutrients.

Diffusion Coefficient

The only diffusion coefficients reported for soil manganese diffusion are those of Halstead and Barber (1968), who measured self-diffusion coefficients in six Indiana soils. Values ranged from 0.33×10^{-7} cm²/s for Chelsea sand to 2.2×10^{-7} cm²/s for Cincinnati silt loam. These D_e values are relatively high for a divalent cation and much higher than those reported for zinc (see Chapter 19). Manganese is possibly held rather loosely by soil-exchange sites. On the basis of b values given in the preceding section, we might expect D_e to vary from 5×10^{-7} to 1×10^{-9} cm²/s.

Adsorption Isotherms

Curtin et al. (1980) measured Mn^{2+} adsorption isotherms for 20 calcareous soils, fitting them to the Langmuir adsorption equation. Adsorption maxima varied from 1712 to 4317 mg/kg and were correlated ($r = 0.96$) with soil cation-exchange capacities. Many of the soils had high affinities for initial amount of manganese added, but bonding energy decreased as the amount of manganese added increased. For four soils, amounts of adsorbed manganese that could be displaced with 0.1 mol/L $CaCl_2$ was measured. At the lowest rate, 125 mg/kg, 6 to 65% of the adsorbed manganese was exchangeable after shaking for 24 h. Although in these experiments Mn^{2+} may have remained as Mn^{2+}, in field situations much of the manganese may be oxidized to Mn^{4+}, precipitated, and become unavailable for absorption by plant roots. Shuman (1977) investigated Mn^{2+} adsorption by four acid soils and correlated adsorption maxima with soil cation-exchange capacities. The Langmuir B values (see Equation 2.17) for the four soils were 810, 450, 50, and 50 mg/g. The Langmuir a values, which are related to bonding energy, were 0.043, 0.162, 0.308, and 0.385 L/mg, since there was a wide range between clay loam at 0.043 L/mg and loamy sand at 0.385 L/mg.

MANGANESE-UPTAKE KINETICS

Maas et al. (1968) measured manganese uptake by excised barley and wheat roots and found uptake to be metabolically mediated, since it was decreased by respiration inhibitors. Uptake increased asymptotically with manganese concentration in solution, reaching a maximum at 2.5 mmol/L. Solution pH also affected manganese uptake; uptake increased from pH 4 to 6, and above pH 6, oxidation of Mn^{2+} reduced uptake rate. While their data were not fit to a Michaelis–Menten equation, inspecting the data suggests a K_m value of approximately 0.1 mmol/L. Since this value is high relative to usual manganese concentrations in solution, manganese uptake should show an approximately linear relation between soil solution Mn^{2+} concentration and uptake. Results obtained by Halstead et al. (1968) confirm this prediction. They grew four plant species in four soils with two transpiration rates. Manganese in the soil solution varied from 0.18 to 16 μmol/L. The authors calculated manganese supplied to the root by mass flow and root interception, comparing these values with manganese uptake by the plant. Results shown in Figure 17.1 indicate that manganese uptake increases with manganese supply over a wide range, with no consistent differences between the four plant species. While the relation in Figure 17.1 might indi-

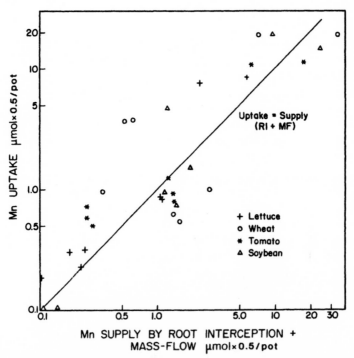

FIGURE   17.1   Relation between manganese uptake and calculated supply of manganese by root interception and mass flow for four plant species growing on four soils, with two rates of transpiration. Reproduced from Halstead et al. (1968) by permission of Soil Science Society of America.

cate manganese uptake to be nonenergy dependent, the high K_m values would give the same results with energy-dependent uptake.

A large fraction of the solution manganese may be complexed with soluble organic materials. The results of Halstead et al. (1968) indicate that manganese is either absorbed directly as the complex or readily dissociates and then is absorbed as Mn^{2+}.

SOIL MANGANESE SUPPLY

The supply of manganese to plant roots by mass flow and diffusion is greatly affected by the concentration in the soil solution, which, in turn is influenced by soil pH and *pe*. In aerated soils, the Mn^{2+} concentration in the soil solution would theoretically decrease by a factor of 100 for every unit increase in pH. If this were the situation in all soils, all soils with a high pH would be deficient in manganese. Since this is not the case, either this solubility relation does not control Mn^{2+} concentration or some other factors increase manganese availability in soils. One factor may be the increased manganese level in solution due to soluble organic complexes, such as siderophores.

The amount of manganese that must be sustained in the soil solution so that mass flow can supply plant manganese requirements can be estimated from plant analysis and water use. Assuming that a plant has 30 mg manganese/kg dry weight and water use is 400 kg/kg of dry weight produced, the soil solution should be 30/400 or 0.075 mg/L (1.4 μmol/L) of manganese. Only soils with pH above 6.5 have soil solution manganese concentrations this low. Oliver and Barber (1966) grew soybeans on a subsoil having a soil solution concentration of 0.36 μmol/L. They found that mass flow supplied less than 15% of the manganese taken up by young soybean plants.

Manganese deficiency in plants occurs in some soils when the pH is increased above 6.2. For other soils, manganese is adequate even though the pH is 7.5 or higher. The manganese chemistry of manganese-deficient soils has not been well defined.

Rule and Graham (1976) used ^{54}Mn to measure the labile pool of manganese in a soil as measured by ladino clover (*Trifolium repens*) and fescue (*Festuca elatior*) uptake of manganese. They obtained values varying from 17.6 to 35 mg/kg. As pH was increased from 4.3 to 6.5 the labile pool for ladino clover increased from 30 to 35 mg/kg, while that measured by fescue decreased from 23.7 to 17.6 mg/kg. These values indicate that levels of labile manganese were relatively insensitive to pH in these soils, which is contrary to what would be expected from solubility calculations.

Plant roots may affect C_{li} values for manganese. Godo and Reisenauer (1980) provided evidence supporting the hypothesis that root exudates reduce Mn^{4+} to Mn^{2+} just as occurs for iron. They suggested reduced manganese may be complexed with soluble organic exudates and moved to the root surface by mass flow. The effect was most evident at a soil pH below

5.5. Just as with other nutrients manganese supply to the root and uptake may be influenced by exudates from the root (as discussed in Chapter 6) so the supply to the root would be greater than that calculated from C_{li}, b, and D_e.

REFERENCES

Barber, S. A., E. H. Halstead, and R. E Follett. 1967. Significant mechanisms controlling the movement of manganese and molybdenum to plant roots growing in soil. *Trans. Joint Meeting Commissions II and IV.* International Soil Science Society, Aberdeen, Scotland. Pp. 299–304.

Curtin, D., J. Ryan, and R. A. Chaudry. 1980. Manganese adsorption and desorption in calcareous Lebanese soils. *Soil Sci. Soc. Am. J.* **44**:947–950.

Duangpatra, P., J. L. Sims, and J. H. Ellis. 1979. Estimating plant available manganese in selected Kentucky soil. *Soil Sci.* **127**:35–40.

Geering, H. R., J. F. Hodgson, and C. Sdano. 1969. Micronutrient cation complexes in soil solution: IV. The chemical state of manganese in soil solution. *Soil Sci. Soc. Am. Proc.* **33**:81–85.

Godo, G. H., and H. M. Reisenauer. 1980. Plant effects on soil manganese availability. *Soil Sci. Soc. Am. J.* **44**:993–995.

Halstead. E. H., and S. A. Barber. 1968. Manganese uptake attributed to diffusion from soil. *Soil Sci. Soc. Am. Proc.* **32**:540–542.

Halstead, E. H., S. A. Barber, D. D. Warncke, and J. Bole. 1968. Supply of Ca, Sr, Mn, and Zn to plant roots. *Soil Sci. Soc. Am. Proc.* **32**:69–72.

Heintz, S. G., and P. J. G. Mann. 1951. A study of various fractions of the manganese of neutral and alkaline soils. *J. Soil Sci.* **2**:234–242.

Lindsay, W. L. 1979. *Chemical Equilibria in Soils.* Wiley-Interscience, New York.

Linehan, D. J., A. H. Sinclair and M. C. Mitchell. 1989. Seasonal changes in Cu, Mn, Zn, and Co concentrations in soil in the root zone of barley (*Hordeum vulgare* L.). *J. Soil Sci.* **40**:103–115.

Maas, E. V., D. P. Moore, and B. J. Mason. 1968. Manganese absorption by excised barley roots. *Plant Physiol.* **43**:527–530.

Oliver, S., and S. A. Barber. 1966. Mechanisms for the movement of Mn, Fe, B, Cu, Zn. Al, and Sr from one soil to the surface of soybean roots (*Glycine max*). *Soil Sci. Soc. Am. Proc.* **30**:468–470.

Rule, J. H., and E. R. Graham. 1976. Soil labile pools of manganese, iron, and zinc as measured by plant uptake and DTPA equilibrium. *Soil Sci. Soc. Am. J.* **40**:853–857.

Shuman, L. M. 1977. Effect of soil properties on manganese isotherms for four soils. *Soil Sci.* **124**:77–81.

Sinclair, A. H., L. A. Mackie-Dawson, and D. J. Linehan. 1990. Micronutrient inflow rates and mobilization into soil solution in the root zone of winter wheat (*Triticum aestivum* L.). *Plant Soil* **122**:143–146.

Walker, J. M., and S. A. Barber. 1960. The availability of chelated Mn to millet and its equilibria with other forms of Mn in the soil. *Soil Sci. Soc. Am. Proc.* **24**:485–288.

CHAPTER **18**

Molybdenum

Molybdenum is present in smaller amounts in the lithosphere than most other micronutrients; the lithosphere averages only 2.3 mg/kg. Concentrations are higher in shales than in sandstones or limestones. Igneous rocks contain 1 to 2 mg/kg. Molybdenum contents in soils range from 0.2 to 36 mg/kg. Lavy et al. (1961) found that Carlisle muck, Limnic Medesaprists, contained 23 mg/kg, while eight mineral soils averaged 2.4 mg/kg. Vlek and Lindsay (1977) found a range of 2.5 to 36.3 mg/kg in 13 Colorado soils. Total molybdenum contents approximate 2 to 4 mg/kg in more weathered soils, with higher values in less weathered soils.

FORMS OF SOIL MOLYBDENUM

Total soil molybdenum can be subdivided into mineral, organic, adsorbed or labile, and solution forms.

Mineral and Organic Molybdenum

Because of low concentrations of molybdenum in soil, few molybdenum-containing minerals have been identified. Three possible minerals are ferrimolybdate, $Fe_2(MoO_4)_3 \cdot 8H_2O$; wulfenite, or lead molybdate, $PbMoO_4$; and powellite, or calcium molybdate, $CaMoO_4$; the latter is the more soluble of the three, with a K_{sp} of 7.2. Follett and Barber (1967a) investigated the possibilities of these minerals controlling the level of molybdenum in a soil solution from Raub silt loam. The solution was undersaturated with respect to both calcium molybdate and ferrous molybdate; data were not available to predict solubility of ferrimolybdate in this soil.

Soil organic matter has a molybdenum content several times higher than the mineral fraction. Plant material varies in molybdenum content from 0.5 to 14 mg/kg. The higher levels in organic matter would have to result from concentration during microbial decomposition.

Soil Solution Molybdenum

Amounts of molybdenum in the soil solution are small (less than 10 μg/L); however, the *Aspergillus niger* procedure is a sensitive molybdenum-measurement technique in this concentration range, which has been used to measure molybdenum concentrations in displaced soil solutions. Lavy and Barber (1964) reported concentrations of 2.2 to 8.1 μg/L in displaced solutions from 11 Indiana soils, and Follett (1966) obtained values of 1 to 13 μg/L on nine different Indiana soils. Follett and Barber (1967b) displaced the soil solution from a Raub silt loam daily for 19 days. When done at 26°C, the molybdenum concentration in the displaced solution did not vary appreciably, indicating that the molybdenum associated with the solid phase rapidly reequilibrated, maintaining the molybdenum level in solution; the level in solution varied with temperature, however. At 65°C, the level was several times higher, and at 4°C, it was about one-half the concentration at 26°C, indicating a temperature-dependent solubility relationship.

The form of molybdenum in soil solution is MoO_4^{2-}, since this is the major form above pH 4.2. Concentrations are not high enough to support the presence of molybdenum complexes in solution.

Adsorbed Molybdenum

Adsorbed molybdenum has been measured in several ways, including determining labile molybdenum with ^{99}Mo (Lavy and Barber, 1964), measuring molybdenum uptake with *Aspergillus niger* M. (Lavy and Barber, 1964), and extracting molybdenum with anion-exchange resin (Bhella and Dawson, 1972; Jackson and Meglen, 1975; Olsen and Watanabe, 1979; Karimian and Cox, 1978). The relative amounts extracted by these three procedures were similar. Amounts extracted by anion-exchange resin from 30 western Oregon hill soils derived from different parent materials ranged from 4 to 21 μg/kg. Values were positively correlated with soil pH ($r = 0.80$). Liming acid members of these soils more than doubled the amount of molybdenum extracted.

Olsen and Watanabe (1979) found much higher anion-exchangeable molybdenum contents in Colorado soils with pH values above 7.5; values ranged from 90 to 1350 µg/kg. Karimian and Cox (1978) measured values for southeastern U.S. soils that ranged from 0.6 to 11.1 µg/kg. For six Indiana soils, Lavy and Barber (1964) found 41 to 211 µg/kg of labile molybdenum, and measurements with *Aspergillus niger* on 15 Indiana soils gave values of 20 to 530 µg/kg. Hence, there is a range in adsorbed molybdenum of 0.6 to 1350 µg/kg. The level increases with increasing pH, being highest for unleached soils with pH values above 7.5.

Adsorbed or available molybdenum can be displaced with water. Follett and Barber (1967b) were able to show (Table 18.1) that displacing the soil solution with water at daily intervals reduced available molybdenum as measured by *Aspergillus niger* bioassay by an amount equal to the amount removed in the displaced solution. Nineteen displacements at 65°C displaced almost all of the available molybdenum, and incubation of the soil subsequently for 6 months did not release additional measureable molybdenum. On this Raub silt loam, there was rapid equilibrium between dissolved and adsorbed molybdenum but little release of mineral molybdenum, which suggests the potential for depleting molybdenum from this soil.

The possibility that the solubility of molybdate minerals can govern molybdenum concentrations in soil solutions was investigated by Follett and Barber (1967a) and Vlek and Lindsay (1977). Calculations indicated that calcium and ferrous molybdate were too soluble for their solid phases to exist in soils; wulfenite, $PbMoO_4$, had a solubility near the value found for one of the 13 soils. For the other soils, it appeared that an adsorption reaction was controlling molybdenum concentration in solution.

For some Colorado soils, the relation between pH and MoO_4^{2-} activity in solution indicated a 10-fold decrease in MoO_4^{2-} per unit decrease in pH. Soils

TABLE 18.1   Effect of Temperature and Daily Water Displacement on Molybdenum Removal and Reduction in Available Soil Molybdenum for Raub Silt Loam

Molybdenum form	Mo (mg/kg)
Total soil Mo	4.50
Initial level of available soil Mo	0.128
Mo removed by 12 displacements at 26°C	0.022
Available soil Mo after 12 displacements at 26°C	0.100
Mo removed by 19 displacements at 65°C	0.221
Available soil Mo after 19 displacements at 65°C and seven months at 26°C	0.002

Source: Reproduced from Follett and Barber (1976b) by permission of the Soil Science Society of America.

derived from volcanic ash had a high adsorption capacity for molybdate (Gonzalez et al., 1974). The adsorption maxima varied from 25.1 to 169.8 mmol/kg, with higher adsorption occurring in C-horizon material. Adsorption was believed mainly due to the presence of allophane and amorphous aluminum, silicon, and iron compounds.

Buffer Power

Buffering of dissolved soil molybdenum covers a wide range. Lavy and Barber (1964) obtained values ranging from 9 on a sandy loam to 83 on a silty clay loam. Little additional information is available, although comparing concentrations of adsorbed molybdenum, varying from 0.6 to 1350 µg/kg, with solution molybdenum concentrations varying from 1 to 13 µg/L indicates that buffer power values range from 1 to 100. This range was indicated by data from Lavy and Barber (1964). Adsorption curves for Australian soils (Barrow, 1970) indicated b values of C_s adsorbed/C_l ranging from 10 to 2000.

Adsorption curves for molybdenum adsorption on volcanic soils were obtained by Gonzalez et al. (1974). They followed the Langmuir equation, with adsorption maxima of 25 to 193 mmol/kg. Adsorption on Australian soils gave maxima of 0.10 to 3.0 mmol/kg (Barrow, 1970). The amounts added to obtain these adsorption curves greatly exceeded amounts of molybdenum normally present in soils, however.

Adsorption by allophanic soils was found to be pH-dependent, dropping rapidly as pH increased above pH 5.5. The pK_2 value for MoO_4^{2-} is 4.5, which is related to the effect of pH on adsorption, as described by the theories of Hingston et al. (1968) for anion adsorption on goethite. These theories were discussed in Chapter 2.

Diffusion Coefficient

Lavy and Barber (1964) measured D_e values for 11 Indiana soils and they obtained values for self-diffusion of molybdate ranging from 4.6×10^{-8} cm^2/s on a Histisol to 8.4×10^{-7} cm^2/s on an Alfisol. Molybdenum diffusion rates were similar to those for potassium and much greater than those for phosphate, with which its characteristics are often compared. Since b values were estimated to vary between 1 and 100, soils on which diffusion was measured must have had consistently low b values. Where b values are high, D_e will probably be smaller than the values given here.

MOLYBDENUM-UPTAKE KINETICS

Uptake from Solution

Few studies have been made on the kinetics of molybdenum uptake by plant roots. Kannan and Ramoni (1978) studied molybdenum absorption and transport in intact and excised bean (*Phaseolus vulgaris* L.) and rice (*Oryza sativa* L.) roots; they found absorption to be energy-dependent. Uptake rate by rice roots was 0.04 mol/$g_{F.W.}$ · h, which could be taken as an I_{max} value, because the solution was 10 μmol molybdenum/L. This is about 100 times the concentration usually found in soil solutions. Adding $FeSO_4$ increased uptake, but adding FeEDDHA did not, and adding zinc depressed uptake. While slower than for rubidium, transport in the plant was still relatively rapid.

Molybdenum concentrations in plants are usually less than 2 mg/kg; hence, a low uptake rate should supply the plant requirement. The molybdenum requirement is much greater for legumes, because it is used in the nodule for nitrogen fixation, and levels needed for plant growth are much lower than levels needed to supply the root nodule with sufficient molybdenum for nitrogen fixation. Hence, most plant species showing responses to molybdenum fertilization are legumes.

Molybdenum concentrations in plants that are adequately supplied with molybdenum range from 1 to 2 mg/kg; with higher rates of molybdenum supply, however, plant concentrations can be as high as 20 mg/kg. This increase without a corresponding increase in plant growth indicates that there is a wide range of molybdenum concentrations over which molybdenum absorption increases concentration of molybdenum in the plant. Hence the influx versus concentration relation may be approximately linear in soil systems.

Stout et al. (1951) found that adding sulfate depressed molybdenum uptake, whereas adding phosphate stimulated molybdenum uptake in solution culture, possibly because sulfate and molybdenum compete for the same absorption sites on the root. Enhancement with phosphate was not explained. The same researchers also found that increasing solution pH depressed molybdenum uptake according to the relation shown in Figure 18.1. This effect on molybdenum absorption is the opposite of that occurring in soil systems. In soil, the effect of pH on molybdenum levels in the soil solution overshadows the effect of solution pH in depressing molybdenum-absorption rate.

Uptake from Soil

Mass flow and diffusion each supplies molybdenum to roots growing in the soil. When the level of molybdenum in the soil solution can be sustained above 0.04 μmol/L, mass flow can usually supply the amount of molybdenum required. Lavy and Barber (1964) made autoradiographs of corn and soybean roots growing in soil with the molybdenum labeled with ^{99}Mo. When the saturation extract contained less than 4 μg/kg (0.04 μmol/L), depletion patterns around the root indicated that diffusion was supplying some of the molybdenum to the plant. At higher soil solution levels, accumulation around the root due to mass flow was evident. The supply of

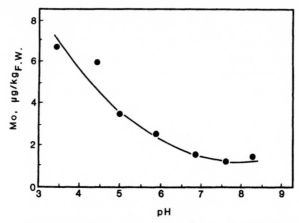

FIGURE 18.1   Effect of solution pH on absorption of molybdenum by tomato from solution culture. Reprinted from Stout et al. (1951) by permission of Martinus Nijhoff Publishers B.V.

molybdenum to the root will be greatly influenced by the soil-solution level of molybdenum, and soil-solution level is affected by adsorbed molybdenum and soil pH.

Soil pH and Molybdenum Supply

For a given soil, molybdenum uptake by the plant increases as soil pH increases. The molybdenum content of a crop grown in a soil of pH 5.0 will, on the average, double if the soil is limed to pH 6.0 and increase six-fold if the soil is limed to pH 7.0. The amount of labile molybdenum measured by extraction with an anion exchange resin also increases (Bhella and Dawson, 1972). Plant absorption of molybdenum decreases with increased pH, though this effect is evidently overshadowed by the increase in molybdenum levels in the soil solution as soil pH is increased. Soil solution molybdenum levels may increase 10-fold for each unit increase in soil pH. Where molybdenum has been added to the soil, the amount adsorbed will decrease as pH increases.

Interactions with Sulfate and Phosphate

Adding gypsum to alkaline soils decreased the molybdenum content of tomato (*Lycopersicon esculentum*) from 2.33 to 1.26 mg/kg (Olsen and Watanabe, 1979); this reduction in molybdenum was also associated with an increase in iron. Karimian and Cox (1978) added molybdenum to 32 southeastern U.S. soils and found that molybdenum adsorption was positively correlated with organic matter and amorphous iron oxide contents of the soil. It was negatively correlated with pH and phosphate levels.

Adding sulfate to soils depresses molybdenum uptake by the root. Increasing pH increases the amount of adsorbed and solution molybdenum, though increasing pH, actually depresses the ability of the root to absorb molybdenum. Adding phosphate increases molybdenum levels in solution and also increases the rate of molybdenum uptake by the root. The chemistry of these reactions has not been completely explained.

Molybdenum concentration in the plant is a measure of the level of available soil molybdenum. Molybdenum concentration in the seed is usually much higher than in the leaf. Lavy and Barber (1963) used the molybdenum content of soybean seed to determine the level of available molybdenum in field soils. They could thus predict where soybeans would respond to molybdenum applications.

In some soils of high pH, plants may adsorb enough molybdenum so that ruminants eating the plants develop molybdenosis. This can happen when forage molybdenum contents are above 5 mg/kg. Hence, methods of reducing molybdenum uptake are of interest for these soils. On acid soils, low in available molybdenum, yield of legumes may be reduced because of lack of molybdenum. The most common soil-extraction procedure for measuring available molybdenum is extraction with acidic ammonium oxalate (Reisenauer, 1965). Since increasing soil pH increases the soil's supply of molybdenum, one of the benefits of liming soils is increasing available molybdenum. On some soils this is the main effect that allows growth of legumes. When this is the case, using 50 g of molybdenum per hectare may be more economical than applying lime.

REFERENCES

Barrow, N. J. 1970. Comparison of the adsorption of molybdate, sulfate, and phosphate by soils. *Soil Sci.* **109**:282–288.

Bhella, H. S., and M. D. Dawson. 1972. The use of anion exchange resin for determining available soil molybdenum. *Soil Sci. Soc. Am. Proc.* **36**:177–179.

Follett, R. F. 1966. Mechanisms for the movement of molybdenum from the soil to the plant root. Ph.D. dissertation Purdue Univ.

Follett, R. F., and S. A. Barber. 1967a. Molybdate phase equilibria in soils. *Soil Sci. Soc. Am. Proc.* **31**:26–29.

Follett, R. F., and S. A. Barber. 1967b. Properties of the available and the soluble molybdenum fractions in a Raub silt loam. *Soil Sci. Soc. Am. Proc.* **31**:191–192.

Gonzalez, R., H. Appelt, E. B. Schalscha, and F. T. Bingham. 1974. Molybdate adsorption characteristics of volcanic-ash-derived soils in Chile. *Soil Sci. Soc. Am. Proc.* **38**:903–906.

Hingston, F. J., A. M. Posner, R. J. Atkinson, and J. P. Quirk. 1968. Specific adsorption of anions on goethite. *Int. Cong. Soil Sci. Trans. Ninth,* Adelaide, Australia **1**:669–678.

Jackson, D. R., and R. R. Meglen. 1975. A procedure for extraction of molybdenum from soil with anion-exchange resin. *Soil Sci. Soc. Am. Proc.* **39**:373–374.

Kannan, S., and S. Ramoni. 1978. Studies on molybdenum absorption and transport in beans. and rice. *Plant Physiol.* **62**:179–181.

Karimian, N., and F. R. Cox. 1978. Adsorption and extractability of molybdenum in relation to some chemical properties of soils. *Soil Sci. Soc. Am. J.* **42**:757–761.

Lavy, T. L., G. Sands, and S. A. Barber. 1961. The molybdenum status of some Indiana soils. *Ind. Acad. Sci. Proc.* **70**:238–242.

Lavy, T. L., and S. A. Barber. 1963. A relationship between the yield response of soybeans to molybdenum applications and the molybdenum content of the seed produced. *Agron. J.* **55**:154–155.

Lavy, T. L., and S. A. Barber. 1964. Movement of molybdenum in the soil and its effect on availability to the plant. *Soil Sci. Soc. Am. Proc.* **28**:93–97.

Olsen, S. R., and F. S. Watanabe. 1979. Interaction of added gypsum in alkaline soils with uptake of iron, molybdenum, manganese, and zinc by sorghum. *Soil Sci. Soc. Am. J.* **43**:125–130.

Reisenauer, H. M. 1965. Molybdenum. In C. A. Black, D. D. Evans, J. L. White, L. E. Ensminger, F. E. Clark, and R. C. Dinauer, Eds. *Methods of Soil Analysis.* Soil Science Society of America, Madison, WI. Pp. 1050–1058.

Stout, P. R., W. R. Meagher, G. A. Pearson, and C. M. Johnson. 1951. Molybdenum nutrition of plant crops. I. The influence of phosphate and sulfate on the absorption of molybdenum from soils and solution cultures. *Plant Soil* **3**:51–87.

Vlek, P. L. G., and W. L. Lindsay. 1977. Thermodynamic stability and solubility of molybdenum minerals in soils. *Soil Sci. Soc. Am. J.* **41**:42–46.

CHAPTER **19**

Zinc

The mean zinc concentration in the lithosphere is 80 mg/kg (Krauskopf, 1972). Total soil zinc ranges from 10 to 300 mg/kg (Lindsay, 1979). Since soil values are in the same range as for the lithosphere, zinc minerals apparently do not weather more readily than average. Dolar and Keeney (1971) reported total zinc values for 36 cultivated soils from Wisconsin; values ranged from 7.4 to 90.2 mg/kg, with an average of 35 mg/kg. Total zinc in 10 calcareous Arizona soils ranged from 16 to 52 mg/kg; higher zinc levels occurred with increased clay and organic matter levels (Udo et al., 1970). For 19 West Virginia soils, total zinc ranged from 19 to 160 ppm (Iyengar et al., 1981).

FORMS OF SOIL ZINC

Relatively insoluble mineral forms account for more than 90% of the zinc in soils. Zinc minerals that may occur in the soil include sphalerite (ZnS), smithsonite ($ZnCO_3$), and hemimorphite [$Zn_4(OH)_2Si_2O_7 \cdot H_2O$]. Zinc also substitutes for magnesium in montmorillonite-type clay minerals (Krauskopf, 1972). Zinc is present in the soil in only the divalent form.

Organic matter forms coordination complexes with zinc; they may be present in both the soil organic matter and soluble organic complexes in soil solution (Hodgson et al., 1966). Zinc that may become available for plant uptake is present as Zn^{2+} in the soil solution, exchangeable zinc on the cation-exchange sites, organically complexed zinc in solution, and organically complexed zinc in the soil solid phase.

Exchangeable Zinc

Exchangeable zinc levels (displaced with 1 mol/L ammonium acetate, pH 7.0) for 60 Indiana soils were measured by Warncke (1967). These soils gave values ranging from 2.5 to 205 μmol/kg, with 18 samples between 23 and 30 μmol/kg (1.5 to 2.0 mg/kg). Values measured for 10 Indian soils (Iyengar and Deb, 1977) ranged from 1.0 to 5.5 μmol/kg; they included alluvial, red, and lateritic soils, with a wide range in soil organic matter contents and pH values. Dolar and Keeney (1971) extracted zinc with 0.5 mol/L $MgCl_2$, pH 5.9 and obtained a range of values from 0.1 to 0.8 mg/kg, with an average of 0.6 mg/kg. Hence, values for Zn exchangeable with ammonium or magnesium appear to be in the range of 0.1 to 2 mg/kg (0.75 to 30.5 μmol/kg). However, lower values probably occur for soils where zinc deficiency in plants occurs.

   Dolar and Keeney (1971) found that exchangeable zinc decreased as soil pH increased. The soil pH ranges and corresponding mean exchangeable zinc values were pH 5.0 to 6.0, 1.2 mg/kg; pH 6.1 to 6.5, 0.5 mg/kg; and pH 6.6 to 7.0, 0.4 mg/kg.

Soil Solution Zinc

Zinc in displaced saturation extracts from 60 Indiana soils was measured by Warncke, who obtained a range of 0.025 to 0.25 mg/L, with 21 samples ranging between 0.025 and 0.05 mg/L (0.4 to 0.8 μmol/L). Hodgson et al. (1966) measured zinc levels in soil solutions displaced from four New York and 20 Colorado soils. In the Colorado soils, concentrations were less than 2 μg/L, while for the more acidic New York soils, concentrations averaged 20 μg/L.

Complexed Solution Zinc

Hodgson et al. (1965, 1966) separated soil solution zinc into ionic and complexed forms. For Colorado soils. the percent complexed varied from 28 to 99% of that present in solution, with an average of 60 ± 15%, New York soils averaged 37 ± 23% complexed. The amount that was complexed increased with an increase in soluble organic matter. McBride and Blasiak (1979) showed that the proportion of the zinc complexed in solution increased as soil pH increased. The soil solution was passed through a cation-exchange column with the zinc passing through the column considered complexed. The relation between pH and the fraction of the metal remaining in solution after passage through the cation-exchange column is shown in Figure 19.1. At soil pH values above 6.5, much of the zinc in solution was complexed with soluble organic materials. The relation shown in Figure 19.1 may explain differences between New York arid Colorado soils as observed by Hodgson et al. (1965, 1966), since Colorado soils had much higher average pH levels than New York soils.

Complexed Soil Zinc

Soil organic matter can hold metal ions such as zinc in coordination complexes from which it is not displaced by ammonium. Zinc may be displaced by copper, however, which has a higher stability constant. Complexed zinc can also be removed by treating the soil with a soluble complexing compound, such as ethylenediamine tetraacetic acid (EDTA) or diethylenetriaminepentaacetic acid (DTPA), since the added complexes are present in larger amounts and usually have higher stability constants than do soil-derived solutes. Some zinc that is tightly held by the mineral fraction of the soil may also be displaced by such treatment. Little information is available on the kinetics of zinc movement between complexed and exchangeable or solution forms.

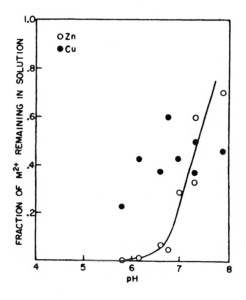

FIGURE 19.1 Fraction of copper and zinc in the soil solution that passed a cation-exchange column as a function of solution pH. Reproduced from McBride and Blasiak (1979) by permission of Soil Science Society of America.

Zinc Fixation

Iyengar and Deb (1977) applied zinc to alluvial, red, and lateritic soil groups from India. They found that only 20 to 60% of the applied zinc could be recovered by complexing agents such as dithizone or DTPA. Nelson and Melsted (1955) found that most of the zinc added to soils was displaceable with ammonium acetate 30 min after addition. With additional time, however, 10 to 20% of the added zinc reverted to forms not removable with ammonium acetate but removable with HCl. Warncke and Barber (1973) added zinc to acid soils (pH 4.8 to 5.1) from Indiana and found that zinc adsorbed by the soil was displaceable by shaking with 0.01 mol/L $CaCl_2$ for 24 h. Since more zinc is adsorbed at higher pH levels, adsorption of zinc into positions from which it is not readily displaceable may occur at higher pH levels.

Zinc Adsorption

Warncke and Barber (1973) determined zinc-adsorption isotherms for four soils varying in clay and organic matter content. All soils had a pH between 4.8 and 5.1, so that zinc-complex formation was probably minimal. Adsorption isotherms were believed to represent mainly adsorption of zinc onto soil cation-exchange sites. Cation-exchange capacities varied from 12.3 to 31.7 $cmol(p^+)/kg$. The adsorption isotherm was determined by shaking 1-g soil samples with 5 mL of zinc containing solutions of 0.01 mol/L $CaCl_2$ for 24 h. Adsorption isotherms for the four soils are shown in Figure 19.2. Slopes of the adsorption isotherms varied among soils. Properties of

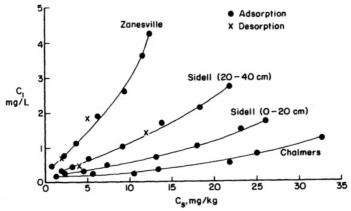

FIGURE   19.2   Relation between solution zinc and adsorbed zinc as the zinc concentration is increased for four soils. Reproduced from Warncke and Barber (1973) by permission of Soil Science Society of America.

the soil and the buffer power at exchangeable zinc levels of 10 µg/g are given in Table 19.1; buffer power for zinc varied from 5 to 63. In general, buffer power increased as cation-exchange capacity increased and decreased as C_{li} increased.

On uniform, well-defined adsorbing surfaces with uniform charge density, adsorption can be predicted from theory (Bowden et al., 1977). However, soils have heterogeneous surfaces. Hence, simpler models, such as the Langmuir adsorption isotherm, have been used to describe the nature of the adsorption.

The adsorption of zinc by soils is increased by increased cation-exchange capacity, clay content, organic matter content, pH, and the presence of calcium carbonate. Shuman (1977) found that aluminum and iron oxides adsorbed zinc in accordance with the Langmuir equation. As described in Chapter 5, the availability of soil zinc to the plant depends upon C_{li} b, and D_e. The Langmuir equation gives the relation between b and C_l as

$$\frac{C_l}{(x/m)} = \left(\frac{1}{aB}\right) + \left(\frac{C_l}{B}\right) \qquad (19.1)$$

where C_l is the concentration in soil solution, x/m is the amount of zinc adsorbed per unit of soil, B is the adsorption maximum, and a is a constant related to the soil's bonding energy for zinc. A plot of $C_l/(x/m)$ versus C_l gives a straight line with slope of $1/B$ and intercept $1/aB$. The inverse of

TABLE 19.1   Characteristics of Four Samples of Indiana Soils as Related to Zinc Adsorption

Soil characteristic	Zanesville silt loam	Sidell silt loam (0-20 cm)	Sidell silt loam (20-40 cm)	Chalmers silty clay loam
pH (water)	4.8	5.0	5.1	5.0
Solution zinc (µmol/L)	12.6	7.1	0.4	0.98
Exchangeable zinc (µmol/kg)	10.1	34.0	17.0	19.7
Cation-exchange capacity [cmol(p⁺)/kg]	12.3	20.7	20.6	31.7
Organic matter (%)	1.55	2.50	1.68	4.69
Buffer power	5	12	21	63

Source: Reproduced from Warncke and Barber (1972a) by permission of the Soil Science Society of America.

$C_l/x(m)$ is equal to b, the buffer power of the soil when values for C_l and x/m are both expressed in volume units. In addition, the reciprocal of b will be proportional to C_l and inversely proportional to the maximum adsorption capacity and size of the bonding energy.

The Langmuir equation has been used to evaluate zinc adsorption on soils by Udo et al. (1970), Shuman (1975), Shukla and Mittal (1979), and Bar-Yosef (1979). For the range of zinc concentrations they used, Shuman and Skukla and Mittal found that it was necessary to divide the adsorption curve into two parts. The value of C_l at the boundary between the two parts was approximately 50 μmol/L for the Indian soils studied by Skukla and Mittal (1979) and in the range of 4 to 10 mg/kg (60 to 153 μmol/L) for the Georgia soils investigated by Shuman (1975). Udo et al. (1970) similarly studied 10 calcareous soils from Arizona and reported a close fit to the Langmuir equation for values of C_{li} below 0.8 mg/kg (12 μmol/L). Bar-Yosef (1979) developed a relationship between adsorbed zinc and zinc in solution as influenced by pH; this relationship was derived using the competitive Langmuir isotherm. The equation related zinc adsorption to its solution concentration and solution pH. The characteristics that affect the distribution of zinc between solution and the exchange phase included clay content, level of zinc, soil pH, and cation-exchange capacity.

Effect of Soil pH on Zinc Adsorption

Soil pH has a strong effect on zinc adsorption. McBride and Blasiak (1979) found the relationship shown in Figure 19.3 for zinc added to a Mardin silt loam soil. The minimal zinc concentration in solution occurred between pH 7 and 8. Concentration of zinc in solution decreased 30-fold for every unit of pH increase in the pH range 5 to 7. The reduction in solution Zn^{2+} concentration for the Mardin soil was believed due to adsorption on hydrous oxide surfaces. Bar-Yosef (1979) plotted pZn in solution versus soil pH for two Israeli soils and two of the Georgia soils reported by Shuman (1975); results are shown in Figure 19.4. The solubility relation suggested by Lindsay (1972) for zinc in soil solution versus pH is shown by the steeper straight line; it was based on other studies of the effect of pH on soil-solution zinc levels.

Saeed and Fox (1977) investigated the effect of suspension pH on the concentration of zinc in solution for four acid Hawaiian soils and three calcareous soils from Pakistan. Solution zinc reached a minimum at pH 7 for the Hawaiian soils but decreased further with increased pH for the calcareous soils. Between pH 5 and 7, the authors obtained a linear relation between pZn and pH. They postulated that reduction of solution zinc levels as pH increased from 5 to 7 was partly due to increases in the number of cation-exchange sites on soils with pH-dependent charge. The increase in solution zinc above 7 may have been due to solubilization of organic matter and an increase in complexed zinc. For calcareous soils, organic matter was not sol-

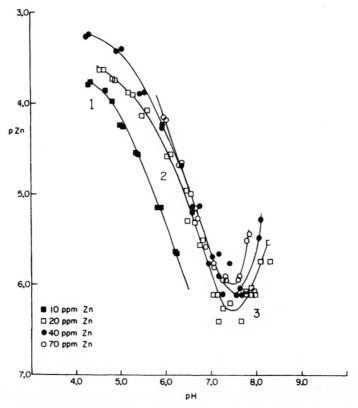

FIGURE   19.3   Relation between Zn^{2+} concentration in the soil solution and pH at four levels of added zinc for Mardin silt loam. Reproduced from McBride and Blasiak (1979) by permission of Soil Science Society of America.

ubilized at higher pH, so that solution zinc concentration continued to decrease.

The level of zinc in solution at a specific soil pH depends on the nature of the soil surfaces and the level of zinc in the soil. Where hydrous oxide surfaces are present, the zinc level in solution is usually lower. At high pH levels, if calcium carbonate is present, the solution zinc level will also be lower. At pH values above 7.5, the level of complexed zinc in solution will depend on the solubility of the organic matter, which, in turn depends on the presence of calcium and other cations that may suppress its solubility.

Buffer Power

The distribution of zinc between the solution and solid phases can be described by the soil buffer power b. The slopes of the adsorption curve described in the last section are a measure of buffer power. The buffer

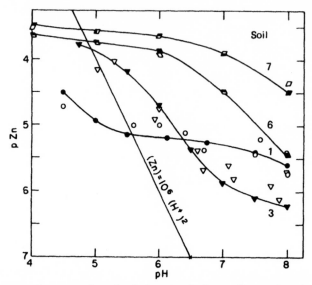

FIGURE   19.4   Concentration of solution zinc in four soils as a function of pH. Also included is the soil zinc solubility line proposed by Lindsay (1979). Reproduced from Bar-Yosef (1979) by permission of Soil Science Society of America.

power was greater the lower the zinc concentration in the soil solution and the larger the cation-exchange capacity. Values for b have been approximated from adsorption data shown graphically in the literature. Warncke and Barber (1973) found values varying from 2.4 to 76; Shuman (1975) estimated b values of 5 to 100 and Elgawhary et al. (1970a) obtained a buffer power of 571 for a Platner loam from Colorado that had a pH of 7.2.

The general relationships from published papers indicate that increasing soil pH increases buffer power greatly, since the distribution of zinc between solid and solution phases is shifted toward the solid phase. Increasing the level of exchangeable zinc by adding zinc reduces b rather rapidly. In soils with a pH above 7.0, part of the zinc in solution may be complexed with organic matter and thus not be in rapid equilibrium with exchangeable zinc.

Diffusion Coefficient

Warncke and Barber (1972a,b) measured the effect of soil water, soil bulk density, and buffer power on zinc diffusion in soils. Effect of soil moisture on zinc movement by diffusion in six Indiana soils (Figure 19.5) varied greatly among soils. Solution and exchangeable zinc levels, and soil pH, are shown in Table 19.2. As zinc levels in solution increased, D_e increased and the effect of soil water level was reduced; the impedance factor was determined by measuring diffusion of chloride, a nonadsorbed ion. This

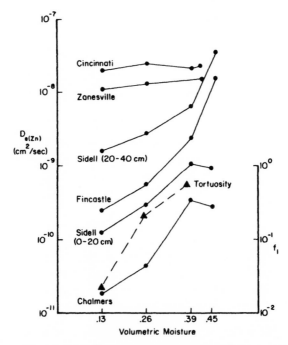

FIGURE   19.5   Influence of soil water on the zinc diffusion of coefficient and impedance factor. Soil bulk density = 1.3. Reproduced from Warncke and Barber (1972a) by permission of Soil Science Society of America.

TABLE 19.2   Solution and Exchangeable Zinc and Soil pH for Six Soils Used by Warncke and Barber in Zinc-Diffusion Studies

| | Zinc | | |
Solution	Displaced solution (μmol/L)	Exchangeable (μmol/kg)	Water pH
Cincinnati sil	20.7	320	4.9
Zanesville sil	1.3	10	4.8
Sidell sil (0–20 cm)	0.7	34	5.0
Sidell sil (20–40 cm)	0.04	17	5.1
Fincastle sil	0.14	27	5.0
Chalmers sil	0.10	20	5.1

Source: Reproduced from Warncke and Barber (1972a) by permission of the Soil Science Society of America.

factor measures the reduction in diffusive flow primarily because of the tortuous nature of the diffusion path through soils. At low levels of soil zinc, the impedance factor appeared to be the main reason for reduction in the D_e value for zinc with reduction in volumetric water content.

Bulk density of the soil affects the tortuosity of the diffusion path and the size of the diffusion coefficient; the effect of bulk density on D_e is shown in Figure 19.6. The diffusion coefficient tended to increase with an increase in bulk density from 1.1 to 1.5, because the water films through the pores were made less tortuous with reduced air-filled porosity. Also, a change in bulk density probably changed pore-size distribution, which subsequently affected D_e. However, when soil bulk density was increased above 1.5, compressing the solid phase of the soil began to make the diffusion path more tortuous once more and D_e decreased.

Soil zinc D_e can be increased by increasing either the zinc level or the soil water content. Since soil pH also plays a large part in determining the distribution of zinc between solution and solid phases, a low pH gives a higher D_e value. Clarke and Graham (1968) found that increasing soil pH reduced the effective diffusion coefficient for zinc. The reduction was related to a decrease in the proportion of zinc in solution. The rate of zinc diffusion can

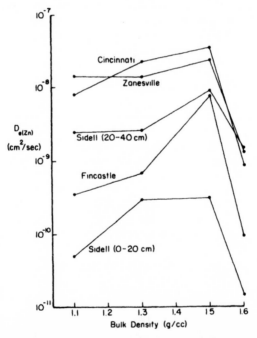

FIGURE 19.6 Influence of soil bulk density on the diffusion coefficient for zinc diffusion in soils at 20% moisture (w/w). Reproduced from Warncke and Barber (1972a) by permission of Soil Science Society of America.

also be increased by adding a soluble complexing agent to the soil (Elgawhary et al., 1970a), which shifts a larger proportion of the zinc into solution.

ZINC-UPTAKE KINETICS

Zinc uptake by intact plants from solutions of less than 10 μmol/L appears to be active, since it is inhibited by respiration inhibitors (Giordano et al., 1974). Uptake rate is also reduced by reducing temperature (Chaudhry and Loneragan, 1972a). Schmid et al. (1965) found that zinc uptake by excised barley roots was active.

The kinetics of zinc uptake have been investigated with intact plants grown in flowing solution culture by Carroll and Loneragan (1969), who grew eight species for 46 days in solutions varying in concentration from 0.01 to 6.25 μmol/L. Log absorption rate gave a linear relation with log zinc concentration, as shown in Figure 19.7. Schmid et al. (1965) also measured zinc uptake by excised barley roots, obtaining the higher rate of uptake shown in Figure 19.7. Chaudhry and Loneragan (1972a) investigated reasons for this difference. When the same species and conditions were used,

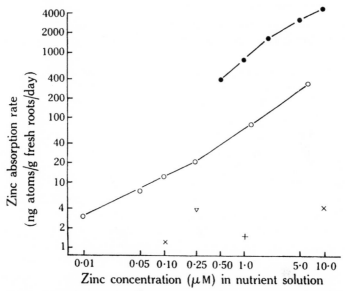

FIGURE   19.7   Relation between zinc concentration in solution and rate of zinc absorption per unit fresh weight of root for different experimental systems. (o) Whole plants in flowing culture (Carroll and Loneragan, 1969); (●) Excised barley roots (Schmid et al., 1965); (×, +, Δ) Adsorption from nonflowing culture. Reprinted from Carroll and Loneragan (1969) by permission of CSIRO.

uptake rate from short-term experiments agreed closely with uptake rates from the long-term experiment of Carroll and Loneragan (1969). The higher uptake rate of Schmid et al. (1965) may have been due to different experimental conditions. However, the rate they report would, if continued, provide the shoot with excessive quantities of zinc. The lower points in Figure 19.7 are from experiments involving whole plants, using nonflowing solution conditions. Uptake was lower because concentrations presumably were not maintained at the root surface.

A K_m value of 3.9 μmol/L was obtained by Hassan and van Hai (1976) for phase-1 zinc uptake by citrus seedlings. Veltrup (1978) obtained a value of 3.18 μmol/L for phase-1 uptake by intact barley roots; K_m values in this case were larger than Zn concentrations usually observed in soil solutions that are mostly below 1 μmol/L. Because soil-solution values are so low relative to K_m, the relation between zinc influx and zinc concentration at the root is essentially linear for zinc uptake by soil-grown plants. There does not appear to be information available on the minimum concentration of zinc required in solution before uptake begins. Carroll and Loneragan (1969), for example, observed uptake at 0.01 μmol/L of solution zinc concentration.

Ion Competition

There is little evidence for the effect of anions on zinc uptake; however cations *do* influence zinc uptake, with evidence of such inhibition existing for a number of cations (Chaudhry and Loneragan, 1972c). Copper has generally been found to inhibit uptake: Hawf and Schmid (1967) found, for example, that copper inhibited zinc uptake but not its translocation; Schmid et al. (1965) found that copper competed for zinc uptake by excised barley roots but that manganese did not. Chaudhry and Loneragan (1972a) also reported calcium inhibition of zinc uptake. Some investigators report inhibition, while others do not. Giordano et al. (1974) found no effect of the copper level on zinc uptake by intact rice plants; this may be related to concentrations of both the competing ions and zinc that are used. Chaudhry and Loneragan (1972a,c) conducted the most extensive study, using a zinc concentration of 1 μmol/L, which is close to that found in soil solution. They found that the reduction in zinc uptake caused by cation addition was in the order $NH_4^+ > Rb^+ > K^+ > Cs^+ > Na^+ > Li^+$. With alkaline earth cations, the order was $Mg^{2+} > Ba^{2+} > Sr^{2+} = Ca^{2+}$. Copper inhibited zinc absorption strongly, while iron and manganese had no effect on zinc uptake, and increasing hydrogen (decreasing pH) inhibited zinc uptake (Chaudhry arid Loneragan, 1972b). Evidence indicates the alkaline-earth cations inhibit zinc absorption noncompetitively. Bowen (1969) observed that hydrogen competitively reduced zinc absorption, with absorption at pH 4.0 only 53.4% of absorption at pH 5.7. Schmid et al. (1965) found that increasing copper levels competitively reduced zinc absorption by excised barley roots. Phosphate has also been shown to inhibit zinc uptake, though this appears to be a translocation effect rather than an effect on zinc uptake via competition at the plasma membrane (Olsen, 1972).

MEASURING BIOAVAILABLE SOIL ZINC

Soil zinc bioavailability has been measured using empirical methods; the more common procedures used are (1) removal with neutral normal ammonium acetate or a similar salt; (2) removal with a soluble chelate, such as DTPA or EDTA; (3) removal with a dilute acid solution, such as 0.1 mol/L HCl; (4) extraction with aqueous ammonium acetate and then concentration with dithizone in CCl_4. Cox and Kamprath (1972) discuss different extraction procedures used for measuring zinc supply to the plant. Because of the effect of soil pH on zinc supply, using both soil pH and amounts of zinc extracted increases the reliability of the method for predicting plant response to zinc. On calcareous soils, adjusting for the level of calcium carbonate increases the value of some methods. Complexing agents have usually been most reliable on soils above pH 7.0, while HCl extraction is only useful on soils below pH 7.0. The levels extracted by these procedures vary with actual method. The level below which zinc deficiency may occur varies from a high of 1.0 to 7.5 mg/kg for 0.1 mol/L HCl extraction to 0.5–1.0 mg/kg for the DTPA + $CaCl_2$ (pH 7.3) procedure.

Lindsay and Norvell (1978) developed a soil test procedure for zinc, iron, manganese, and copper. The procedure has been frequently used to evaluate available soil zinc.

MODELING ZINC UPTAKE FROM SOIL

Levels of zinc in soil solution in the range of 0.025 to 0.05 mg/L will normally supply sufficient zinc to the root by mass flow to produce a plant containing 5 to 10 mg of zinc/kg if we assume that 200 gs of water is transpired for each gram of plant dry matter produced. Since plants commonly contain in excess of 20 mg of zinc/kg mass flow usually supplies only a portion of the zinc for soil-grown plants; diffusion supplies the remainder. Wilkinson et al. (1968) demonstrated zinc depletion around wheat (*Triticum vulgare*) roots, using autoradiographs of root-soil systems where the zinc was labeled with ^{65}Zn. In their system, the soil solution contained 2.3 µmol/L (0.15 mg/kg) zinc, the plant used 173 ml of water/g of dry weight produced, and a final zinc concentration of 50 mg/kg was produced in the plant tops. This was double what would have been supplied to the root by mass flow. Hence, zinc uptake reduced the concentration at the root surface and caused a zinc concentration gradient perpendicular to the root as diffusion occurred. The autoradiograph showed depletion around the root and verified that diffusion was an important supply mechanism.

Since mass flow and diffusion each supply zinc to the root, the process can be described mathematically using the model of Barber and Cushman (1981). Relevant soil parameters are C_{li}, concentration in the soil solution initially; b, buffer power of the diffusible soil zinc for C_{li}; and D_e, effective-diffusion coefficient for zinc movement in the soil. Values for these parameters for zinc, and their variation with soil parameters, are discussed in this section.

Factors Affecting Zinc Uptake

Soil-Solution Zinc

The level of solution zinc can be determined by measuring zinc concentrations in displaced soil solutions. At pH values above 6.2, part of the solution zinc may be combined with dissolved organic complexes. The quantities of both Zn^{2+} and complexed zinc in solution should be measured to produce values for predicting the zinc supply by mass flow.

Buffer Power

The levels of exchangeable or diffusible zinc, C_s, can be measured by displacing zinc with cations such as ammonium and magnesium. The value needed is the amount of zinc associated with the solid phase that readily equilibrates with Zn^{2+} in solution. The relation of $\Delta C_s/\Delta C_l$ for zinc has commonly been found to vary from 5 to 100.

Effective Diffusion Coefficient

The effective diffusion coefficient may be determined by measuring the rate of zinc diffusion in the soil (Warncke and Barber, 1972a); it may also be estimated using Equation 3.2. Values for D_e vary from 10^{-7} to 10^{-10} cm²/s. D_l in water at 20°C is 7.20×10^{-6} cm²/s (Parsons, 1959).

Calculations

An untested example of the effect of changing the zinc level in the soil on the amount of zinc absorbed by a corn plant growing from 4 to 17 days is illustrated in Table 19.3. Only the soil zinc parameters of the model were varied; all three soil parameters were varied simultaneously, because the size of D_e is affected by the size of both C_{li} and b. These are typical effects what might occur if zinc were added to one soil and then several rates of zinc addition compared.

Values for root size, growth rate (k, L_0, and r_0) and water flux (v_0) were taken from experiments by Claassen and Barber (1976). The zinc-uptake characteristics of the roots were estimated from values for barley given by Schmid et al. (1965) and for several additional crops studies by Carroll and

TABLE 19.3   Effect of Zinc Levels in the Soil on the Predicted Uptake by Corn Growing in Chalmers Silt Loam[a]

Soil zinc parameters				Predicted zinc zinc uptake (μmol/plant)	Predicted zinc concentration in the plant (mg/kg)
D_e (10^9cm²/s)	C_{li} (mmol/L)	b ($\Delta C_v/\Delta C_l$)	C_v (mg/kg)		
1.0	0.0001	200	1.0	0.169	4.3
2.0	0.0003	180	2.66	0.681	17.4
5.0	0.001	140	6.9	3.069	78.6
6.0	0.002	130	12.7	6.269	160
7.0	0.003	120	17.7	9.23	236
10.0	0.005	80	19.69	12.44	318
50.0	0.10	40	157	16.85	431
100	1.0	5	246	16.86	432

Source: Barber and Claassen (1977).
[a]Values of plant parameters in the model were v_0, 3.2×10^{-6} ml/cm²; I_{max}, 1.5 nmol/m² · s; E, 0.01 nmol/m² · s; K_m, 3.0 μmol/L; k, 2.0×10^{-6}/s; L_0, 500 cm; r_0, 1.5×10^{-2} cm; t, 5.5 $\times 10^5$ s.

Loneragan (1969). Values used for D_e, C_{li}, and b were estimated from experimental data of Warncke and Barber (1972a) for a Chalmers silty clay loam. Data for soil and plant parameters shown in Table 19.3 were used in the mathematical model to compute predicted uptake of zinc by the corn plant at eight different levels of soil zinc. The concentration in the plant was based on an average plant dry weight of 2.5 g at the end of the experiment. Zinc was not assumed to have affected plant weight.

As the zinc level in the soil was increased, zinc uptake increased until a level was reached where further increases in D_e and C_{li} had little effect on uptake; the limiting factor then became the ability of the root to absorb zinc. At the last two zinc levels, supply to the root by mass flow and diffusion was faster than uptake, so that concentrations at the root surface increased to a maximum of 1.05 and 1.25 of the initial level, respectively, for C_s values of 157 and 246 mg of zinc/kg. For the first six zinc levels, the concentration at the root C_{l0} decreased to 0.03, 0.03, 0.04, 0.06, 0.09, and 0.38, respectively, of the original level as uptake progressed.

The zinc level in the plant increased to a maximum of 432 mg/kg, which is at the low end of the range of values (429 to 980) obtained by Carroll and Loneragan (1969) for several species grown for 18 days in 6.25 μmol/L zinc solution. The maximum level depends on the parameters used to describe zinc uptake by the plant. Increasing I_{max} to double the value used would essentially double predicted zinc concentration in the plant at the highest zinc levels. Increasing the rate of root growth in relation to shoot growth could also increase predicted plant zinc concentration in proportion to the number of new roots added.

Values in Table 19.3 illustrate how a mathematical model for metal uptake can be used to evaluate the effect of changing soil zinc levels. Predicted values should then be compared with values determined experimentally, to verify the adequacy of the model and the magnitude of the parameters used.

Gladstone and Loneragan (1967) evaluated the uptake of zinc by 25 annual crop and pasture plants; their results suggested that species differed in their abilities to obtain zinc from the soil. Measuring the parameters used in the mathematical model for these species, and using the measured values in the model, is one method of determining the reason for differences in these species.

In addition to the model of Barber and Claassen (1977), Bar-Yosef et al. (1980) developed a model describing zinc movement to single roots in soil. In their model, they were able to evaluate the effect of hydrogen efflux from the root on increased zinc influx. When roots exude hydrogen and reduce soil pH near the root, zinc supply to the root may be increased, which could be one reason for differences between species in the absorption of soil zinc. Results of root-simulation experiments by Elgawhary et al. (1970b) indicate that release of hydrogen from a simulated root increases zinc movement to the root. However, in the more usual situation, little hydrogen is exuded from the root.

Application of high rates of phosphorus (160 kg/ha) induced zinc deficiency in wheat (Singh et al., 1986), which was corrected, and yield increased by adding zinc. Phosphorus applications reduced VA mycorrhizae infection, hence there was a correlation between VA mycorrhizae infection and grain and straw zinc concentration. The reduction in VA mycorrhizae infection may have been part of the reason for reduced zinc uptake with phosphate application.

REFERENCES

Barber, S. A., and N. Claassen. 1977. A mathematical model to simulate metal uptake by plants growing in soil. *Biological Implications of Metals in the Environment. Proc. Fifteenth Hanford Life Sciences Symposium.* Pp. 358–364.

Barber, S. A., and J. H. Cushman. 1981. Nitrogen uptake model for agronomic crops. In J. K. Iskandar, ed., *Modeling Waste Water Renovation—Land Treatment.* John Wiley, New York.

Bar-Yosef, B. 1979. pH-dependent Zn adsorption by soils. *Soil Sci. Soc. Am. J.* **43**:1095–1099.

Bar-Yosef, B., S. Fishman, and H. Talpaz. 1980. A model of zinc movement to single roots in soil. *Soil Sci. Soc. Am. J.* **44**:1272–1279.

Bowden, J. W., A. M. Posner, and J. P. Quirk. 1977. Ionic adsorption on variable charge mineral surfaces. Theoretical charge development and titration curves. *Aust. J. Soil Res.* **15**:121–136.

Bowen, J. E. 1969. Absorption of copper, zinc, and manganese by sugar-cane leaf tissue. *Plant Physiol.* **44**:255–261.

Carroll, M. D., and J. F. Loneragan. 1969. Response of plant species to concentrations of zinc in solution: II. Rates of zinc absorption and their relation to growth. *Aust. J. Agric. Res.* **20**:457–463.

Chaudhry, F. M., and J. F. Loneragan. 1972a. Zinc absorption by wheat seedlings: I. Inhibition by macronutrient ions in short-term experiments and its relevance to long-term zinc nutrition. *Soil Sci. Soc. Am. Proc.* **36**:323–327.

Chaudhry, F. M., and J. F. Loneragan. 1972b. Zinc absorption by wheat seedlings: II. Inhibition by hydrogen ions and by micronutrient cations. *Soil Sci. Soc. Am. Proc.* **36**:327–331.

Chaudhry, F. M., and J. F. Loneragan. 1972c. Zinc absorption by wheat seedlings and the nature of its inhibition by alkaline earth cations. *J. Exp. Bot.* **23**:552–560.

Claassen, N., and S. A. Barber. 1976. Simulation model for nutrient uptake from soil by a growing plant root system. *Agron. J.* **68**:961–964.

Clarke. A. L., and E. R. Graham. 1968. Zinc diffusion and distribution coefficients in soil as affected by soil texture, zinc concentration, and pH. *Soil Sci.* **105**:409–419.

Cox, F. R., and E. J. Kamprath. 1972. Micronutrient soil tests. In J. J. Mortvedt, P. M. Giordano, and W. L. Lindsay, Eds. *Micronutrients in Agriculture.* Soil Science Society of America, Madison, WI. Pp. 289–318.

Dolar, S. G., and D. R. Keeney. 1971. Availability of Cu, Zn, and Mn in soils. I. Influence of soil pH, organic matter, and extractable phosphorus. *J. Sci. Fd. Agric.* **22**:273–282.

Elgawhary, S. M., W. L. Lindsay, and W. D. Kemper. 1970a. Effect of EDTA on the self-diffusion of zinc in aqueous solution and in soil. *Soil Sci. Soc. Am. Proc.* **34**:66–70.

Elgawhary, S. M., W. L. Lindsay, and W. D. Kemper. 1970b. Effect of complexing agents and acids on the diffusion of zinc to a simulated root. *Soil Sci. Soc. Am. Proc.* **34**:211–214.

Giordano, P. M., J. C. Noggle, and J. J. Mortvedt. 1974. Zinc uptake by rice as affected by metabolic inhibitors and competing cations. *Plant Soil* **41**:637–646.

Gladstone, J. S., and J. F. Loneragan. 1967. Mineral elements in temperate crops and pasture plants. I. Zinc. *Aust. J. Agric. Res.* **18**:427–446.

Hassan, M. M., and T. van Hai. 1976. Kinetics of zinc uptake by citrus roots. *Z. Pfanzenphysiol Bd.* **79**:177–181.

Hawf, L. R., and W. E. Schmid. 1967. Uptake and translocation of zinc by intact plants. *Plant Soil* **17**:249–260.

Hodgson, J. F., H. R. Geering, and W. A. Norvell. 1965. Micronutrient cation complexes in soil solution: Partition between complexed and uncomplexed forms by solvent extraction. *Soil Sci. Soc. Am. Proc.* **29**:665–669.

Hodgson, J. F., W. L. Lindsay, and J. F. Trierweiler. 1966. Micronutrient cation complexing in soil solution: II. Complexing of zinc and copper in displaced solution from calcareous soils. *Soil Sci. Soc. Am. Proc.* **30**:723–726.

Iyengar, B. R. V., and D. L. Deb. 1977. Contribution of soil zinc fractions to plant uptake and fate of zinc applied to the soil. *J. Ind. Soc. Soil Sci.* **25**:426–432.

Iyengar, S. S., D. C. Martens, and W. P. Miller. 1981. Distribution and plant availability of soil zinc fractions. *Soil Sci. Soc. Am. J.* **45**:735–739.

Krauskopf, K. B. 1972. Geochemistry of micronutrients. In. J. J. Monvedt, P. M. Giordano, and W. L. Lindsay, Eds. *Micronutrients in Agriculture.* Soil Science Society of America, Madison, WI. Pp. 31–33.

Lindsay, W. L. 1972. Zinc in soils and plant nutrition. *Adv. Agron.* **24**:147–186.

Lindsay, W. L. 1979. *Chemical Equilibria in Soils.* Wiley-Interscience, New York.

Lindsay, W. L., and W. A. Norvell. 1978. Development of a DPTA soil test for zinc, iron, manganese and copper. *Soil Sci. Soc. Am. J.* **42**:421–428.

McBride, M. B., and J. J. Blasiak. 1979. Zinc and copper solubility as a function of pH in an acid soil. *Soil Sci. Soc. Am. J.* **43**:866–870.

Nelson, J. L., and S. W. Melsted. 1955. The chemistry of zinc added to soils and clays. *Soil Sci. Soc. Am. Proc.* **19**:439–443.

Olsen, S. R. 1972. Micronutrient interactions. In J. J. Mortvedt, P. M. Giordano, and W. L. Lindsay, Eds. *Micronutrients in Agricultures.* Soil Science Society of America, Madison, WI. Pp. 243–264.

Parsons, R. 1959. *Handbook of Electrochemical Constants.* Academic Press, New York.

Saeed, M., and R. L. Fox. 1977. Relations between suspension pH and zinc solubility in acid and calcareous soils. *Soil Sci.* **124**:199–204.

Schmid, W. E., H. P. Haag, and E. Epstein. 1965. Absorption of zinc by excised barley roots. *Physiol. Plant.* **18**:860–869.

Shukla, U. C., and S. B. Mittal. 1979. Characterization of zinc adsorption in some soils of India. *Soil Sci. Soc. Am. J.* **43**:905–908.

Shuman, L. M. 1975. The effect of soil properties on zinc adsorption by soils. *Soil Sci. Soc. Am. Proc.* **39**:454–458.

Shuman, L. M. 1977. Adsorption of Zn by Fe and Al hydrous oxides as influenced by aging and pH. *Soil Sci. Soc. Am. J.* **41**:703–706.

Singh, J. P., R. E. Karamanos, and J. W. B. Stewart. 1986. Phosphorus-induced zinc deficiency in wheat on residual phosphorus plots. *Agron. J.* **78**:668–675.

Udo, E. J., H. L. Bohn, and T. C. Tucker. 1970. Zinc adsorption by calcareous soils. *Soil Sci. Soc. Am. Proc.* **34**:405–407.

Veltrup, W. 1978. Characteristics of zinc uptake by barley roots. *Physiol. Plant.* **42**:190–194.

Warncke, D. D. 1967. Mechanisms for zinc supply to plant roots growing in soil and alteration of these mechanisms by phosphorus application and soil pH. M.Sc. thesis, Purdue Univ.

Warncke, D. D., and S. A. Barber. 1972a. Diffusion of zinc in soils: I. The influence of soil moisture. *Soil Sci. Soc. Am. Proc.* **36**:39–42.

Warncke, D. D., and S. A. Barber. 1972b. Diffusion of zinc in soils: II. The influence of soil bulk density and its interaction with soil moisture. *Soil Sci. Soc. Am. Proc.* **36**:42–46.

Warncke, D. D., and S. A. Barber. 1973. Diffusion of zinc in soil: III. Relation to zinc adsorption isotherms. *Soil Sci. Soc. Am. Proc.* **37**:355–358.

Wilkinson. H. F., J. F. Loneragan, and J. P. Quirk. 1968. The movement of zinc to plant roots. *Soil Sci. Soc. Am. Proc.* **32**:831–833.

CHAPTER **20**

Water

One of the most significant factors in crop production is the water supply to the plant. Many investigations focus on developing plant root systems that will supply additional water to the plant, making it less susceptible to stress during drought periods. The supply of water to the plant, and its absorption, are treated in this chapter as if water were a nutrient supplied by the soil. Only a very brief description of the role of water in soil and plant growth is giiven here; detailed descriptions are found in textbooks by Hillel (1980), Nobel (1974), and Hanks and Ashcroft (1980).

SOIL WATER

An important part of the soil system is the water present in the soil pore spaces. Water in the soil–plant system is important as (1) a medium for diffusion of solutes, (2) a temperature-regulating liquid, (3) a solvent for biochemical reactions, (4) an aid to plant support, (5) a medium for supplying nutrients to the plant by mass flow, (6) a medium for moving nutrients throughout the plant, and (7) a source of hydrogen in photosynthesis.

Properties of Water

Water consists of two hydrogen ions bonded to one oxygen; this arrangement is illustrated in Figure 20.1. The H–O–H angle is 105°, and the O–H distance 0.099 nm. The oxygen atom is strongly electronegative and hence draws electrons away from the hydrogen atoms; this results in a polar molecule. The oxygen has a partial negative charge, and the hydrogen a partial positive charge. Because of this separation of charge, hydrogen bonding (as illustrated in Figure 20.1) can occur among water molecules. The hydrogen bonding has an energy of 20.1 kJ/mol of hydrogen bonds.

Ice has essentially all water molecules joined by hydrogen bonds. Adding heat breaks some of the hydrogen bonds, so liquid is formed. About 15% of the hydrogen bonds are broken during ice melting. The heat of fusion of ice is 6.0 kJ/mol of water. When water changes from the liquid state to the vapor state, all remaining hydrogen bonds are broken. The heat of vaporization of water is 40.7 kJ/mol of water, which is the highest heat of vaporization of any known liquid. This is important in plant physiology, because vaporization of water at the leaf surface cools the leaf surface and permits control of leaf temperature. Heat loss during evaporation of water at both the leaf and soil surface dissipates heat gained by net radiation from the sun.

The strong attraction among water molecules makes water very cohesive, because of this, long, intact columns of water can exist in the seive tubes of plants. In addition, water has a strong adhesive property. There is a strong attraction between water and the walls of capillary tubes, which results in the capillary rise of water. Water also has tensile strength, which allows it to be pulled up in long columns. The tensile strength of water corresponds to a stress of 1800 MPa.

Because of its highly polar nature, water is a good solvent for polar substances. Its small molecular size also make it a good general-purpose solvent.

Two important characteristics of soil water are its water content and the soil water potential. Soil water content can be expressed as fractional volumetric water m^3/m^3. Hence, it is dimensionless and designated by the symbol θ. Soil water content also can be expressed as cm of water per m^2 of soil surface, which is the depth of water that would occur if it were all ponded on the surface of the soil. The soil depth to which soil water is

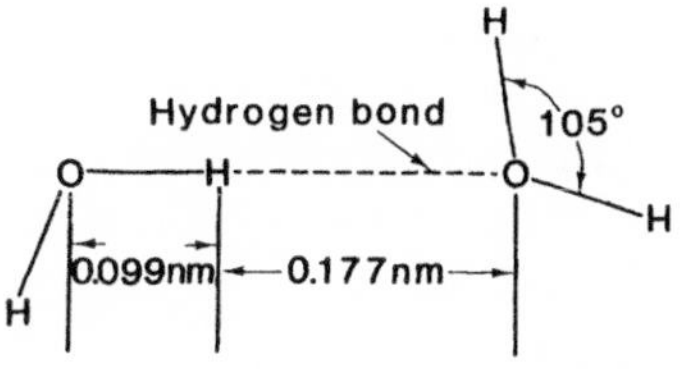

FIGURE 20.1 Schematic structure of water molecules, showing H–O–H angle and H-bonding between molecules.

measured should correspond to the depth to which plant roots can remove water.

Soil Water Potential

Soil water potential is the difference in partial specific free energy between soil water and standard free water. Total soil water potential is the sum of potentials due to a number of force fields; they include the gravitational potential, the pressure or matrix potential, and the osmotic potential. The amount of energy in the form of gravitational potential depends on the water's elevation compared to standard free water. Water at a free-water surface has zero pressure potential. Water under this surface has a positive pressure potential. The capillary and adsorptive forces of the soil matrix reduce the matrix potential below that of bulk water. The capillary effect predominates in sandy soils, whereas the adsorptive forces predominate in determining the matrix potential in clay soils. The osmotic potential is the effect of solutes on lowering the vapor pressure of water. This is important both for soil water and water within the plant.

Soil water potential has frequently been expressed in terms of atmospheres or bars. The SI unit designation is kilopascal (kPa). One bar is equal to 100 kPa. The relation between soil water content and soil water potential for a particular soil can be described by the soil-water characteristic curve.

Soil-Water Characteristic Curve

When a saturated soil is in equilibrium with free water at the same elevation, soil water potential is zero. If suction is applied to the soil, the large pores empty, and soil water potential decreases. (Hence, there is an increase in magnitudes of the negative values.) A comparison of two soils with respect to their relations between matrix potential and volumetric water content is shown in Table 20.1. There is an approximate amount of water to which saturated soil will be drained due to gravitational forces; this moisture level has often been referred to as the field capacity. A water tension of 30 kPa (0.3 bar) can often be used to define this value. On the other end of the soil-water characteristic curve is the quantity of water remaining in the soil after permanent wilting has occurred for plants growing on the soil. This is called the permanent wilting point. A tension of 1500 kPa (15 bars) is usually used to approximate this value. The amount of available water present in a soil at field capacity is the difference between θ at field capacity and θ at the permanent wilting point. Soils high in clay content usually have a much higher θ value at the permanent wilting point than sandy soils. For example, Raub silt loam, Aquic Argiudoll, had 12.7% volumetric moisture at 1500 kPa, while Aubbeenaubbee sandy loam, Aric Ochraqualfs, had only 3.2% (Table 20.1). Soils high in silt content usually have a higher pro-

TABLE 20.1   Soil Water Characteristic for Raub Silt Loam and Aubbeenaubbee Sandy Loam

Suction head		Volumetric wetness	
kilopascals	(cm)	Raub	Aubbeenaubbee
		(%)	
0	0	50.5	40.5
1	10	48.2	28.6
3	30	43.6	24.8
5	50	43.4	22.7
10	100	31.9	10.1
33	330	27.1	9.7
100	1000	24.2	8.1
300	3000	19.4	5.3
500	5000	17.0	4.6
1000	10,000	14.0	3.6
1500	15,000	12.7	3.2

Source: Barber and Mackay (unpublished)

portion of their water that is regarded as available. In Table 20.1, the volumetric amount of available water in Raub silt loam was 14.4% while that in Aubbenaubbee sandy loam was only 6.5%. The soil water characteristic curve can be used to calculate the amount of available water in a soil profile that has been fully charged with water.

PLANT WATER USE

The amount of water used by a crop in the field depends on the amount of radient energy reaching the field, relative humidity, and rate of water supply to the root by the soil. In humid environments, most of the net radiation reaching the earth is used to evaporate water. As conditions become drier, as in a desert climate, more of the energy is used to heat the soil and the air above it. Evaporation from an open pan has been used to determine the potential for evapotranspiration that may occur each day.

It is desirable to know the amount of evapotranspiration from a crop; to obtain this value, crop coefficients are determined. The crop coefficient is the relation between evapotranspiration by the crop and open-pan evaporation; the relation varies with both crop and climate. Values for a well-watered crop may vary from 0.6 to 0.8. However, when soil water potential decreases, there is a decrease in evapotranspiration; the relation between these variables can be complex. The reader is referred to books such as Hanks and Ashcroft (1980) for further information.

The amount of water transpired per gram of dry matter produced varies with the plant species. Sorghum may be as low as 200 and alfalfa as high as 800. Crops that are efficient users of water are important in the drier areas of the world. Water-use efficiency is also directly affected by crop yield, and since water use is directly related to net radiation reaching the crop, a small crop will use almost as much water per hectare as a large crop. Hence, other practices that increase yield also increase water-use efficiency. Consequently, high-yielding crops will usually have a higher water-use efficiency than crops producing less dry matter.

Water Uptake

While there is evidence that applying large amounts of respiration inhibitors can decrease water uptake, it is not certain that respiration energy is used in water uptake. Because of the large amounts of inhibitor required, it may be that the inhibitor eventually decreases the root's permeability to diffusion (Shone and Flood, 1981). The wide fluctuations in water influx, and its ready efflux, suggest that there is little direct relation between respiration energy and water uptake. The energy for transpiration comes from the sun's energy evaporating water at the leaf surface. Water appears to follow the apoplastic pathway through the root. Suberization of the endodermis has been shown to restrict water uptake by barley and marrow, but not by corn. In studies with corn, Stephens and Clarkson (1981) found that rate of water uptake and hydraulic conductivity both increased along the root toward the base of the root. In studies with barley roots, Sanderson (1982) found that the rate of uptake was much less as the root matured. Temporary disruption of the Casparian strip where lateral roots emerge provided an apoplastic pathway.

For corn seedlings growing in solution culture, the leaf water potential, which is believed to be the drawing force for water movement through the plant, is about 400 kPa (4 bars). Hydraulic conductivity of water through the root system of corn was calculated at 6 to 15×10^{-2} cm^3/m · s MP$_a$. Water uptake ranged from 1.3 to 3.3×10^{-2} cm^3/m · s with the lower rate 0–1 cm and the higher rate 20–21 cm from the tip. Results for corn were different in this respect than results with barley (Stephens and Clarkson, 1981).

While there is an influx of water into the root and flow of water through the xylem to the shoot, there is also efflux of water from the root; uptake is the difference between influx and efflux. Shone and Wood (1977) applied tritiated water to the apical segments of corn roots, while passing the remainder of the root through nontritiated water. Much of the water absorbed by the apical portion had moved back into solution before the water had reached the base of the root.

Taylor and Klepper (1978) described conceptual models of water flow through plants. Since water only flows during evaporation at the leaf surface, no water flow occurs at night. In the morning, impingement of sunlight energy on the leaf causes water to vaporize and move out through open stomata, which causes a reduction in water potential and a flow of water from adjacent plant tissue. This ultimately leads to a flow of water through the soil to the root and then through the plant to the leaf. The rate of flow depends on the resistance to flow in each part of the system.

Conceptual models describe water flow through the plant's root system (Taylor and Klepper, 1978); flow through the plant is believed to be the source of greatest resistance to water flow, and there are differences between species in the amount of this resistance. Dunham and Nye (1973) found experimentally that steep water-potential gradients do not normally exist in the soil near plant roots. Little information is available on the relation between rooting density in the soil and water uptake, but it is generally believed that plants have ample roots to provide for water uptake under moist soil conditions. Removal of soil water is believed to occur more rapidly from surface layers, as soils have been observed to dry out progressively from surface layers downward. Root density is also higher in surface layers.

Water influx into the root can be varied over a five-fold range by changing the relative humidity of the air around the shoot from 95 to 5%. The mean rate of water flux into roots observed for plants growing in controlled climate chambers varies between 0.2 and 1.0×10^{-6} cm^3/cm$^2 \cdot$ s. Higher rates occur for plants having fewer coarser roots.

Water and Ion Uptake

Water affects nutrient availability through its influences on supply by mass flow and diffusion. Supply by mass flow is directly related to the rate of water influx. If nutrient flux to the root is less than I_{max}, increasing water flux will increase uptake.

Rate of ion diffusion is also directly related to θ. Increasing θ reduces the tortuosity of the diffusion path and increases f_i; hence, there is often a linear relation between θ and D_e. An example of this is shown in Figure 4.10.

Rate of water diffusivity in soil ranges from 1 to 10×10^{-6} cm^2/s, which is rapid enough to supply adequate water to the root. Values will decrease as moisture level in the soil decreases. Values of 1 to 10×10^{-6} cm^2/s would indicate that there should be a relatively flat water gradient around the plant root. This being the case, it should be sufficiently accurate to adjust D_e for θ by using changes in mean θ in the soil.

Soil water level, θ, affects the D_e value for ions diffusing toward the root surface. Cox and Barber (1992) measured phosphorus uptake by maize growing in four soils that varied in soil θ from 0.13 to 0.40 at -33 kPa soil water potential. Since D_e increased with increases in θ, the level of available soil phosphorus as measured by C_{si} required to obtain the same phosphorus supply to the root decreased as θ increased.

REFERENCES

Cox, M. S., and S. A. Barber. 1992. Soil phosphorus levels needed for equal P uptake from four soils with different water contents at the same water potential. *Plant Soil* **143**:93–98.

Dunham, R. J., and P. H. Nye. 1973. The influence of soil water content on the uptake of ions by roots. I. Soil water content gradients near a plane of onion roots. *J. Appl. Ecol.* **10**:585–598.

Hanks, R. J., and G. L. Ashcroft. 1980. *Applied Soil Physics.* Springer-Verlag, New York.

Hillel, D. 1980. *Fundamentals of Soil Physics.* Academic Press, New York.

Nobel, P. S. 1974. *Introduction to Biophysical Plant Physiology.* W. H. Freeman and Co., San Francisco.

Place, G. A., and S. A. Barber. 1964. The effect of soil moisture and Rb concentration on diffusion and uptake of Rb[86]. *Soil Sci. Soc. Am. Proc.* **28**:239–243.

Sanderson, J. 1982. The possible significance of the tertiary development of the endodermis for the pathways of radial water flow in the barley root. *Annual Report 1981.* Letcombe Laboratory, Agricultural Research Council, Wantage, England. Pp. 58–59.

Shone, M. G. T., and A. V. Wood. 1977. Further studies on the longitudinal movement and loss of water in barley roots. *Annual Report, 1976.* Letcombe Laboratory, Agricultural Research Council, Wantage, England. Pp 12–15.

Shone, M, G. T., and A. V. Flood. 1981. Effect of sodium azide on radial diffusion of labeled water in barley roots. *Annual Report 1980.* Letcombe Laboratory, Agricultural Research Council, Wantage, England. Pp. 66–67.

Stephens, J. S., and D. T. Clarkson. 1981. Water uptake and hydraulic conductivity in various zones of maize roots. *Annual Report 1980.* Letcombe Laboratory, Agricultural Research Council, Wantage, England. Pp. 62–63.

Taylor, H. M., and B. Klepper. 1978. The role of rooting characteristics in the supply of water to plants. *Adv. Agron.* **30**:99–128.

CHAPTER **21**

Application of the Mechanistic Uptake Model

Since the mechanistic uptake model predicts uptake of several nutrients that agrees with observed uptake, the model can be used in several research applications. Using the model in place of conducting plant growth experiments greatly reduces the time and expense required in studying various aspects of soil nutrient availability. The applications described in this chapter are from our research and are examples of several types of applications. Others can be used for research on specific soil nutrient availability problems. The applications are described briefly. See the publications referred to for details.

EVALUATING PHOSPHATE FERTILIZER

As an example of the use of the uptake model, Barber and Ernani (1990) evaluated relative predicted phosphorus uptake after application of four different phosphate fertilizers to one soil, Raredon silt loam (clayed, mixed mesic Aquic Hapludalfs) at two soil pH levels, 5.0 and 5.7. Monocalcium phosphate (MCP), diammonium phosphate (DAP), partially acidulated rock phosphate (PARP), and rock phosphate (RP) were the materials compared. The first two had a particle diameter less than 0.5 mm and the latter two had particle diameters less than 0.1 mm. All were mixed with the soil at a rate of 232 mg phosphorus/kg of soil and incubated at 25°C for 18 days before the soil measurements of C_{li}, C_{si}, and θ were made. These measurements were used to obtain C_{li}, b, and D_e (see Chapter 2 and Equation 4.6) for use in the model. The remaining model parameters were the same for all treatments and are the same as those given in Table 9.11. Greater detail as well as the results with other soils are given in Barber and Ernani (1990) and Ernani and Barber (1990). Uptake for 10 days growth of maize after transplanting 4-day-old seedlings into the pots was used in the calculations. At harvest, root density in the soil was similar to that found with maize in the field at tasseling. The results are given in Table 21.1.

Predicted P uptake was greater for RP and PARP at pH 5.0 than at pH 5.7 as would be expected. Predicted P uptake for RP at pH 5.7 was not significantly higher than for unfertilized soil at pH 5.0. Partially acidulating the rock phosphate approximately doubled P availability but the predicted P

TABLE 21.1   Comparison of Four Fertilizer P Materials at Two Soil pH Levels

Phosphorus source	C_{si} (mmol/L)	C_{li} (mmol/L)	Predicted P uptake (μmol/kg soil)
pH 5.0			
No phosphate	0.08	0.0009	6
Rock phosphate	4.98	0.0016	23
Partially acidulated			
rock phosphate	3.74	0.0041	37
MCP	7.98	0.020	131
DAP	9.28	0.015	118
pH 5.7			
Rock phosphate	4.20	0.0005	9
Partially acidulated			
rock phosphate	3.55	0.0018	21
MCP	7.62	0.015	108
DAP	8.62	0.025	154
SE	2.06	0.0014	10

uptake from MCP was 3.5- to 5-fold greater than that for PARP. The differences between MCP and DAP switched with the change in soil pH level.

The mathematical model predicts P uptake where we assume root growth is the same for all materials so that we are evaluating only soil chemistry. When we compare these materials in pot experiments, root growth could differ among materials. Differences would usually be greater the greater the availability of phosphorus, since plants would be larger on soils with higher levels of available phosphorus. Using the model along with a pot experiment allows you to determine the reasons for differences between materials in observed phosphorus uptake for plants growing in soils fertilized with the phosphorus materials, since soil phosphorus supply and root growth effects can be separated. Using only the soil measurements as was done here allows the rapid screening of many fertilized materials and also the screening with a range of soils and time of incubation of the phosphate with the soil. It also shows the differences that would be due only to the soil supply of phosphorus to the root.

NUTRIENT PLACEMENT IN THE SOIL

In previous chapters, nutrients were assumed to exist in uniform concentrations throughout the volume of soil penetrated by plant roots. Under field conditions, however, nutrient levels may vary with location in the soil. Surface horizons are usually higher in available nutrients than subsoils. When nutrients are applied to supplement the soil level, either they may be applied broadcast, and more or less uniformly mixed with the soil, or their application may be localized, which occurs where nutrients are banded. Nutrients are usually applied to only the tilled surface layer. There are many possible soil distributions for placing added nutrients, and placement often influences the extent of the plant's nutrient recovery.

Early studies of the effect of nutrient placement on crop yield usually involved low rates of application (less than that removed by the crop). Comparisons have usually been between fertilizer banded near, or in, the seed row at planting versus fertilizer broadcast on the surface and mixed by plowing or surface tillage. Frequently banding, particularly of phosphorus, gave greater yield increases than the same amount of nutrient broadcast and then mixed with the soil (Stanford and Pierre, 1953; Widdowson and Cooke, 1958: Welch et al. 1966). Research involving a wide range of application rates (Barber, 1958, 1959) has shown that rate influences the effect of treatment on yield response. At high fertilization rates, broadcast applications may give higher yields than band applications.

In field research, the effect of placement on efficiency of nutrient uptake may be influenced by soil, weather conditions, crop, and nutrient. In this section we use the mechanistic uptake model to aid in understanding the effect of nutrient placement on nutrient uptake. Placement affects root distribution and phosphorus addition stimulates root growth in the fertilized volume. In addition, the initial level of nutrient in the soil and the rate of nutrient addition affect influx and nutrient-supply characteristics of the soil. At a given fertilization rate, the efficiency of phosphate use will be affected by the relative soil volume fertilized. Total uptake includes the sum of uptake by roots growing in treated and untreated soil. We will discuss the effect of placement on root growth, uptake kinetics, and rate of nutrient supply to the root surface by the soil. Since fertilizer placement is frequently used with phosphorus, we will use phosphorus as an example, then make briefer comments about nitrogen and potassium.

PHOSPHORUS PLACEMENT EFFECT ON ROOT GROWTH

Solution Experiments

Stryker et al. (1974) grew corn in split-root systems where half of the roots were supplied with a nutrient solution containing phosphorus and half maintained without phosphorus. They found greater root growth on the side containing the phosphorus. However, maximal dry matter accumulation occurred only where all the roots were supplied with phosphorus. Drew and Saker (1978) supplied phosphorus to a segment of the root length of barley and not to the remainder. There was stimulation of both the number and length of lateral roots in the segment supplied with phosphorus. Removing the apical meristem caused similar stimulation of lateral roots, indicating that morphogenic effects or localized phosphorus supply may be controlled by the apical meristem and availability of carbon assimilates.

Anghinoni and Barber (1980a) grew corn in solution culture using containers divided into two parts. Corn seedlings were placed on a divider between the two parts, with roots distributed between compartments of phosphorus-containing solution and phosphorus-lacking solution in ratios of 1:1, 1:3, 3:1, and 1:7. Seedlings with eight roots were used in all cases. The ratio of solution volumes for the two parts was the same as the ratio of the number of roots placed in the phosphorus-containing and the phosphorus-lacking compartments. Plants were placed in the solutions at 6 days and harvested at 18 days. The effect of the percentage of roots exposed to phosphorus on shoot and root growth and on shoot phosphorus concentration is given in Table 21.2. Reducing the proportion of roots contacted by phosphorus reduced phosphorus uptake by the plant. Root growth was greater in the phosphorus-containing solution, though total root growth for the plant was not affected by phosphorus distribution. Phosphorus stimulated root growth (Table 21.2) in the phosphorus-containing solution at the expense of growth in the phosphorus-lacking solu-

TABLE 21.2   Effect of Supplying Different Portions of the Root System with Phosphorus on Shoot and Root Growth of 18-Day-Old Corn Plants Grown in Solution Culture

Roots exposed to P (%)	Shoot			Root length		
	Dry weight (g/pot)	P (%)	Total P uptake (mmol/pot)	Dry weight (g/pot)	Total (cm/pot)	In P solution (%)
100	7.20[a]	1.01	2.484	1.55	25,287	100
50	7.07	0.69	1.654	1.59	24,537	52
25	6.01	0.52	1.025	1.35	24,509	36
12.5	5.92	0.47	0.747	1.44	26,807	32
L.S.D.						
0.05	1.06	0.09	230	0.19	n.s.	

Source: Reproduced from Anghinoni and Barber (1980a) by permission of the American Society of Agronomy.
[a] Mean values.

tion. There was also an increase in phosphorus influx to the roots for the side containing phosphorus, as shown by data for I_{max} and K_m in Table 21.3. The increase in I_{max} was correlated with a decrease in phosphorus concentration in the roots. The reduced phosphorus status in the plant stimulated an increase in I_{max} for phosphorus uptake.

TABLE 21.3   Effect of Supplying Different Portions of the Root System with Phosphorus on Phosphorus-Uptake Parameters and Percent Phosphorus in the Shoot and Root of 18-Day-Old Corn Plants Grown in Solution Culture

Initial root exposed to P (%)	P uptake parameters			P concentration	
	I_{max} (nmol/m$^2 \cdot$ s)	K_m (μmol/L)	C_{min}	Shoot (%)	Root +P side (%)
100	22	1.1	0.04	1.01	0.47
50	30	1.6	0.03	0.69	0.28
25	36	2.8	0.06	0.52	0.23
12.5	36	2.9	0.17	0.47	0.18
L.S.D.					
0.05	3	0.6	n.s.	0.09	0.05

Source: Reproduced from Anghinoni and Barber (1980a) by permission of the American Society of Agronomy.

Soil Experiments

Anghinoni and Barber (1980b) mixed phosphorus with portions ranging from 12.5 to 100% of the soil in pots; they also used two soils varying in phosphorus-adsorption properties. The soils were from the Ap layer of Wellston silt loam (Ultic Hapludalf) and Raub silt loam (Aquic Argiudoll). Six-day-old corn seedlings were transplanted into the pots and then grown for 6 and 12 days. The lengths of roots in phosphorus-treated and untreated soil were measured separately. The effect of soil phosphorus treatment on root distribution is shown in Figure 21.1, where the proportion of roots in the phosphorus-treated soil is compared with the proportion in treated soil. The values were described by the relation $y = x^{0.68}$, where y is the proportion of total root length in phosphorus-treated soil and x is the volumetric proportion of soil treated with phosphorus. Placement effects were similar for the two soils. The effect was presumably a plant-root response to the effects of phosphorus distribution. Borkert (1983) conducted a similar experiment using soybeans and found that phosphorus increased root growth in the treated zone just as it had for corn.

Yao and Barber (1986) conducted a similar experiment with wheat (*Triticum vugare* L.) and found the degree of root stimulation from placed phosphate was similar to that obtained for corn by Anghinoni and Barber (1980a) and soybean (*Glycine max*) by Borkert and Barber (1985b), hence these three crops reacted similarly. As shown in Figure 21.2, the relation of roots in the P-fertilized soil volume, y, with the fraction of soil volume P-fertilized, x, could be described by the relation $y = x^{0.7}$. These experiments

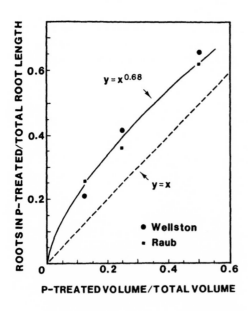

FIGURE 21.1 Effect of phosphorus treatment of part of the soil on the distribution of roots of 18-day-old corn plants between treated and untreated soil. Reproduced from Anghinoni and Barber (1980b) by permission of Soil Science Society of America.

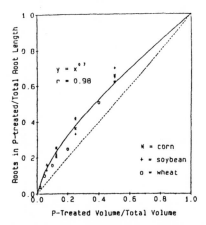

FIGURE   21.2   Effect of fraction of soil treated with P on the fraction of the roots in P-treated soil. Data are given for separate experiments with corn, soybean, and wheat. Reproduced from Yao and Barber (1986) by permission of Marcel Dekker, Inc.

did not investigate the effect of a wide range of C_{si} or of P rates added on root distribution.

Zhang and Barber (1992) studied a wide range of C_{si} and of phosphorus added to 20% of the soil volume in pot experiments in the growth chamber using corn grown on three soils varying in C_{si} levels and with three rates of applied phosphorus varying from 50 to 300 mg/kg. Root density, cm/cm³, in the P-fertilized soil volume, RDF, and a comparable 20% volume of unfertilized soil, RDU, was measured (the distributions of the sampled soil volumes are shown in Figure 21.3) and compared with C_{si} in the P-fertilized soil, $C_{si}F$, and unfertilized soil, $C_{si}U$. The F, fertilized, and U, unfertilized, considered had equal opportunity for root growth from the plants. There was a curvilinear relation of the results described by the equation

$$y = 1.20 + 2.74 \log x \qquad (r^2 = 0.97) \qquad (21.1)$$

where y is RDF/RDU and x is $C_{si}F/C_{si}U$ (see Figure 21.4). Hence as the $C_{si}U$ level increased RDF/RDU decreased and as the rate of phosphorus applied increased RDF/RDU increased. The relation between $C_{si}F/C_{si}U$ and RDF/RDU can be used to predict root growth rates to use in the fertilized and unfertilized soil when using a mechanistic nutrient uptake model to calculate effect of phosphorus placement on phosphorus uptake.

The large effect of phosphorus on root growth observed by Drew (1975), where phosphorus was added to a portion of a root system grown in solution culture while the remaining root system was in solution devoid of phosphorus, can be explained by Equation 21.1. Since $C_{si}U$ would be almost zero,

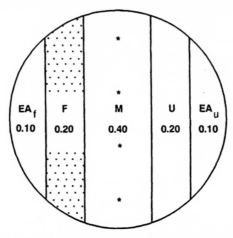

FIGURE   21.3   Top view of the arrangement of vertical screens separating the different parts of the pot into the middle, M, showing the location of the maize plants, the fertilized portion, F, the unfertilized portion, U, and the contiguous parts EA_f and EA_u. Reproduced from Zhang and Barber (1992) by permission of the Soil Science Society of America.

$C_{si}F/C_{is}U$ would be infinitely large and this would give a large RDF/RDU value. In soils $C_{si}F/C_{si}U$ is almost never this large.

Equation 21.1 can be used to calculate RDF/RDU where the same amount of phosphorus is added per pot, the fractional volume of soil fertilized varies from 0.01 to 1.0, and the relation between C_{si} and phosphorus added is known. This is the relation used by Anghinoni and Barber (1980a), Borkert and Barber (1985a), and Yao and Barber (1986) giving $y = x^{0.7}$, where y was the fraction of roots in fertilized soil and x was the fraction of soil fertilized. The results of these calculations are shown in Figure 21.5. The curve $y = x^{0.71}$ was close to the relation obtained by Anghinini and Barber (1980a), hence Equation 21.1 can be used to calculate root distributions for many fractional soil volumes receiving phosphate fertilization where rate of phosphate per unit of total soil volume can be varied and soil can be varied.

To test the effect of soil volume fertilized, additional calculations were made where C_{si} of the fertilized soil was the same for all fractions of the soil fertilized, as fraction of soil fertilized was reduced, phosphorus applied per unit of total soil volume decreased. This relation had little curvature; the effect due to differences in soil volume fertilized is shown as $C_{si}F/C_{si}$ constant in Figure 21.5. Both fractional volume fertilized and differences in $C_{si}F$ contributed to the relation $y = x^{0.7}$, however, differences in $C_{si}F$ dominated.

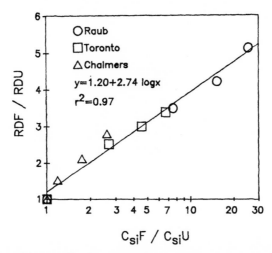

FIGURE 21.4   Relation of ratio of root density in P-fertilized soil to that in unfertilized soil (RDF/RDU) with ratio of initial resin-exchangeable P in fertilized soil to that in unfertilized soil ($C_{si}F/C_{si}U$). Reproduced from Zhang and Barber (1992) by permission of the Soil Science Society of America.

Effect of Varying RDF/RDU on Influx

In a split-pot solution experiment with phosphorus, Anghinoni and Barber (1980a) found that as the fraction of roots was reduced the I_{max} value after 18 days for the roots in phosphate increased (Table 21.3). This is the same experiment as that shown in Table 21.2 for the effect on root growth. I_{max} values increased from 22 to 36 μmol/m^2 · s as phosphorus concentration of the roots decreased from 0.47 to 0.18%. Phosphorus starvation of whole plants caused similar increases in I_{max}. Similar relations were shown for I_{max} and shoot potassium concentration in Figure 3.8 and for nitrogen uptake kinetics of corn in Table 3.3.

While starvation of the plant for a particular nutrient may increase I_{max}, I_{max} also increases when C_{li} is increased to high levels. Borkert and Barber (1985b) doing research on phosphorus uptake by soybean found that when grown in soils with C_{li} values above 25 μmol and D_e values above 1.5×10^{-8} cm^2/s, I_{max} values more than doubled. While increased I_{max} may occur in placement experiments, the fraction of the roots in the high phosphate soil is very low so that increase in uptake is limited because of the low fraction of roots that is absorbing at a high rate.

In summary, I_{max} may increase due to plant starvation for a nutrient or it may increase because of a change in influx kinetics as solution concentration at the root increases to high levels. These kinetics need to be investigated if predicted nutrient uptake does not agree with observed uptake for research experiments on nutrient placement.

Placement Effect on Soil Supply Characteristics

The effect of rate of phosphate and potassium applied to the soil on C_{li}, b, and D_e, the soil supply parameters, was studied for 33 diverse soils by Kovar and Barber (1987, 1988, 1989). These results have been discussed for phosphorus in Chapter 9 and for potassium in Chapter 10. For both nutrients there was a linear relation between C_{si} and rate of application and a curvilinear relation between C_{li} and rate of applied nutrient. Since the sensitivity analysis (Figure 9.9) indicated that C_{li} had the greatest affect on uptake and that C_{li} is an important part of the calculation of b ($\Delta C_{si}/\Delta C_{li}$) as well as D_e (it was calculated $D_l f_l \theta/b$), C_{li} plays a large part in determining soil supply.

The relation between C_{li} and phosphate added to the soil (Kovar and Barber, 1988) is shown in Figure 9.2 for four soils, illustrating the range of values. A plot of curves for all 33 soils was made by Caldwell et al. (1992) that shows the rather uniform distribution of the curves over the range found for the 33 soils. Since placing phosphate increases the rate applied on the fertilized soil, it is important to learn how the increase in rate affects the supply characteristics for the soils considered in order to determine the effect of phosphate placement.

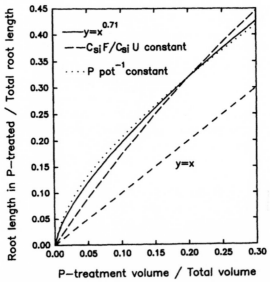

FIGURE 21.5   The relation between fraction of root length in fertilized soil and fraction of soil fertilized, where the relation was calculated using Equation 21.1 assuming the ratio of initial resin-exchangeable P for fertilized to unfertilized soil remained constant and where $C_{si}F/C_{si}U$ changed to give the same total P per pot. These are compared with the curve $y = x^{0.71}$. Reproduced from Zhang and Barber (1992) by permission of the Soil Science Society of America.

Fractional Placement Effect on Nutrient Uptake

The uptake model can be used to calculate predicted uptake for a series of treatments application of the same rate of nutrient per unit area of land in placements varying from applying the nutrient to 1% of the soil volume to mixing it uniformly with all the soil. As volume of soil fertilized increases the phosphorus concentration in the fertilized soil decreases. The soil supply values of C_{li}, b, and D_e for a wide range of rates of application to samples of 1 soil are determined by experimentally measuring C_{li}, and C_{si} for each rate, b is calculated from $\Delta C_{si}/\Delta C_{li}$; θ, D_l, and f_l will be common for all rates and these values are used to calculate D_e from $D_e = D_l \theta f_l / b$. The difference in root density in the fertilized and unfertilized volumes of soil for each fraction fertilized can be calculated from Equation 21.1. The same kinetics common for the plant species may be used with adjustment for soil supply levels. When these values are used to calculate predicted uptake a curvilinear relation similar to that shown in Figure 21.6 is obtained. When a low fractional volume is used, predicted uptake increases as volume fertilized increases because more roots are in the fertilized zone. Uptake was limited by amount of root-fertilized soil contact. As the fractional volume of soil fertilized increases, maximum predicted uptake is obtained, and then begins to decrease as the volume fertilized increases further toward mixing the fertilizer uniformly with all the soil.

The question arises, does the predicted data agree with experimental data? Borkert and Barber (1985b) conducted a pot experiment with phosphate placement for soybean. One rate of phosphorus (60 mg P/kg of soil in the pot) was applied to volume fractions of 1.0, 0.5, 0.25, 0.125, and 0.0625. Values used in calculating the predictions were the C_{li}, b, and D_e measured for soil fertilized at 0, 60, 120, 240, and 480 mg P/kg of soil; root density measurements and root radii were made for the fertilized and unfer-

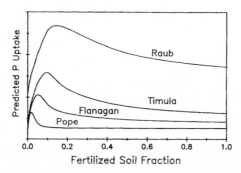

FIGURE 21.6   Using the Barber–Cushman model to calculate the effect of placing phosphorus in varying soil fractions on predicted phosphorus uptake from four widely differing soils as described in Kovar and Barber (1987).

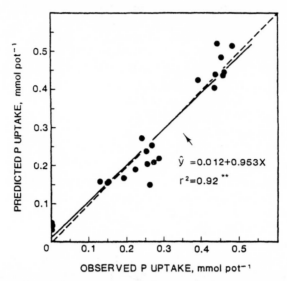

FIGURE 21.7 Correlation between predicted and observed P uptake for 24-day-old soybean plants grown in a Raub silt loam in an experiment where one P rate/pot was applied to decreasing volumes of soil. There were six treatments replicated four times. Reproduced from Borkert and Barber (1985) by permission of the Soil Science Society of America.

tilized portions of each pot, and uptake kinetics determined experimentally for a range of C_{li} values in solution culture. The relation between the observed phosphorus uptake and predicted phosphorus uptake for the four replications of each treatment is given in Figure 21.7. The predicted curvilinear relation, obtained using more application rates for the phosphorus-treated fraction, is shown in Figure 21.8. The curve labeled 60 represents the one measured with the experimental data. The additional curves are given to show the relation between rate of phosphorus used and the position of the curve. The curve for zero indicated the phosphorus taken up from the soil not receiving phosphorus.

In an earlier experiment with corn, Anghinoni and Barber (1980a) obtained a linear relation of $y = 12.2 + 0.89x$ ($r = 0.85$) where y is predicted uptake and x is observed uptake. The greater variation in this experiment was probably due to our laboratory having less experience in accurately measuring the many values used in the calculations.

Kovar and Barber (1989) investigated the reason for differences among soils in the volume placement of phosphorus that gave maximum predicted uptake. The increase in solution phosphorus with added phosphorus is curvilinear. It can be explained by the equation $y = ax^c + d$ where y is solution phosphorus, x is rate of phosphorus added, a is the linear coefficient, c is the curvilinear coefficient, and d is the intercept with the y axis.

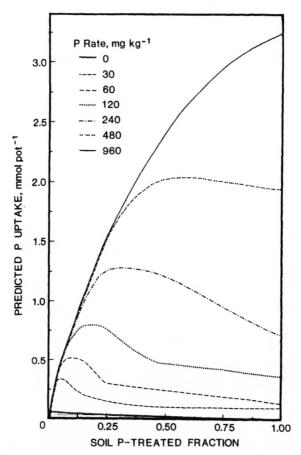

FIGURE   21.8   Predicted P uptake for 24-day-old soybean plants grown in a Raub silt loam as a function of P rate (mg/kg, values shown on lines) and fractional volume of the soil fertilized with P. Reproduced from Borkert and Barber (1985) by permission of the Soil Science Society of America.

Changes in the size of c were investigated (Kovar and Barber, 1989). Since c was correlated with a, a was changed as c changed. Data for Toronto silt foam [fine silty mixed mesic (Udollic Ochraqualf)] from the study of 33 soils were used. Its c value was 2.01 and was near the mean of the 33 soils. Values of c of 2.51, 25% higher, 1.51, 25% lower, and 1.0 were used in calculations of C_{li} increase with added phosphorus. Values of C_{li} and C_{si} were taken from regression equations of C_{li} vs. P added and C_{si} vs. P added and used to calculate b and D_e for each rate of P added. The relation of C_{li} vs. P added is shown in Figure 21.9. Predicted phosphorus uptake was determined after 18 days for 18 placements of P. Rates of root growth var-

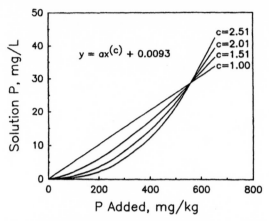

FIGURE   21.9   The effect of changing (c) and (a) constants from the equation C_{li} = $ax^c + d$, on the relation of C_{li} to P added (x). Reproduced from Kovar and Barber (1989) by permission of the Soil Science Society of America.

ied according to differences in C_{si} values of the fertilized and unfertilized soil.

The results for predicted uptake vs. soil fraction fertilized as affected by the value of c are shown in Figure 21.10. The c value has a large effect on the fraction of the soil to fertilize to maximize P uptake as the P rate is increased.

A sensitivity analysis of the effect of h in the relation $C_{si} = g + hx$ was conducted where h is the slope of the linear relation between C_{si} and the rate of

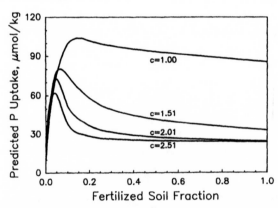

FIGURE   21.10   The effect of changing constants (c) and (a) from the equation C_{li} = $ax^c + d$ on the relation of volume of soil fertilized with P to predicted P uptake. Reproduced from Kovar and Barber (1989) by permission of the Soil Science Society of America.

P added, x. The value of h did not appreciably affect the relation between predicted P uptake and fraction of soil fertilized.

While the calculations here were based on using a uniform volume of soil, and use phosphorus as an example, they show the principles involved in using fertilizer placement for increasing the efficiency of fertilizer application. They show how the difference in curvilinearity (size of c values) among soils can influence the results of phosphate placement.

SUMMARY

Placement of phosphate is most important in soils low in available phosphorus that adsorb or fix large quantities of added phosphorus. Phosphorus placement is also more important the lower the rate of application used. As rate of application increases, the volume percent of the soil that has to be fertilized in order to maximize uptake will also increase. Localizing phosphorus increases root growth in the fertilized soil volume relative io the unfertilized soil. Localizing potassium does not affect root growth as far as is known. Hence, potassium localization is not likely to be as important unless it is accompanied by nitrogen or phosphorus, which accentuates root growth in the fertilized zone.

Evaluation of Soil pH Effect on Phosphorus and Potassium Availability

Soil pH influences nutrient levels in the soil and their uptake by plants. Soil pH may range from 4.0 or less to above 8.5. The nutrient uptake model can be useful for evaluating the effect of soil pH on nutrient uptake since parameters of the model describe the uptake process.

Chen and Barber (1990) investigated the effect of a range of pHs from 3.8 to 8.3 on phosphorus and potassium uptake using three replicates of six soil pH values. Chalmers silt loam (fine silty, mixed, mesic Typic Haplaquoll) initially at pH 4.7 was acidified with sulfuric acid to a pH of 3.8. Soil pHs of 5.7, 6.5, 7.6, and 8.3 were obtained by adding CaO. The soils were incubated for 10 days at 28°C and field capacity moisture content. No nutrients were added since the soil was relatively fertile. The experimental uptake was determined by transplanting four 6-day-old corn seedlings into each pot containing 3 kg of soil compressed to 1.2 Mg/m³ bulk density. Plants were grown for 12.5 days in a controlled-climate chamber with 16 h of light and a temperature of 25°C. Pots were rerandomized, weighed, and watered daily to 25% (w/w).

The mechanistic model parameters were measured and used in the model to predict phosphorus and potassium uptake. The results for phosphorus are plotted in Figure 21.11 to show the relation between uptake from plant analysis and uptake predicted by the model for each soil pH.

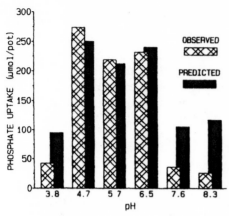

FIGURE 21.11 Relation between observed and predicted P uptake calculated with the Barber–Cushman model for maize plants grown for 18.5 days. Correlation between observed (x) and predicted (y) was $y = 92.8 + 0.61x$ ($r^2 = 0.95$) where data for pH 3.8 were omitted. Reproduced from Chen and Barber (1990) by permission of the Soil Science Society of America.

For pH values 4.7, 5.7, and 6.5 agreement was relatively close. The other three pH values showed a wide divergence between observed and predicted values.

The $H_2PO_4^-$ form of solution phosphorus is believed to be the principal form absorbed by the root. When soil pH is above 5.0 part of the solution phosphorus is in the HPO_4^{2-} form, the proportion increasing with pH. Hendrix (1967) found that the rate of phosphorus uptake from HPO_4^{2-} was about one-tenth that from $H_2PO_4^-$ at pH 8.7. Hence separate calculations were made for the uptake of each phosphate form, totalled and used to predict phosphorus uptake; when this was done as shown in Table 21.4, the values for pH 7.6 and 8.3 agreed much more closely with observed uptake. The difference at pH 3.8 is believed to be due to the treatment used to get the low pH inhibited phosphorus uptake kinetics. Omitting the pH 3.8 treatment, observed uptake agreed with predicted uptake ($y = 0.93x + 0.44$, $x^2 = 0.99$).

TABLE 21.4   The Calculated Fraction and Predicted Uptake of H_2PO_4 and HPO_4 at Varying pH Values

	Fraction		Predicted uptake (μmol/pot)			Observed
pH	H_2PO_4	HPO_4	H_2PO_4	HPO_4^{a}	Total	uptake (μmol/pot)
3.8	0.991	0.001	94.3	Trace	94.3	42.1
4.7	0.997	0.003	250.4	Trace	250.4	273.7
5.7	0.971	0.029	201.0	0.5	201.5	218.5
6.5	0.840	0.160	219.8	1.5	221.3	231.7
7.6	0.294	0.706	31.0	7.5	38.5	36.1
8.3	0.077	0.923	8.7	10.9	19.6	26.1

[a]HPO_4 uptake = one-tenth of the predicted uptake rate of H_2PO_4 (Hendrix, 1967).

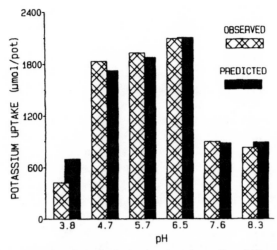

FIGURE   21.12   Relation between observed and predicted K uptake calculated with the Barber–Cushman model for plants grown for 18.5 days. Correlation was $y = 67 + 0.94x$ ($r^2 = 0.99$), where y is predicted uptake and x is observed uptake and data for pH 3.8 were omitted. Reproduced from Chen and Barber (1990) by permission of the Soil Science Society of America.

Comparison of Observed with Predicted Potassium Uptake

Unlike P, K species do not change with soil pH. Calculations of predicted K uptake were affected primarily by differences in soil supply and root growth parameters. Predicted K uptake closely agreed with observed K uptake (Figure 21.12), indicating that the nutrient uptake model satisfactorily predicted the effect of soil pH on K uptake by corn. However, K uptake was also overestimated at pH 3.8, which supports the assumption that inhibition of uptake kinetics was the cause of overpredicted P uptake at pH 3.8.

This comparison of phosphorus and potassium serves to determine the reason for the initial lack of agreement between observed and predicted uptake for phosphorus in this study. Errors in measuring root length can also cause lack of agreement. If this were the case, it would occur for both phosphorus and potassium to the same degree.

Since many soils have pH values above 7.5, the information given here gives a potential reason for phosphate deficiency where chemical measurements on the soil, without adjusting for phosphate species, indicate adequate phosphate.

Calculating Changes of Legume Rhizosphere Soil pH and Soil Solution Phosphorus from Phosphorus Uptake

Li and Barber (1991) used the uptake model in a study of phosphorus uptake by three legumes grown in pot culture. Corn, a nonlegume, was included since with the soil used it had been shown that it did not appreciably change rhizosphere pH.

The legumes used were alfalfa (*Medicago sativa* L.), faba bean (*Vicia faba* L.), and Austrian winter pea (*Lathyrus hirsutus*). The species were grown in Chalmers silt loam limed to pH levels of 5.72, 6.30, 7.22, and 8.30. The soil was equilibrated for 3 weeks at 25°C and -33 kPa water tension. Seedlings were transplanted into 3-L pots containing 2.5 kg soil at 1.2 Mg/m^3 density and placed in a growth chamber at 25°C and a 16-h day with 6000 MW/cm^2 irradiance. Plants were harvested after 12 to 28 days depending on species. Model parameters were measured as described previously in this book and by Li and Barber (1991). Since $H_2PO_4^-$ and HPO_4^{2-} are both present at different ratios depending on pH level (log $[HPO_4^{2-}/H_2PO_4^-]$ = pH − 7.22 for solution pHs between 6.0 and 9.0), amount of each form was calculated and I_{max} for HPO_4^{2-} was assumed to be one-tenth that for $H_2PO_4^-$. Separate calculations of uptake were made for each phosphate form and totalled. Comparisions were made between predicted and observed phosphorus uptake at each initial soil pH. The results are shown in Table 21.5.

Predicted and observed values were similar for corn at each initial soil pH. Predicted values for all legumes were less than observed, with the differences becoming greater as the initial soil pH increased. Observed potassium uptake was measured and predicted potassium uptake was calculated. Comparison of predicted and observed uptake for all species at all pH levels gave a linear relation of $y = 0.98x − 36.3$ ($r^2 = 0.99$), as shown in Figure 5.2. Since soil pH has little affect on potassium uptake, the observed difference between phosphate and potassium appears to be due to changes in rhizosphere soil pH.

The amount of rhizosphere soil pH change required to make the observed difference was calculated from the observed and predicted data. The procedure used was to take the equations for initial soil pH and predicted phosphorus uptake, which were as follows:

Alfalfa: $\quad\quad\quad\quad\quad y = 8.4 − 4.3 \times 10^{-3}x − 5.1 \times 10^{-6}x^2$

Faba bean: $\quad\quad\quad\quad y = 8.1 − 6.3 \times 10^{-3}x − 1.6 \times 10^{-4}x^2$

Austrian winter pea: $\quad y = 8.6 − 0.024x − 2.5 \times 10^{-5}x^2$

where y is soil pH and x is predicted uptake. Then by substituting the observed phosphorus uptake into the equation at each pH level, solve for the rhizoshpere pH. Figure 21.13 shows the curves: the lower curve shows the relation between bulk soil pH (open circles) and predicted phosphorus uptake (solid line); the upper curve shows the relation between bulk soil pH (solid circles) and observed phosphorus uptake (dashed line). The rhizosphere pH is obtained by dropping a line from upper dashed line to the lower solid line where the solid diamonds give the rhizosphere soil pH necessary to give the observed uptake. For alfalfa, where the bulk soil pH is 8.30, the

TABLE 21.5   Fraction of Different P Species and P Uptake by Plants at Varying pH Values

pH	Fraction (%)		Predicted uptake (μmol/pot)			Observed uptake (μmol/pot)
	$H_2PO_4^-$	HPO_4^{2-}	$H_2PO_4^-$	HPO_4^{2-}	Total	
Corn						
5.8	96.3	3.7	742.7	2.9	745.6	718.4
7.3	45.4	54.6	321.3	39.1	366.4	381.2
7.6	29.4	70.6	168.0	39.5	207.5	211.6
8.3	7.7	92.3	8.8	12.1	20.9	22.9
Alfalfa						
5.8	96.3	3.7	372.6	12.9	385.5	430.5
7.3	45.4	54.6	180.8	20.1	200.9	265.0
7.6	29.4	70.6	113.4	19.7	133.1	240.2
8.3	7.7	92.3	7.0	18.5	25.5	156.8
Faba bean						
5.72	96.9	3.1	95.2	9.1	104.3	114.4
6.30	89.3	10.7	82.0	4.7	86.7	104.7
7.22	50.0	50.0	53.5	5.3	58.8	80.8
7.86	18.6	81.4	19.4	4.7	24.1	60.5
Austria winter pea						
5.72	96.9	3.1	119.5	8.5	128.0	141.0
6.30	89.3	10.7	91.9	5.7	91.6	121.7
7.22	50.0	50.0	64.7	6.5	71.2	92.7
7.86	18.6	81.4	28.8	6.3	35.1	70.0

rhizosphere pH is 7.53; where the bulk soil pH is 7.60, the rhizosphere pH is 6.96; where the bulk soil pH is 7.30, the rhizosphere pH is 6.78; where the bulk soil pH is 5.80, the rhizosphere pH is 5.41. Figure 21.13 is plotted with pH on the y axis and P uptake on the x axis because we are using P uptake as the determinate factor to calculate rhizosphere pH.

Changes in C_{li} for phosphate can be determined from a curve of the relation of C_{li} vs. soil pH for the soil. In this study legumes reduced rhizosphere soil pH 0.39 to 0.77 units and increased phosphorus availability 21 to 242%.

Evaluating Whether Mycorrhizae, Root Hairs, or Change of Rhizosphere pH Affect Phosphorus Uptake

The mechanistic model can be divided into root size and morphology, root kinetics, and soil supply to the root. When an experiment is conducted to

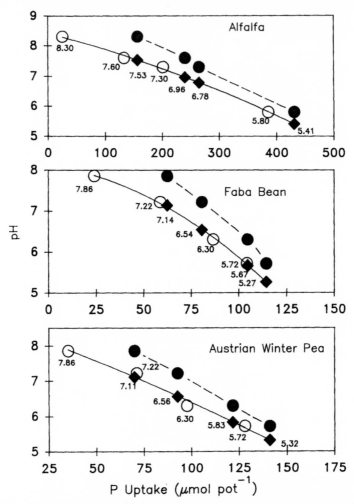

FIGURE 21.13 The relation between bulk soil pH (open circles) and predicted P uptake (solid line), between bulk soil pH (solid circles) and observed P uptake (dashed line), and predicted rhizosphere pH (solid diamonds on solid line) for each bulk soil pH (open circles) for three legume species. Reproduced from Li and Barber (1991) with permission of Marcel Dekkar, Inc.

verify the model for phosphorus uptake, and predicted uptake does not agree with the observed uptake, the reason for the lack of agreement should be investigated. One reason for lack of agreement may be because parameters for the effect of mycorrhizae, root hairs, or change of rhizosphere pH are not included in the Barber–Cushman model. Comparing predicted and observed potassium uptake on the same experiment may indicate the reason. Mycorrhizae, root hairs, or acidification of the rhizosphere seldom affect potassium uptake. The reason mycorrhizae and root hairs have little effect on potassium uptake is because D_e for potassium may be 10 to 100 times that for phosphorus and the depletion zone near the root is much greater than for phosphorus (see Figure 4.6). Root hairs and mycorrhizae provide a greater uptake surface, however, especially for root hairs, absorption is within the rhizosphere zone being deleted due to potassium uptake by the root but not of phosphorus by phosphorus uptake. If predicted potassium uptake agrees with observed uptake, such as shown in Figure 5.2, then the predictions with the model are being calculated correctly. The size and geometry of the root system (except for root hairs) and its rate of increase are being calculated correctly for potassium uptake. If predicted potassium uptake still does not agree with observed uptake, incorrect values used for one or more parameters may be at fault and they should be investigated. If the lack of agreement for both phosphorus and potassium is similar, values used for L_0, k, or r_0 may be incorrect, since they would be the same for both nutrients.

Assuming predicted potassium uptake agrees with observed uptake, then the difference between phosphorus and potassium uptake is usually due to root hairs, mycorrhizae, or rhizosphere soil acidification by the root. Acidification of the rhizosphere frequently increases C_{li} for phosphorus. At low C_{li} values for phosphorus (below 10 μmol/L), root hairs are more likely to have an effect, because low soil phosphorus stimulates root hair growth and, in addition, low soil phosphorus usually has a low soil D_e so phosphorus does not diffuse far. If the plant species used has long root hairs as shown in Table 7.2, root hairs will have a greater effect on phosphorus uptake.

Mycorrhizae increase phosphorus uptake. At low phosphorus C_{li} more mycorrhizae will be present. However, mycorrhizae do not become effective until plants have been growing for 2 weeks, except where grown on undisturbed soil (see Chapter 7). Mycorrhizae can also be reduced by use of fumigation or heating of the soil to 70°C. We have frequently used the latter after application of phosphate to moist soil to speed equilibration.

Acidification of the rhizosphere occurs with legumes or with use of NH_4^+ fertilizer to nonlegumes. It occurs because of greater cation uptake than anion uptake and H^+ is exuded to balance the reaction.

Selection of maize (corn) as a test crop reduces the effects of root hairs, since they are shorter and less dense on the roots. Also unless NH_4^+ fertilizer is added the rhizosphere pH is not affected appreciably where maize is grown in 3-L pots of soil. Plant should be harvested after about 2 weeks of growth in disturbed soil, hence mycorrhizae will not be effective. Root density at 2 weeks will be about that found in the field at maximum growth. Values will be affected by soil temperature. Considering these factors, experiments can be designed to evaluate the influence of root hairs, mycorrhizae, and rhizosphere acidification.

A Soil Supply Standard to Use for Soil Test Calibration

Since the Barber–Cushman model predicts nutrient uptake that agrees close-ly with observed uptake, the predicted values of the model can be used as a standard to correlate with various empirical soil tests. This does not involve using plant growth as a measure of the standard. A common set of values can be used for everything except C_{li}, b, and D_e, hence all that is needed is to determine C_{li}, C_{si}, and θ on each soil since b is C_{si}/C_{li} and D_e can be approxi-mated from $D_e = D_l\, \theta f_l/b$ (see Equation 4.6 in Chapter 4). Then calculate the predicted nutrient uptake using a constant set of values for the remaining model parameters of I_{max}, K_m, C_{min}, v_0, L_0, k, r_0, r, and t from an experiment with the plant species being considered such as shown in Table 5.2. The pre-dicted uptake values can be used as the standard against which various soil tests for the nutrient can be evaluated. This will give initial information in developing tests using plant growth experiments.

REFERENCES

Anghinoni, I., and S. A. Barber. 1980a. Phosphorus influx and growth characteris-tics of corn roots as influenced by phosphorus supply. *Agron. J.* **72**:685–688.

Anghinoni, I., and S. A. Barber. 1980b. Phosphorus application rate and distribution in the soil and phosphorus uptake by corn. *Soil Sci. Soc. Am. J.* **44**:1041–1044.

Barber, S. A. 1958. Relation of fertilizer placement to nutrient uptake and crop yield. I. Interaction of row phosphorus and the soil level of phosphorus. *Agron. J.* **50**:535–539.

Barber, S. A. 1959. Relation of fertilizer placement to nutrient uptake and crop yield. II. Effects of row potassium, potassium soil-level, and precipitation. *Agron. J.* **51**:97–99.

Barber, S. A. 1977. Placement of phosphate and potassium for increased efficiency. *Solutions* **21**:24–25.

Barber, S. A., and P. R. Ernani. 1990. Use of a mechanistic model to evaluate phos-phate fertilizer. *Int. Soil Sci. Soc. Proc.* **II**:136–139.

Borkert, C. 1983. Mechanistic modeling of phosphate placement to maximize phos-phate recovery by soybeans. Ph.D. dissertation, Purdue Univ.

Borkert, C. M., and S. A. Barber. 1985a. Soybean shoot and root growth and phos-phorus concentration as affected by phosphorus placement. *Soil Sci. Soc. Am. J.* **49**:152–155.

Borkert, C. M., and S. A. Barber. 1985b. Predicting the most efficient phosphorus placement for soybeans. *Soil Sci. Soc. Am. J.* **49**:901–904.

Caldwell, M. M., L. M. Dudley, and B. Lilieholm. 1992. Soil solution phosphate, root uptake kinetics and nutrient acquisition: Implications for a patchy environ-ment. *Oecologia* **89**:305–309.

Chen, J. H., and S. A. Barber. 1990. Effect of soil pH and P and K uptake by maize as evaluated with a mechanistic uptake model. *Soil Sci. Soc. Am. J.* **54**:1032–1036.

Drew, M. C. 1975. Comparison of the effects of localized supply of phosphate, nitrate, ammonium, and potassium on the growth of the seminal root system and shoot of barley. *New Phytol.* **75**:479–490.

Drew, M. C., and L. R Saker. 1978. Nutrient supply and the growth of the seminal root system in barley. 3. Compensatory increases in growth of lateral roots and in rates of phosphorus uptake, in response to a localized supply of phosphate. *J. Exp. Bot.* **29**:435–451.

Ernani, P. R., and S. A. Barber. 1990. Comparison of P availability from monocalcium and diammonium phosphates using a mechanistic nutrient uptake model. *Fert. Res.* **22**:15–20.

Hendrix, J. E. 1967. The effect of pH on the uptake and accumulation of phosphate and sulfate ions by bean plants. *Am. J. Bot.* **54**:560–564.

Kovar, J. L., and S. A. Barber. 1987. Placing phosphorus and potassium for greatest recovery. *J. Fert. Issues* **4**:1–6.

Kovar, J. L., and S. A. Barber. 1988. Phosphorus supply characteristics of 33 soils as influenced by seven rates of phosphorus addition. *Soil Sci. Soc. Am. J.* **52**:160–165.

Kovar, J. L., and S. A. Barber. 1989. Reasons for differences among soils in placement of phosphorus for maximum predicted uptake. *Soil Sci. Soc. Am. J.* **53**:1733–1736.

Li, Y., and S. A. Barber. 1991. Calculating changes of legume rhizosphere soil pH and soil solution phosphorus from phosphorus uptake. *Commun. Soil Sci. Plant Anal.* **22**:955–973.

Stanford, G., and W. H. Pierre. 1953. Soil management practices in relation to phosphorus availability and use. In W. H. Pierre and A. G. Norman, Eds. *Soil and Fertilizer Phosphorus in Crop Nutrition.* Academic Press, New York. Pp. 243–280.

Stryker, R. B., J. W. Gilliam, and W. A. Jackson. 1974. Nonuniform phosphorus distribution in the root zone of corn: Growth and phosphorus uptake. *Soil Sci. Soc. Am. Proc.* **38**:334–340.

Welch, L. F., D. L. Mulvaney, L. V. Boone, G. F. McKibben, and J. W. Pendleton. 1966. Relative efficiency of broadcast versus band phosphorus for corn. *Agron. J.* **58**:283–287.

Widdowson, F. V., and G. W. Cooke. 1958. Comparisons between placing and broadcasting nitrogen, phosphorus, and potassium fertilizers for potatoes, peas, beans, kale, and maize. *J. Agric. Sci.* **51**:53–61.

Yao, J., and S. A. Barber. 1986. Effect of one phosphorus rate placed in different soil volumes on P uptake and growth of wheat. *Commun. Soil Sci. Plant Anal.* **17**:819–827.

Zhang, J., and S. A. Barber. 1992. Maize root distribution between phosphorus-fertilized and unfertilized soil. *Soil Sci. Soc. Am. J.* **56**:819–822.

Index

Printed in the United Kingdom
by Lightning Source UK Ltd.
134328UK00001B/214/A